AF845621

VOLUME ONE HUNDRED AND FIVE

VITAMINS AND HORMONES

Erythropoietin

VOLUME ONE HUNDRED AND FIVE

VITAMINS AND HORMONES

Erythropoietin

Series Editor

GERALD LITWACK, PhD

Toluca Lake, North Hollywood, California

ELSEVIER

ACADEMIC PRESS

An imprint of Elsevier

Cover photo credit:
Frank Guarnieri
Analysis of the Asymmetry of Activated EPO Receptor Enables Designing Small Molecule Agonists
Vitamins and Hormones (2017) **105**, pp. 19–38.

Academic Press is an imprint of Elsevier
50 Hampshire Street, 5th Floor, Cambridge, MA 02139, United States
525 B Street, Suite 1800, San Diego, CA 92101-4495, United States
The Boulevard, Langford Lane, Kidlington, Oxford OX5 1GB, United Kingdom
125 London Wall, London, EC2Y 5AS, United Kingdom

First edition 2017

Copyright © 2017 Elsevier Inc. All rights reserved.

No part of this publication may be reproduced or transmitted in any form or by any means, electronic or mechanical, including photocopying, recording, or any information storage and retrieval system, without permission in writing from the publisher. Details on how to seek permission, further information about the Publisher's permissions policies and our arrangements with organizations such as the Copyright Clearance Center and the Copyright Licensing Agency, can be found at our website: www.elsevier.com/permissions.

This book and the individual contributions contained in it are protected under copyright by the Publisher (other than as may be noted herein).

Notices
Knowledge and best practice in this field are constantly changing. As new research and experience broaden our understanding, changes in research methods, professional practices, or medical treatment may become necessary.

Practitioners and researchers must always rely on their own experience and knowledge in evaluating and using any information, methods, compounds, or experiments described herein. In using such information or methods they should be mindful of their own safety and the safety of others, including parties for whom they have a professional responsibility.

To the fullest extent of the law, neither the Publisher nor the authors, contributors, or editors, assume any liability for any injury and/or damage to persons or property as a matter of products liability, negligence or otherwise, or from any use or operation of any methods, products, instructions, or ideas contained in the material herein.

ISBN: 978-0-12-812265-5
ISSN: 0083-6729

For information on all Academic Press publications
visit our website at https://www.elsevier.com/books-and-journals

ELSEVIER
Book Aid International
Working together to grow libraries in developing countries

www.elsevier.com • www.bookaid.org

Publisher: Zoe Kruze
Acquisition Editor: Alex White
Editorial Project Manager: Helene Kabes
Production Project Manager: Vignesh Tamil
Cover Designer: Greg Harris

Typeset by SPi Global, India

Former Editors

ROBERT S. HARRIS

Newton, Massachusetts

JOHN A. LORRAINE

University of Edinburgh
Edinburgh, Scotland

PAUL L. MUNSON

University of North Carolina
Chapel Hill, North Carolina

JOHN GLOVER

University of Liverpool
Liverpool, England

GERALD D. AURBACH

Metabolic Diseases Branch
National Institute of
Diabetes and Digestive and
Kidney Diseases
National Institutes of Health
Bethesda, Maryland

KENNETH V. THIMANN

University of California
Santa Cruz, California

IRA G. WOOL

University of Chicago
Chicago, Illinois

EGON DICZFALUSY

Karolinska Sjukhuset
Stockholm, Sweden

ROBERT OLSEN

School of Medicine
State University of New York
at Stony Brook
Stony Brook, New York

DONALD B. MCCORMICK

Department of Biochemistry
Emory University School of
Medicine, Atlanta, Georgia

CONTENTS

CONTRIBUTORS

Begum Alural
Izmir International Biomedicine and Genome Institute; Health Science Institute, Dokuz Eylul University, Izmir, Turkey

Zeynep O. Ayyildiz
Izmir International Biomedicine and Genome Institute; Health Science Institute, Dokuz Eylul University, Izmir, Turkey

Gaetano Cairo
University of Milan, Milano, Italy

Céline Caravagna
Institut de Neurosciences de la Timone, Marseille, France

Regina M. Day
Uniformed Services University of the Health Sciences, Bethesda, MD, United States

Soumyadeep Dey
National Institute of Diabetes and Digestive and Kidney Diseases, National Institutes of Health, Bethesda, MD, United States

Mitsugu Fujita
Kindai University Faculty of Medicine, Osaka-Sayama, Japan

Yankel Gabet
Sackler Faculty of Medicine, Tel-Aviv University, Tel-Aviv, Israel

Kursad Genc
Health Science Institute, Dokuz Eylul University, Izmir, Turkey

Sermin Genc
Izmir International Biomedicine and Genome Institute; Health Science Institute, Dokuz Eylul University, Izmir, Turkey

Frank Guarnieri
Virginia Commonwealth University School of Medicine, Richmond, VA; PAKA Pulmonary Pharmaceuticals, Acton, MA, United States

Verena Günther
Institute for Zoology and Anthropology, Georg-August-University Goettingen, Goettingen, Germany

Ralf Heinrich
Institute for Zoology and Anthropology, Georg-August-University Goettingen, Goettingen, Germany

Sahar Hiram-Bab
Sackler Faculty of Medicine, Tel-Aviv University, Tel-Aviv, Israel

CongBao Kang
Experimental Therapeutics Centre, Agency for Science, Technology and Research (A*STAR), Nanos, Singapore

Yong-Chul Kim
Uniformed Services University of the Health Sciences, Bethesda, MD, United States

Steven J. Korzeniewski
Wayne State University School of Medicine, Detroit, United States

Qingxin Li
Institute of Chemical & Engineering Sciences, Agency for Science, Technology and Research (A*STAR), Jurong Island, Singapore, Singapore

Alan List
Division of Clinical Sciences, H. Lee Moffitt Cancer Center & Research Institute, Tampa, FL, United States

Robert T. Mallet
Institute for Cardiovascular and Metabolic Diseases, University of North Texas Health Science Center, Fort Worth, TX, United States

Kathy McGraw
Division of Clinical Sciences, H. Lee Moffitt Cancer Center & Research Institute, Tampa, FL, United States

Natasa Miljus
Institute for Zoology and Anthropology, Georg-August-University Goettingen, Goettingen, Germany

Victor Monjaras-Embriz
Facultad de Ciencias de la Salud, Universidad Panamericana, Ciudad de Mexico, Mexico

Ognoon Mungunsukh
Uniformed Services University of the Health Sciences, Bethesda, MD, United States

Drorit Neumann
Sackler Faculty of Medicine, Tel-Aviv University, Tel-Aviv, Israel

Constance T. Noguchi
National Institute of Diabetes and Digestive and Kidney Diseases, National Institutes of Health, Bethesda, MD, United States

Athina Pappas
Wayne State University School of Medicine, Detroit, United States

Leticia Rocha-Zavaleta
Instituto de Investigaciones Biomédicas, Universidad Nacional Autónoma de México; Facultad de Ciencias de la Salud, Universidad Panamericana, Ciudad de Mexico, Mexico

Ilona Rybinska
University of Milan, Milano, Italy

Myoung-Gwi Ryou
Institute for Cardiovascular and Metabolic Diseases, University of North Texas Health Science Center; Tarleton State University, Fort Worth, TX, United States

Tommy Seaborn
Faculty of Medicine, Université Laval, Quebec City, QC, Canada

Kemal U. Tufekci
Izmir International Biomedicine and Genome Institute; Health Science Institute, Dokuz Eylul University, Izmir, Turkey

Maria J. Vazquez-Mellado
Instituto de Investigaciones Biomédicas, Universidad Nacional Autónoma de México, Ciudad de Mexico, Mexico

Yoshiko Yasuda
Cancer Center, Kyoto Prefectural University of Medicine, Kyoto, Japan

Maria P. Zafeiriou
Institute of Pharmacology and Toxicology, Heart Research Center Goettingen, University Medical Center Goettingen, Georg-August-University Goettingen, Goettingen, Germany

PREFACE

Erythropoietin is a glycoprotein cytokine-like hormone that is mainly synthesized in the kidney and to a lesser extent in the liver. Erythropoietin regulates the formation of red blood cells by stimulating the proliferation and differentiation of red blood cell precursors in bone marrow. In response to anemia or hypoxia, the kidney releases erythropoietin into the bloodstream. The oxygen level of blood is sensed in the kidney by the hypoxia-inducible factor, a transactivating factor that binds to hypoxia response elements in nuclear DNA. Although the early discoveries of erythropoietin characterized its effects on producing new red blood cells, it is currently known to have many other physiological functions. Among these are: interaction with iron metabolism, central respiration control, regulation by angiotensin II, effects on bone, neuroprotection, effects on ischemic brain, relationship with lipid rafts, regulation of metabolism, protection of the heart, and effects on cancers. All of these subjects are covered in this volume in addition to structural studies of ligand, small molecule agonists, and the erythropoietin receptor.

The first chapters are centered on structure–function. "Erythropoietin receptor structural domains" is the subject of Q. Li and C. Kang. Then, F. Guarnieri introduces "Analysis of the asymmetry of activated EPO receptor enables designing small molecule agonists." S.J. Korzeniewski and A. Pappas describe "Endogenous erythropoietin." "Erythropoietin regulation by angiotensin II" is the topic of Y.-C. Kim, O. Mungunsukh, and R.M. Day. K. McGraw and A. List report on "Erythropoietin receptor signaling and lipid rafts." "Erythropoietin and hypothalamic–pituitary axis" is the subject discussed by S. Dey and C.T. Noguchi. T. Seaborn and C. Caravagna report on "Erythropoietin-mediated regulation of central respiratory command." "Mutual cross talk between iron homeostasis and erythropoiesis" is the theme of the chapter by I. Rybinska and G. Cairo. S. Hiram-Bab, D. Neumann, and Y. Gabet contribute a chapter entitled "Context-dependent skeletal effects of erythropoietin." R. Heinrich, V. Günther, and N. Miljus discuss "Erythropoietin-mediated neuroprotection in insects suggests a prevertebrate evolution of erythropoietin-like signaling." "Erythropoietin: endogenous protection of ischemic brain" is explored by R.T. Mallet and M.-G. Ryou. A chapter provided by M.P. Zafeiriou is entitled "The erythropoietin system protects the heart upon injury by cardiac progenitor cell activation." The volume is completed by three chapters on cancer, the first of which is "Erythropoietin promotes glioblastoma via miR-451 suppression" by B. Alural, Z.O. Ayyildiz, K.U. Tufekci, S. Genc, and K. Genc. Next, M.J. Vazquez-Mellado, V. Monjaras-Embriz, and L. Rocha-Zavaleta present "Erythropoietin, stem cell factor, and cancer cell migration." Finally, Y. Yasuda and M. Fujita detail the "Significance of erythropoietin receptor antagonist EMP9 in cancers."

The illustration on the cover is reproduced from Fig. 5 of Chapter 2 by F. Guarnieri entitled "Analysis of the asymmetry of activated EPO receptor enables designing small molecule agonists." The figure legend is: "This is the RWJ structure with the EPOR dimer shown in *light* and *dark blue ribbons* and the EMP1 peptide ligand dimer in *gray ribbons*. R178 from one receptor is in VDW contact with L175 of the other receptor forming a homodimer interacting quartet at the center of the receptor. The E176 from both receptors packs against this quartet."

The publication of this volume was assisted by the work of Helene Kabes of Elsevier, Oxford, United Kingdom and by Vignesh Tamilselvvan of Elsevier/S and T Book Production, Chennai, India.

Gerald Litwack
North Hollywood, CA
January 28, 2017

CHAPTER ONE

Erythropoietin Receptor Structural Domains

Qingxin Li*, CongBao Kang†,1
*Institute of Chemical & Engineering Sciences, Agency for Science, Technology and Research (A*STAR), Jurong Island, Singapore, Singapore
†Experimental Therapeutics Centre, Agency for Science, Technology and Research (A*STAR), Nanos, Singapore
[1]Corresponding author: e-mail address: cbkang@etc.a-star.edu.sg

Contents

Abstract

Erythropoietin (EPO) is a hormone that is important for regulating red blood cell production. It is functional through binding to its receptor—EpoR. EpoR is a single-span membrane protein. It contains an extracellular region, a transmembrane domain, and a C-terminus. The extracellular region is important for binding to EPO, and its conformation is critical for signal transduction. The transmembrane domain contains 21 residues forming a helix which plays an important role in transferring ligand-induced conformational changes of the extracellular domain across the cell membrane. The C-terminal region contains the Janus kinase 2-binding sites and eight tyrosine residues that can be phosphorylated to become binding sites for transcription factors to active the downstream pathways. This chapter focuses on structural description of the domains of the EpoR. The recent progress in the structural determination of these domains is summarized, which will be useful for understanding their function in signal transduction.

Vitamins and Hormones, Volume 105
ISSN 0083-6729
http://dx.doi.org/10.1016/bs.vh.2017.02.005
© 2017 Elsevier Inc.
All rights reserved.

1. INTRODUCTION

Red blood cells are abundant in the body and play essential roles in transporting oxygen and carbon dioxide. The number of red blood cells is well managed for sustaining their normal biological functions. Erythropoiesis is a complex physiological process to maintain oxygen level in the body through production of red blood cells (Elliott, Pham, & Macdougall, 2008). Erythropoiesis is regulated by erythropoietin (EPO)—a hormone that is regulated by oxygen concentration in circulation and is critical for stimulating and differentiation of red blood cell precursors in the hemopoietic tissues such as bone marrow that are responsible for erythropoiesis. EPO is a 30-kDa protein comprising 165 amino acids and is produced by kidneys (Elliott et al., 2008; Lai, Everett, Wang, Arakawa, & Goldwasser, 1986; Lin et al., 1985). In the absence of EPO, precursor cells of the red blood cell may undergo apoptosis (Kelley et al., 1994). EPO is functional through its receptor—erythropoietin receptor (EpoR). One EPO molecule binds two EpoRs, which can induce conformational changes at the C-terminus of EpoR. Conformational change in the C-terminus will activate the Janus kinase 2 (JAK2) that is a direct signaling molecule at the downstream of EpoR (Lamon & Russell, 2013; Pelletier, Gingras, Funakoshi-Tago, Howell, & Ihle, 2006). Activated JAK2 or other kinases may promote different downstream pathways under various conditions (Miyake et al., 2013). The activated JAK can phosphate several tyrosine residues in the C-terminus of EpoR and itself (Ihle et al., 1998; Waters, Brooks, & Chhabra, 2014). Phosphorylated receptor can bind to Src homology 2 (SH2) domain-containing proteins. SH domains comprise approximately 100 residues and can bind selectively to phosphorylated tyrosine residues (Filippakopoulos, Muller, & Knapp, 2009). Phosphorylated EpoR can interact with SH2-containing transcription factors, signal transducers and activator of transcription (STAT) proteins, which will induce blood cell production (Constantinescu, Ghaffari, & Lodish, 1999; Ihle, 1994; Ihle & Kerr, 1995; Richmond, Chohan, & Barber, 2005).

2. DOMAIN ARCHITECTURE OF THE EpoR

EpoR is a single-span transmembrane protein with 507 residues (human EpoR). It belongs to the cytokine receptor family that also includes growth hormone receptor, prolactin receptor, and thrombopoietin receptor. All

A

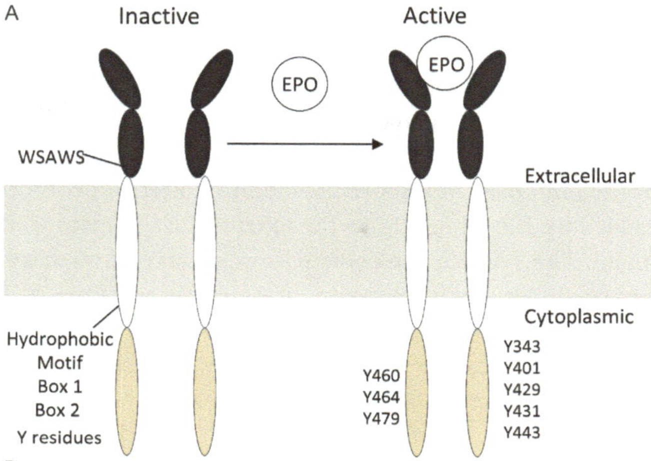

B

```
MDHLGASLWP QVGSLCLLLA GAAWAPPPNL PDPKFESKAA LLAARGPEEL
LCFTERLEDL VCFWEEAASA GVGPGNYSFS YQLEDEPWKL CRLHQAPTAR
GAVRFWCSLP TADTSSFVPL ELRVTAASGA PRYHRVIHIN EVVLLDAPVG
LVARLADESG HVVLRWLPPP ETPMTSHIRY EVDVSAGNGA GSVQRVEILE
GRTECVLSNL RGRTRYTFAV RARMAEPSFG GFWSAWSEPV SLLTPSDLDP
LILTLSLILV VILVLLTVLA LLSHRRALKQ KIWPGIPSPE SEFEGLFTTH
KGNFQLWLYQ NDGCLWWSPC TPFTEDPPAS LEVLSERCWG TMQAVEPGTD
DEGPLLEPVG SEHAQDTYLV LDKWLLPRNP PSEDLPGPGG SVDIVAMDEG
SEASSCSSAL ASKPSPEGAS AASFEYTILD PSSQLLRPWT LCPELPPTPP
HLKYLYLVVS DSGISTDYSS GDSQGAQGGL SDGPYSNPYE NSLIPAAEPL
PPSYVACS
```

Fig. 1 Diagram of EpoR. (A) Diagram of EpoR receptor on the membrane. The membrane is shown as a *gray box*. The extracellular domain, the TM domain, and the cytoplasmic region are shown in *dark gray*, *white*, and *light gray*, respectively. The WSAWS in the extracellular domain, the three-residue hydrophobic motif, box 1 and box 2 motifs, and tyrosine residues are indicated. Receptor was predicted to have inactive (monomeric) and active (dimeric) forms in the absence and presence of EPO (*circle*). (B) Amino acid sequence of the human EpoR. The sequence was from UniProt Knowledgebase (http://www.uniprot.org) with access number P19235. The first 24 residues are shown in *red* and are not counted in most of the publications. Mouse EpoR (UniProtKB ID P14753) contains 507 residues, and human EpoR contains 508 residues. The extracellular domain is shown in *purple*. The TM domain is shown in *cyan*. The tyrosine residues are highlighted in *red*. The WSAWS motif is shown in *green*, and the hydrophobic motif near the TM domain is highlighted in *blue*. The box 1 and box 2 motifs are shown in a box.

these receptors play key roles in many cellular processes. The EpoR consists of a cytoplasmic region, a transmembrane (TM) domain and a cytoplasmic region (Fig. 1). The extracellular domain contains two fibronectin type II domains that bind EPO to induce conformational changes. Structural study has demonstrated that the extracellular domain can form a dimer even in the absence of EPO (Livnah et al., 1999). In the presence of EPO, conformational changes are induced in the fibronectin type II domains within the extracellular domain to affect JAK2 activity (Richmond et al., 2005). The TM domain of the EpoR contains 21 residues (Fig. 1). The TM domain can form dimers and play an important role in transferring conformational changes induced by EPO binding to the extracellular domain to the cytoplasmic domain. The TM domain contains important residues that are essential for receptor dimerization, rotation, and activation (Ebie & Fleming, 2007). The C-terminus of the EpoR contains both JAK2-binding site and several residues that can be phosphorylated. Structural studies on EpoR have been conducted using both X-ray crystallography and nuclear magnetic resonance (NMR) spectroscopy (Livnah et al., 1998, 1996, 1999; Moraga et al., 2015). The structure of the full-length EpoR is not available due to its large size and difficulty in crystallization. Nevertheless, accumulated data have provided insight into the structure and function of the individual domain of the receptor.

3. THE EXTRACELLULAR DOMAIN OF THE EpoR

The extracellular region of the EpoR contains approximately 220 amino acids (Fig. 1). This region is responsible for interacting with EPO, and the resulting conformational changes can induce downstream signalings. This region is characterized with two discrete domains named as D1 and D2 (Bazan, 1990; Livnah et al., 1996). Crystal structures of this region in the absence (Fig. 2A) and presence of EPO, EPO mimics, and diabodies have been solved (Fig. 2; Livnah et al., 1996), providing insight into its role in signal transduction. The first crystal structure of the extracellular domain of the human EpoR is determined when it forms a complex with EPO mimetic peptide-EMP1 that is a 20-residue cyclic peptide (Livnah et al., 1996; Fig. 2B). The structure of EMP1 is different from that of EPO, but it can mimic EPO's function (Wrighton et al., 1996). Crystal structure of the EpoR reveals that each D domain comprises approximately 100 residues that form antiparallel β strands (Livnah et al., 1996). Both D1 (residues 10–114) and D2 (residues 119–220) have a similar topology to that of fibronectin type III (FBN III) domains. D1 contains a short α-helix preceding the first β sandwich and its fold is considered as a hybrid of the FBN III fold with an Ig fold (Livnah et al., 1996). There is an irregular coil structure instead of a short helix preceding the first β sandwich in D2. The folding of D2 is similar

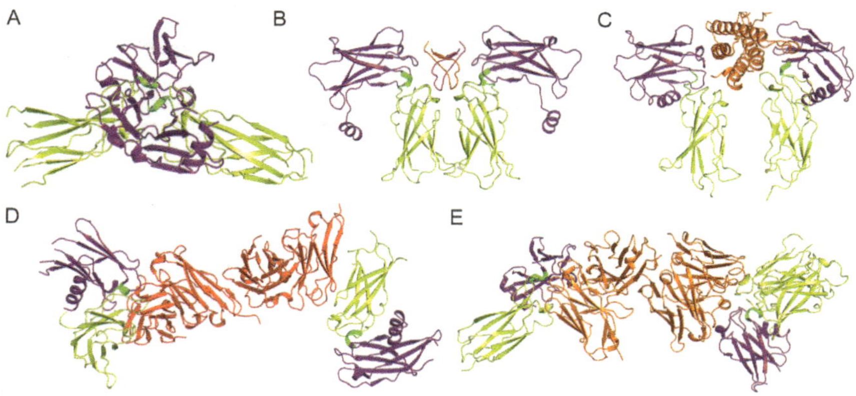

Fig. 2 Structures of the extracellular domain of EpoR. (A) Free extracellular domain of EpoR. The free form of extracellular domain of EpoR (PDB ID 1ERN) is shown. The D1, D2, and loop between D1 and D2 are shown in *blue, lime, and green*, respectively. (B) Structure of the EpoR in a complex with peptide agonist. The structure of EpoR/peptide complex (PDB ID 1EBP) is shown. Two peptides are interacting with two EpoR molecules. The two peptides are shown in *brown* and *magenta*, respectively. (C) Structure of the EpoR–EPO complex. The structure of EpoR in complex with EPO (PDB ID 1CN4) is shown. The EPO peptide is shown in *brown*. This structure is the active form of the receptor. (D and E) Structures of EpoR with different diabodies. Structures of EpoR in complex with full agonist (PDB ID 4Y5V) and nonagonist (PDB ID 4Y5X) are shown. The antibodies are shown in *red* (D) and *brown* (E), respectively.

to that of FBN III. D1 and D2 are connected by a short helical linker formed by residues 115–118 (Fig. 2). D1 contains a pair of conserved disulfide bridges, namely, Cys28 to Cys38 and Cys67 to Cys83 (Livnah et al., 1996). D2 only contains a free cysteine residue without any disulfide bridges (Fig. 2B). There is a WSxWS (x is any residue and it is Ala in the human EpoR) box in D2 forming a β bulge between two β sheets (Fig. 2B; Livnah et al., 1996). This WSxWS box contains tryptophan and serine residues, which is shown to be important for receptor signaling. This box region adopts a polyproline type II helix conformation and does not play any role in ligand binding (Livnah et al., 1996). Mutation in this box region affects receptor expression on cell surface, suggesting that this box may play an important role in receptor folding (Hilton, Watowich, Katz, & Lodish, 1996).

In the structure of the EpoR–EMP1 complex, there are two peptides interacting with two EpoR molecules (Fig. 2B) to generate a symmetric 2:2 assembly. D1 and D2 are aligned at approximately 90 degree to each other, forming a T-shaped structure (Livnah et al., 1996). Four loop regions from EpoR interact with the cyclic peptide, namely, L1 (residues 33 and 34),

L3 (residues 90–94) of D1 and L5 (residues 148–153), L6 (residues 203–205) of D2 (Livnah et al., 1996). In the structure of the EpoR–EPO complex, there is one hormone molecule and two EpoRs forming a 1:2 complex stoichiometry (Syed et al., 1998). In the complex, two EpoR molecules are interacting with one EPO molecules through binding to two regions—a high-affinity interface ($K_d = 1$ nM) and a low-affinity interface ($K_d = 1$ μM) (Fig. 2C). Different from the EpoR–EMP1 complex, six loops (L1, L2, and L3 from D1, L4 of D1 and D2 linker, and L5 and L6 from D2) are interacting with the EPO molecule. The conserved WSxWS box has been shown to have molecular interactions with the helix preceding the first β sheet in D1, suggesting that this helix may affect receptor folding (Syed et al., 1998). Comparison of the EpoR–EPO and EpoR–EMP1 structures reveals that the receptors have different orientations especially for the D1 domain. The two D1 domains in the EpoR–EMP1 complex are positioned ~180 degree, compared with ~120 degree in the EpoR–EPO complex (Fig. 2B). Receptor–receptor interaction is also observed in the EpoR–EPO complex, where residues (S135 and E134) in D2 form hydrogen bond interactions.

Structural studies have demonstrated that the active receptor exists in a dimeric form. Further structural information on receptor in complex with EMP33—an inactive ligand—has demonstrated that the inactive receptor also exists as a dimer (Livnah et al., 1998). EMP33 is a 20-residue peptide which cannot active the receptor at low concentration probably due to its low-binding affinity to the receptor (Livnah et al., 1998; Moraga et al., 2015). The complex structure is similar to that of receptor in complex with EMP1 with difference in the orientation of D1 and D2, suggesting that orientation of the dimer affects receptor activation (Livnah et al., 1998). The free receptor structure suggests that two molecules can form a cross-shaped dimer (Livnah et al., 1999). Residues involving self-dimer interface are similar to those that bind to EPO (Livnah et al., 1999). Ligand binding may affect the orientations of the dimeric structure of the EpoR. In the presence of EMP1 or EPO, the distance of the two D2 C-termini residues (residue 220) in the complexes is 39 and 34 Å, respectively (Livnah et al., 1999; Syed et al., 1998; Wilson & Jolliffe, 1999). The corresponding distance is 73 Å in the free receptor. Based on these structural data, a hypothesis is proposed that the EpoR can form an inactive dimer on the cell surface. Binding of EPO or other ligands would affect the conformation to form an active dimer. This model has been confirmed by a study using dihydrofolate reductase (Remy, Wilson, & Michnick, 1999). A recent structural study reveals that the orientation of the extracellular domain can affect the downstream signaling pathway significantly (Moraga et al., 2015). Modified antibodies–diabodies are shown to have different effects on EpoR signaling (Lim et al., 2010; Moraga et al., 2015). Three diabodies, namely, DA5, DA330, and DA10, are active, partially active, and nonactive ligands against EpoR (Fig. 2D and E). Structures of their complexes with EpoR are determined (Moraga et al., 2015). Although two EpoR molecules and one diabody are present in the complex, their overall architectures are different from that of the EpoR–EPO complex (Moraga et al., 2015). In the structures of the EpoR–EPO and the EpoR–EMP1 complexes, the C-terminal regions of D2 in the complex are in a close proximity. In the structures of EpoR–diabodies complexes, the distance of the C-termini is large, ranging from ~127 Å in the EpoR–DA5 (full agonist) complex to ~148 Å in the EpoR–DA10 (nonagonist) complex (Moraga et al., 2015). The structural data for the diabodies suggest that extracellular topology correlates differently in signaling amplitude, which may arise from its influence on the orientation of JAK molecules that are important for phosphorylation (Moraga et al., 2015). Dimer reorientation may be one of the strategies to fine and tune the signaling output to achieve a maximum effect without causing significant toxicity (Moraga et al., 2015). Despite these differences, the structures of D1 and D2 in the complexes are very similar to those in EpoR–EPO complex.

4. TM DOMAIN OF THE EpoR

Like other cytokine receptors, EpoR contains a single TM span. The TM domain is important for receptor dimerization. Using an immunofluorescence colocalization assay, this domain is shown to be sufficient for maintaining dimerization of the full-length receptor (Constantinescu, Keren, et al., 2001). This domain is also demonstrated to be important for activation of downstream signalings. Orientation of the TM domain dimer may have different effects on signaling, which is evidenced by using a series of coiled-coil-containing mutants that can constrain the TM domain into different orientations (Seubert et al., 2003). The TM domain and the juxtamembrane region are important for both receptor dimerization and EPO signaling (Lu, Gross, & Lodish, 2006). The TM domain alone is shown to be self-associated and two leucine residues (L240 and L241) are proposed to be important for dimerization (Kubatzky et al., 2001). A further thermodynamic study has demonstrated that peptide derived from the transmembrane domain can form dimer in detergent micelles (Ebie & Fleming, 2007). In the study, the TM domain of the mouse EpoR may have a stronger self-dimerization affinity than that of the human EpoR, which is also consistent with the functional analysis (Ebie & Fleming, 2007).

Solution structures of the TM domains of both human and mouse EpoRs in detergent micelles have been solved using solution NMR spectroscopy (Fig. 3). The TM domain forms a standard helix in solution (Li, Wong, Huang, & Kang, 2014; Li, Wong, Yueqi Lee, Li, & Kang, 2015). The C-terminal juxtamembrane region is also helical in solution and is shown

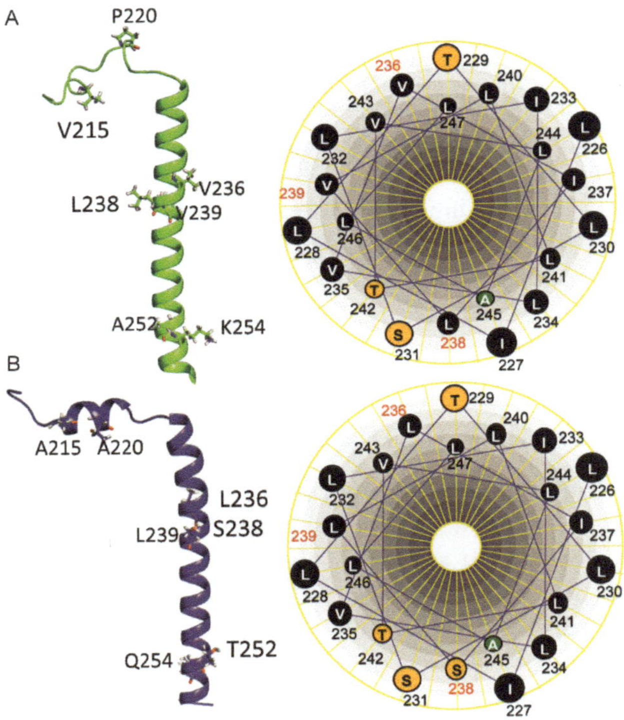

Fig. 3 Structures of the TM domains of EpoRs. (A) Structure of the TM domain of the human EpoR in detergent micelles. (B) Structure of the TM domain of the mouse EpoR in detergent micelles. *Left panel* is the structure of the TM domain, and the *right panel* is helix wheel representation of residues in the helix. Different residues between mouse and human EpoRs are highlighted in *red* (*right panel*). *These figures are from Li, Q., Wong, Y. L., Yueqi Lee, M., Li, Y., & Kang, C. (2015). Solution structure of the transmembrane domain of the mouse erythropoietin receptor in detergent micelles.* Scientific Reports, 5, *13586, with slightly modifications.*

to be exposed to the solvent, which is confirmed by NMR spectroscopy, fluorescence spectroscopy, and paramagnetic relaxation enhancement experiments (Li et al., 2014). Further dynamic studies reveal that the TM domain is rigid in solution. The N-terminal region preceding the TM domain is dynamic, and this region is the linker between the extracellular domain and the TM domain. It may play a role in transforming conformational changes of the extracellular region induced by EPO binding to the C-terminus. Although the TM domains from both mouse and human share very high sequence homology, the residues preceding the N-terminus of the TM domain of the mouse EpoR have a tendency to form a short helix, which may arise from the fact of presence of an alanine residue at position 220 instead of a proline (a helix breaker) (Li et al., 2015). Cysteine-scanning mutagenesis study demonstrated that L223C mutation in the juxtamembrane region results in a construct that can form a disulfide bond, suggesting that L223 is important in receptor activation (Kubatzky et al., 2005). In addition to L223, residues including L226 and L227 are in close contacts, suggesting that these residues may be important for TM dimerization because mutants on those sites form constitutively active receptors (Kubatzky et al., 2005). The C-terminal juxtamembrane region contains several hydrophobic residues including L253, I257, and W258. These residues can form a hydrophobic patch that may be important for their function in signaling (Constantinescu, Huang, Nam, & Lodish, 2001; Li et al., 2014). This juxtamembrane region is stable in solution and not as dynamic as the N-terminal residues preceding the TM domain (Li et al., 2015). The TM domain is important for signal transduction upon EPO binding due to its physical location. TM domain dimerization may be important for its function and the residues involving in dimerization may include several hydrophilic residues including S231, S238, T242, and S248 (Seubert et al., 2003). Studies have also suggested that the TM domain of EpoR or other cytokine receptors may have different dimer conformations representing active or inactive receptors (Seubert et al., 2003). TM domain rotation may be one of the mechanisms during receptor activation (Brooks et al., 2014; Brown et al., 2005; Matthews et al., 2011). Although there is no dimeric structure of the TM domain of EpoR available, a left-hand dimeric model has been proposed based on few restraints from NMR experiments. This model is in consistent with a previous model based on functional data analysis (Li et al., 2015; Seubert et al., 2003). In the previous NMR studies, multiple cross peaks in the spectra were observed, suggesting that the TM domain may have multiple conformations. This is in consistent with previous finding that the TM helix may have active and inactive conformations under some conditions (Seubert et al., 2003).

Although different studies have shown that the TM can form dimers both in vivo and in vitro, recent studies on the structures of the extracellular domain of EpoR–diabodies suggest that the active EpoR dimer may not involve the TM domain (Moraga et al., 2015). Further structural study on a construct containing both the extracellular region and the TM domain will be useful to understand the roles of the TM domain in signal transduction. The challenge for X-ray study on such a construct may be at the crystallization step. Further structural investigation using solution or solid-state NMR spectroscopes will be an alternative way to overcome such challenge.

5. THE C-TERMINUS OF THE EpoR

The cytoplasmic region of the human EpoR contains 235 residues, and this region is responsible for JAK2 binding. There is no structural information available for this region so far probably due to the protein dynamics. Sequence analysis and functional studies revealed that there are two regions, namely, box 1 and box 2 in the cytokine receptors. For the EpoR, box 1 is an eight-residue region containing two conserved prolines. Box 2 is a 10-residue motif following box 1 motif and this motif contains a conserved core formed by four residues (LEVL) (Jiang, He, Miyajima, & Wojchowski, 1996). Mutagenesis studies demonstrate that these regions are important for receptor activation and recruitment of JAK kinase (Chiba et al., 1992; Miura, Cleveland, & Ihle, 1993; Witthuhn et al., 1993). Further study also shows that the box 1 domain is required for the specification, recruitment, and activation of JAK2 (Jiang et al., 1996). An alanine-scanning experiment suggests that two domains close to the TM domain are sufficient for JAK2 binding, activation, and signal transduction (Pelletier et al., 2006). JAK2 is proposed to be the sole molecule at the downstream of EpoR signaling. The two JAK2-binding regions include the box 1, box 2, and a highly conserved glycine residue (G277) among cytokine receptors (Pelletier et al., 2006). The JAK kinases contain seven JAK-homology (JH) domains. The C-terminal region of JAK2 contains an active tyrosine kinase domain (JH1) and an adjacent pseudokinase domain (JH2). JH2 plays a negative regulatory role in kinase activity (Saharinen, Takaluoma, & Silvennoinen, 2000). EPO binding to EpoR can cause some conformational changes, which will affect JH2 of JAK2 to activate its kinase activity. One possible mechanism is to remove the inhibitory effect arisen from JH2 (Brooks et al., 2014).

Tyrosine residues in EpoR are phosphorylated, and the resulting residues can interact with SH domain-containing proteins. There are eight tyrosine residues in the cytoplasmic region of mouse EpoR including Y343, Y401, Y429, Y431, Y443, Y460, Y464, and Y479 that can be phosphorylated by kinases (Montoye, Lemmens, Catteeuw, Eyckerman, & Tavernier, 2005; Richmond et al., 2005). These same residues are present in human EpoR with a different sequence number (Montoye et al., 2005). Phosphorylation on different residues may cause receptor binding to different SH2-containing transcription factors, resulting in different signaling pathways (Wu, Klingmuller, Acurio, Hsiao, & Lodish, 1997). A study has shown that residue Y343 is important for downstream signaling through binding to STAT5 (Zang et al., 2001). The binding is confirmed to be rapid and transient (Chin et al., 1996). Further study indicates that Y343F mutation has no couple to EPO binding. Peptide competition study confirms that phosphorylated Y343 is a binding site for STAT5. Phosphorylated peptide can disrupt STAT5 DNA-binding activity (Quelle et al., 1996). Y432 is critical for activation of STAT1 and STAT3, which is verified by using chimers of human EpoR mutants in its cytoplasmic region fused with the extracellular region of the granulocyte colony-stimulating factor receptor (Kirito et al., 2002). Phosphorylated Y429 of EpoR is important for binding to hematopoietic protein tyrosine phosphatase (SH-PTP1). Mutation in Y429 abolishes the interaction with SH-PTP1 and the cells harboring such mutant are hypersensitive to EPO and exhibit prolonged autophosphorylation of JAK2, suggesting that receptor binding to SH-PTP1 is important in terminating proliferation signals (Klingmuller, Lorenz, Cantley, Neel, & Lodish, 1995). Phosphorylation of Y443, Y460, and Y464 may be important for calcium channel activation (Miller et al., 1999). Phosphorylated Y460 interacts with SH2 domain of CrkL—a member of the Crk family of adaptor proteins, leading to activation of kinases such as ERK (Arai, Kanda, Nosaka, Miyasaka, & Miura, 2001). Phosphorylated Y429, Y431, and Y479 residues in EpoR bind to p85 subunit of PI3 kinase upon EPO stimulation. The binding is sufficient for mediating EPO-dependent EpoR internalization (Sulahian, Cleaver, & Huang, 2009). Y401 is a binding site for several SH2-containing proteins. Y401 also can be the direct binding site for cytokine-inducible SH2 protein-3 (CIS3). Deletion of this residue reduces the inhibitory effect of CIS3 (Sasaki et al., 2000). Both Y401 and the intact SH2 domain of CIS are required for the growth inhibitory effect of CIS on EpoR-dependent cell line (Ketteler et al., 2003). Phosphorylated Y401 binds to CIS3 with a binding constant

(K_d) of 9.5 μM and phosphorylated Y429 and Y431 show a ninefold higher affinity than Y401 (Hortner, Nielsch, Mayr, Heinrich, & Haan, 2002). Y401 is a major binding site for SH2 inositol 5-phosphatase 1 (Ship 1) using a panel of deletion mutants (Mason, Beattie, Liu, Dumont, & Barber, 2000). Y429 and Y431 play a role in recruiting Ship 1 (Mason et al., 2000). Using coimmunoprecipitation and other biochemical assays, phosphorylated Y401 has been shown to be critical for binding to Syp—a SH2-containing phosphatase (Tauchi et al., 1996). Although there is still no structure of SH2 domain-containing protein in complex with a phosphorylated peptide of EpoR available, the structures of SH2 domain in complex with phosphorylated peptides from other proteins have been determined using both X-ray crystallography and NMR spectroscopy (Fig. 4). For example, the structures of mouse SOCS3 in complex with a phosphopeptide from gp130 receptor have been determined (Babon et al., 2006; Bergamin, Wu, & Hubbard, 2006). Based on the structural studies, the phosphorylated tyrosine residues are critical for the binding with the SH2 domain through both hydrophobic and charge–charge interactions (Fig. 4). Some residues close to the phosphorylated tyrosine are also important for interactions with SH2 domain (Babon et al., 2006).

6. CONCLUSION AND FUTURE DIRECTIONS

EpoR is critical for EPO-induced signaling transduction. Many studies have been conducted to understand the mechanism of the signal transduction and the molecules that are involved in downstream signaling transductions. Several assay systems and animal models have been set up to validate the mechanisms. Despite these findings, further structural studies are still needed to understand the mechanisms. EPO binding to the extracellular domain has been well characterized by both biochemical and structural studies which have shown that EPO binding to the extracellular domain will cause conformational changes. It has been suggested that the TM domain is critical for receptor dimerization and activation. A recent structural report using diabodies shows that the TM domain dimerization may not be required for receptor activation, which is different from the previous understanding of EpoR activation. How the conformational changes induced by EPO binding to the receptor are transferred to the cytoplasmic region through the TM domain is still not well understood. Whether the TM domain dimerization is critical for receptor activation needs more

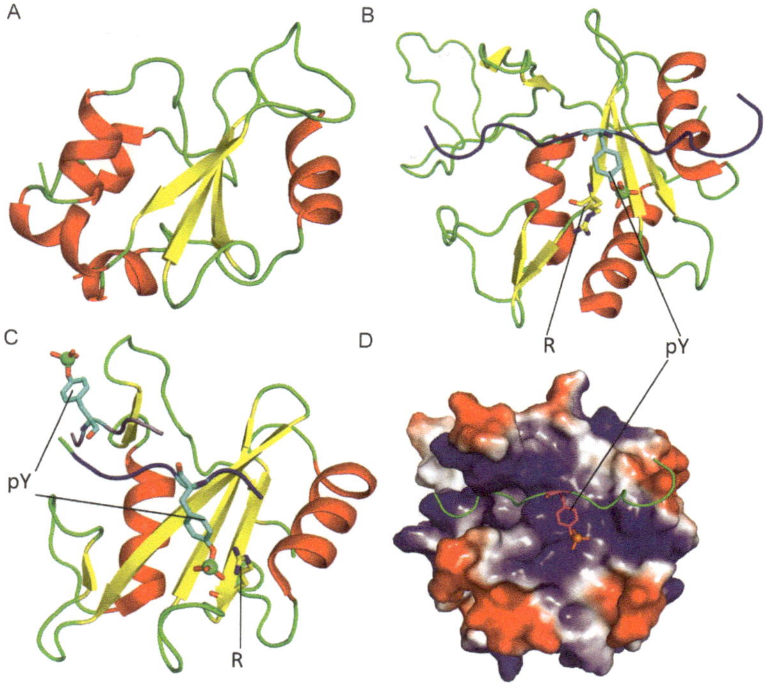

Fig. 4 Structures of the SH2 domain and its complex with phosphorylated tyrosine-containing peptide. (A) Structure of the SH2 domain of STAT5. The structure is from protein data bank (http://www.pdb.org) with access number 1Y1U. Only the SH2 domain is shown. (B) Structure of SH2–peptide complex. The structure of the SOCS3 in a complex with a phosphopeptide from the gp130 receptor (PDB ID 2BBU) is shown. (C) Structure of the SHP-2 domain in a 1:2 complex with RVIpYFVPLNR peptide. The two peptides containing a phosphorylated tyrosine (pY) are shown in *blue* and *light blue*, respectively. The complex structure (PDB ID 3TKZ) is shown. (D) Surface presentation of the SOCS3/peptide complex. The structure is the same as Fig. 4B, and SOCS3 is shown surface-charge representation. Positively and negatively charged surfaces are shown in *blue* and *red*, respectively. The phosphorylated peptides are shown in *sticks*, and the phosphate group is highlighted as a *sphere*. All the figures are made using PyMOL (http://www.pymol.org).

structural information to validate. A structural study on a construct containing both the extracellular domain and the TM domain will be helpful to understand the role of the TM domain in signal transduction. In addition, EpoR has several tyrosine residues that can be phosphorylated by kinases in its cytoplasmic region. Phosphorylation at different sites showed various binding affinity to SH2-containing proteins, which can induce diverse signaling pathways. Further structural studies of SH2-containing proteins in complex with the cytoplasmic region of EpoR will provide valuable insight into signal transduction. As no structure of the cytoplasmic region is available and structural study on this region will be useful to understand its function as well, structural information for different domains of EpoR will provide insight into their roles in signal transduction and provide useful information for structure-based drug discovery.

REFERENCES

Arai, A., Kanda, E., Nosaka, Y., Miyasaka, N., & Miura, O. (2001). CrkL is recruited through its SH2 domain to the erythropoietin receptor and plays a role in Lyn-mediated receptor signaling. *The Journal of Biological Chemistry*, *276*(35), 33282–33290.

Babon, J. J., McManus, E. J., Yao, S., DeSouza, D. P., Mielke, L. A., Sprigg, N. S., et al. (2006). The structure of SOCS3 reveals the basis of the extended SH2 domain function and identifies an unstructured insertion that regulates stability. *Molecular Cell*, *22*(2), 205–216.

Bazan, J. F. (1990). Structural design and molecular evolution of a cytokine receptor superfamily. *Proceedings of the National Academy of Sciences of the United States of America*, *87*(18), 6934–6938.

Bergamin, E., Wu, J., & Hubbard, S. R. (2006). Structural basis for phosphotyrosine recognition by suppressor of cytokine signaling-3. *Structure*, *14*(8), 1285–1292.

Brooks, A. J., Dai, W., O'Mara, M. L., Abankwa, D., Chhabra, Y., Pelekanos, R. A., et al. (2014). Mechanism of activation of protein kinase JAK2 by the growth hormone receptor. *Science*, *344*(6185), 1249783.

Brown, R. J., Adams, J. J., Pelekanos, R. A., Wan, Y., McKinstry, W. J., Palethorpe, K., et al. (2005). Model for growth hormone receptor activation based on subunit rotation within a receptor dimer. *Nature Structural & Molecular Biology*, *12*(9), 814–821.

Chiba, T., Kishi, A., Sugiyama, M., Amanuma, H., Machide, M., Nagata, Y., et al. (1992). Functionally essential cytoplasmic domain of the erythropoietin receptor. *Biochemical and Biophysical Research Communications*, *186*(3), 1236–1241.

Chin, H., Nakamura, N., Kamiyama, R., Miyasaka, N., Ihle, J. N., & Miura, O. (1996). Physical and functional interactions between Stat5 and the tyrosine-phosphorylated receptors for erythropoietin and interleukin-3. *Blood*, *88*(12), 4415–4425.

Constantinescu, S. N., Ghaffari, S., & Lodish, H. F. (1999). The erythropoietin receptor: Structure, activation and intracellular signal transduction. *Trends in Endocrinology and Metabolism*, *10*(1), 18–23.

Constantinescu, S. N., Huang, L. J., Nam, H., & Lodish, H. F. (2001). The erythropoietin receptor cytosolic juxtamembrane domain contains an essential, precisely oriented, hydrophobic motif. *Molecular Cell*, 7(2), 377–385.

Constantinescu, S. N., Keren, T., Socolovsky, M., Nam, H., Henis, Y. I., & Lodish, H. F. (2001). Ligand-independent oligomerization of cell-surface erythropoietin receptor is mediated by the transmembrane domain. *Proceedings of the National Academy of Sciences of the United States of America*, *98*(8), 4379–4384.

Ebie, A. Z., & Fleming, K. G. (2007). Dimerization of the erythropoietin receptor transmembrane domain in micelles. *Journal of Molecular Biology*, *366*(2), 517–524.

Elliott, S., Pham, E., & Macdougall, I. C. (2008). Erythropoietins: A common mechanism of action. *Experimental Hematology*, *36*(12), 1573–1584.

Filippakopoulos, P., Muller, S., & Knapp, S. (2009). SH2 domains: Modulators of nonreceptor tyrosine kinase activity. *Current Opinion in Structural Biology*, *19*(6), 643–649.

Hilton, D. J., Watowich, S. S., Katz, L., & Lodish, H. F. (1996). Saturation mutagenesis of the WSXWS motif of the erythropoietin receptor. *The Journal of Biological Chemistry, 271*(9), 4699–4708.

Hortner, M., Nielsch, U., Mayr, L. M., Heinrich, P. C., & Haan, S. (2002). A new high affinity binding site for suppressor of cytokine signaling-3 on the erythropoietin receptor. *European Journal of Biochemistry, 269*(10), 2516–2526.

Ihle, J. N. (1994). Signaling by the cytokine receptor superfamily just another kinase story. *Trends in Endocrinology and Metabolism, 5*(3), 137–143.

Ihle, J. N., & Kerr, I. M. (1995). Jaks and Stats in signaling by the cytokine receptor superfamily. *Trends in Genetics, 11*(2), 69–74.

Ihle, J. N., Thierfelder, W., Teglund, S., Stravopodis, D., Wang, D., Feng, J., et al. (1998). Signaling by the cytokine receptor superfamily. *The Annals of the New York Academy of Sciences, 865*, 1–9.

Jiang, N., He, T. C., Miyajima, A., & Wojchowski, D. M. (1996). The box1 domain of the erythropoietin receptor specifies Janus kinase 2 activation and functions mitogenically within an interleukin 2 beta-receptor chimera. *The Journal of Biological Chemistry, 271*(28), 16472–16476.

Kelley, L. L., Green, W. F., Hicks, G. G., Bondurant, M. C., Koury, M. J., & Ruley, H. E. (1994). Apoptosis in erythroid progenitors deprived of erythropoietin occurs during the G1 and S phases of the cell cycle without growth arrest or stabilization of wild-type p53. *Molecular and Cellular Biology, 14*(6), 4183–4192.

Ketteler, R., Moghraby, C. S., Hsiao, J. G., Sandra, O., Lodish, H. F., & Klingmuller, U. (2003). The cytokine-inducible Scr homology domain-containing protein negatively regulates signaling by promoting apoptosis in erythroid progenitor cells. *The Journal of Biological Chemistry, 278*(4), 2654–2660.

Kirito, K., Nakajima, K., Watanabe, T., Uchida, M., Tanaka, M., Ozawa, K., et al. (2002). Identification of the human erythropoietin receptor region required for Stat1 and Stat3 activation. *Blood, 99*(1), 102–110.

Klingmuller, U., Lorenz, U., Cantley, L. C., Neel, B. G., & Lodish, H. F. (1995). Specific recruitment of SH-PTP1 to the erythropoietin receptor causes inactivation of JAK2 and termination of proliferative signals. *Cell, 80*(5), 729–738.

Kubatzky, K. F., Liu, W., Goldgraben, K., Simmerling, C., Smith, S. O., & Constantinescu, S. N. (2005). Structural requirements of the extracellular to transmembrane domain junction for erythropoietin receptor function. *The Journal of Biological Chemistry, 280*(15), 14844–14854.

Kubatzky, K. F., Ruan, W., Gurezka, R., Cohen, J., Ketteler, R., Watowich, S. S., et al. (2001). Self assembly of the transmembrane domain promotes signal transduction through the erythropoietin receptor. *Current Biology: CB, 11*(2), 110–115.

Lai, P. H., Everett, R., Wang, F. F., Arakawa, T., & Goldwasser, E. (1986). Structural characterization of human erythropoietin. *The Journal of Biological Chemistry, 261*(7), 3116–3121.

Lamon, S., & Russell, A. P. (2013). The role and regulation of erythropoietin (EPO) and its receptor in skeletal muscle: How much do we really know? *Frontiers in Physiology, 4*, 176.

Li, Q., Wong, Y. L., Huang, Q., & Kang, C. (2014). Structural insight into the transmembrane domain and the juxtamembrane region of the erythropoietin receptor in micelles. *Biophysical Journal, 107*(10), 2325–2336.

Li, Q., Wong, Y. L., Yueqi Lee, M., Li, Y., & Kang, C. (2015). Solution structure of the transmembrane domain of the mouse erythropoietin receptor in detergent micelles. *Scientific Reports, 5*, 13586.

Lim, A. C., Ketchem, R. R., Borges, L., Carabeo, T., Carter, J., Hoover, J. E., et al. (2010). A diversity of antibody epitopes can induce signaling through the erythropoietin receptor. *Biochemistry, 49*(18), 3797–3804.

Lin, F. K., Suggs, S., Lin, C. H., Browne, J. K., Smalling, R., Egrie, J. C., et al. (1985). Cloning and expression of the human erythropoietin gene. *Proceedings of the National Academy of Sciences of the United States of America*, *82*(22), 7580–7584.

Livnah, O., Johnson, D. L., Stura, E. A., Farrell, F. X., Barbone, F. P., You, Y., et al. (1998). An antagonist peptide-EPO receptor complex suggests that receptor dimerization is not sufficient for activation. *Nature Structural Biology*, *5*(11), 993–1004.

Livnah, O., Stura, E. A., Johnson, D. L., Middleton, S. A., Mulcahy, L. S., Wrighton, N. C., et al. (1996). Functional mimicry of a protein hormone by a peptide agonist: The EPO receptor complex at 2.8 A. *Science*, *273*(5274), 464–471.

Livnah, O., Stura, E. A., Middleton, S. A., Johnson, D. L., Jolliffe, L. K., & Wilson, I. A. (1999). Crystallographic evidence for preformed dimers of erythropoietin receptor before ligand activation. *Science*, *283*(5404), 987–990.

Lu, X., Gross, A. W., & Lodish, H. F. (2006). Active conformation of the erythropoietin receptor: Random and cysteine-scanning mutagenesis of the extracellular juxtamembrane and transmembrane domains. *The Journal of Biological Chemistry*, *281*(11), 7002–7011.

Mason, J. M., Beattie, B. K., Liu, Q., Dumont, D. J., & Barber, D. L. (2000). The SH2 inositol 5-phosphatase Ship1 is recruited in an SH2-dependent manner to the erythropoietin receptor. *The Journal of Biological Chemistry*, *275*(6), 4398–4406.

Matthews, E. E., Thevenin, D., Rogers, J. M., Gotow, L., Lira, P. D., Reiter, L. A., et al. (2011). Thrombopoietin receptor activation: Transmembrane helix dimerization, rotation, and allosteric modulation. *The FASEB Journal*, *25*(7), 2234–2244.

Miller, B. A., Barber, D. L., Bell, L. L., Beattie, B. K., Zhang, M. Y., Neel, B. G., et al. (1999). Identification of the erythropoietin receptor domain required for calcium channel activation. *The Journal of Biological Chemistry*, *274*(29), 20465–20472.

Miura, O., Cleveland, J. L., & Ihle, J. N. (1993). Inactivation of erythropoietin receptor function by point mutations in a region having homology with other cytokine receptors. *Molecular and Cellular Biology*, *13*(3), 1788–1795.

Miyake, M., Goodison, S., Lawton, A., Zhang, G., Gomes-Giacoia, E., & Rosser, C. J. (2013). Erythropoietin is a JAK2 and ERK1/2 effector that can promote renal tumor cell proliferation under hypoxic conditions. *Journal of Hematology & Oncology*, *6*, 65.

Montoye, T., Lemmens, I., Catteeuw, D., Eyckerman, S., & Tavernier, J. (2005). A systematic scan of interactions with tyrosine motifs in the erythropoietin receptor using a mammalian 2-hybrid approach. *Blood*, *105*(11), 4264–4271.

Moraga, I., Wernig, G., Wilmes, S., Gryshkova, V., Richter, C. P., Hong, W. J., et al. (2015). Tuning cytokine receptor signaling by re-orienting dimer geometry with surrogate ligands. *Cell*, *160*(6), 1196–1208.

Pelletier, S., Gingras, S., Funakoshi-Tago, M., Howell, S., & Ihle, J. N. (2006). Two domains of the erythropoietin receptor are sufficient for Jak2 binding/activation and function. *Molecular and Cellular Biology*, *26*(22), 8527–8538.

Quelle, F. W., Wang, D., Nosaka, T., Thierfelder, W. E., Stravopodis, D., Weinstein, Y., et al. (1996). Erythropoietin induces activation of Stat5 through association with specific tyrosines on the receptor that are not required for a mitogenic response. *Molecular and Cellular Biology*, *16*(4), 1622–1631.

Remy, I., Wilson, I. A., & Michnick, S. W. (1999). Erythropoietin receptor activation by a ligand-induced conformation change. *Science*, *283*(5404), 990–993.

Richmond, T. D., Chohan, M., & Barber, D. L. (2005). Turning cells red: Signal transduction mediated by erythropoietin. *Trends in Cell Biology*, *15*(3), 146–155.

Saharinen, P., Takaluoma, K., & Silvennoinen, O. (2000). Regulation of the Jak2 tyrosine kinase by its pseudokinase domain. *Molecular and Cellular Biology*, *20*(10), 3387–3395.

Sasaki, A., Yasukawa, H., Shouda, T., Kitamura, T., Dikic, I., & Yoshimura, A. (2000). CIS3/SOCS-3 suppresses erythropoietin (EPO) signaling by binding the EPO receptor and JAK2. *The Journal of Biological Chemistry*, *275*(38), 29338–29347.

Seubert, N., Royer, Y., Staerk, J., Kubatzky, K. F., Moucadel, V., Krishnakumar, S., et al. (2003). Active and inactive orientations of the transmembrane and cytosolic domains of the erythropoietin receptor dimer. *Molecular Cell*, *12*(5), 1239–1250.

Sulahian, R., Cleaver, O., & Huang, L. J. (2009). Ligand-induced EpoR internalization is mediated by JAK2 and p85 and is impaired by mutations responsible for primary familial and congenital polycythemia. *Blood*, *113*(21), 5287–5297.

Syed, R. S., Reid, S. W., Li, C., Cheetham, J. C., Aoki, K. H., Liu, B., et al. (1998). Efficiency of signalling through cytokine receptors depends critically on receptor orientation. *Nature*, *395*(6701), 511–516.

Tauchi, T., Damen, J. E., Toyama, K., Feng, G. S., Broxmeyer, H. E., & Krystal, G. (1996). Tyrosine 425 within the activated erythropoietin receptor binds Syp, reduces the erythropoietin required for Syp tyrosine phosphorylation, and promotes mitogenesis. *Blood*, *87*(11), 4495–4501.

Waters, M. J., Brooks, A. J., & Chhabra, Y. (2014). A new mechanism for growth hormone receptor activation of JAK2, and implications for related cytokine receptors. *JAKSTAT*, *3*, e29569.

Wilson, I. A., & Jolliffe, L. K. (1999). The structure, organization, activation and plasticity of the erythropoietin receptor. *Current Opinion in Structural Biology*, *9*(6), 696–704.

Witthuhn, B. A., Quelle, F. W., Silvennoinen, O., Yi, T., Tang, B., Miura, O., et al. (1993). JAK2 associates with the erythropoietin receptor and is tyrosine phosphorylated and activated following stimulation with erythropoietin. *Cell*, *74*(2), 227–236.

Wrighton, N. C., Farrell, F. X., Chang, R., Kashyap, A. K., Barbone, F. P., Mulcahy, L. S., et al. (1996). Small peptides as potent mimetics of the protein hormone erythropoietin. *Science*, *273*(5274), 458–464.

Wu, H., Klingmuller, U., Acurio, A., Hsiao, J. G., & Lodish, H. F. (1997). Functional interaction of erythropoietin and stem cell factor receptors is essential for erythroid colony formation. *Proceedings of the National Academy of Sciences of the United States of America*, *94*(5), 1806–1810.

Zang, H., Sato, K., Nakajima, H., McKay, C., Ney, P. A., & Ihle, J. N. (2001). The distal region and receptor tyrosines of the Epo receptor are non-essential for in vivo erythropoiesis. *The EMBO Journal*, *20*(12), 3156–3166.

CHAPTER TWO

Analysis of the Asymmetry of Activated EPO Receptor Enables Designing Small Molecule Agonists

Frank Guarnieri[1]
Virginia Commonwealth University School of Medicine, Richmond, VA, United States
PAKA Pulmonary Pharmaceuticals, Acton, MA, United States
[1]Corresponding author: e-mail address: frankguarnieri@yahoo.com

Contents

Abstract

Amgen solved the high-resolution cocrystal structure of erythropoietin (EPO) bound to the extracellular part of the receptor (EPOR) in 1998, which reveals that the EPO–EPOR interaction surface is formed by 11 salt bridges, 17 H-bonds, and 2 hydrophobic clusters centered at a pair of crucial phenylalanines (F93). The EPOR has two domains, one that penetrates the membrane and a second extracellular domain that forms one arm of the binding site for the EPO ligand. The complete competent receptor-binding site is a homodimer of EPOR with the two arms forming a funnel-shaped cup where EPO binds. The two binding arms of the EPOR dimer meet at the membrane at a 120 degree angle, which Amgen characterizes as, "erythropoietin imposes a unique angular relationship and orientation that is responsible for optimal signaling." They come to this conclusion, because the EPOR cocrystallized with 2 equivalents of a 20 residue EPO mimetic peptide created at Robert Wood Johnson (RWJ) activates the receptor with a 3 order of magnitude reduction in potency, and the binding arms are forced to meet at the membrane with an angle of 180 degrees. The vast interaction surface between EPO and EPOR forms a singularly important three-dimensional structure responsible for hematopoietic stem cell proliferation and differentiation—this is Amgen's conclusion. This goal of this work

Vitamins and Hormones, Volume 105
ISSN 0083-6729
http://dx.doi.org/10.1016/bs.vh.2017.03.004

© 2017 Elsevier Inc.
All rights reserved.

is to present experimental and computational evidence that the Amgen structure is a postsignaling off-state and that the RWJ structure with the partially active peptide mimetics is an on-state. A detailed side-by-side comparison of the two structures will be presented along with literature evidence that calls into question the Amgen claim that their structure is a unique on-state. A computational fragment-based drug discovery method applied to the RWJ structure was used to locate and characterize a new predicted small molecule binding site and a fragment analysis was performed based on theories of asymmetry to create a proposed agonist with MW < 300. When this molecule was experimentally tested, it displaced radiolabeled EPO with nanomolar potency and transformed human hematopoietic stem cells into red blood cells with subnanomolar potency. Obviously, this small molecule makes none of the EPO–EPOR interactions that Amgen stated were essential for fully turning on the receptor and provides strong evidence that stabilizing receptor asymmetry, not specific interactions, is the critical factor needed for activating signal transduction. Finally, when the agonist was altered to remove the asymmetric component, it still was able to displace radiolabeled EPO in competition binding experiments, but it no longer activated the receptor.

1. INTRODUCTION

During the first half of the 20th century, it was generally believed that low blood oxygen levels directly stimulated the bone marrow to make new red blood cells. In 1950 Reissmann (1950) carried out surgical experiments to parabiotically join pairs of rats with an anastomosis, which results in a sharing of the circulatory system. When only one partner is subject to hypoxic conditions, both had elevated levels of erythropoiesis. The interpretation of these pioneering and difficult experiments is that hypoxia induces corrective action in the one rodent only, which is shared with the other rodent through the blood, thus providing the first compelling evidence that a circulating factor (erythropoietin or EPO) directly induces the creation of erythrocytes. Erslev (1953) demonstrated that normal rabbits would produce large amounts of reticulocytes 4–6 days after they were transfused with blood drawn from anemic animals, because "anemic" plasma presumably contains the factor required to replenish the red cells. Jacobsen and coworkers (Plzak, Fried, Jacobson, & Bethard, 1955) quantitated this with similar experiments using radiolabeled iron. Jacobsen then eliminated any remaining doubts by showing that the kidneys (Jacobson, Goldwasser, Fried, & Plzak, 1957) are the source of EPO by systemically removing one organ at a time, inducing anemia, and showing that restoration of erythrocytes occurred in all cases except when nephrectomy was performed. This was about 50 years after

Carnot (Carnot & Deflandre, 1906; No-Authors-Listed, 1972) proposed the radical and prescient notion that a hormone was responsible for red blood cell proliferation.

In the 1970s, it became clear that some forms of anemia could be effectively treated with EPO (Radtke et al., 1979; Ward, Kurnick, & Pisarczyk, 1971) instead of transfusions. It took 7 years to develop a protocol for producing high purity EPO (Goldwasser & Kung, 1971; Miyake, Kung, & Goldwasser, 1977) in milligram quantities, which is essential (Browne et al., 1986; Dukes, Egrie, Strickland, Browne, & Lin, 1986; Egrie, Browne, Lai, & Lin, 1985; Egrie et al., 1986; Jacobs et al., 1985; Lai, Everett, Wang, Arakawa, & Goldwasser, 1986; Lin, n.d.; Lin et al., 1986; Lin et al., 1985) for amino acid sequencing, DNA probe construction, genomic library screening, cloning, and protein expression. Clinical trials in 1987 conducted by Amgen (Eschbach, Egrie, Downing, Browne, & Adamson, 1987) showed that recombinant EPO restored hematocrit to normal levels in renoprival patients suffering from anemia with no toxicities. Recombinant EPO has been used to treat cancer chemotherapy-induced anemia and renal failure for decades making it a multibillion dollar per year product (Locatelli & Vecchio, 2001; Pavlou & Reichert, 2004) for Amgen and Johnson & Johnson. Solving the fundamental biomedical research question of how bone marrow hematopoietic stem cells are transformed into new red blood cells, resulted in the creation of the first blockbuster commercial product using recombinant DNA, and thus it also launched the biotechnology revolution. Additional interesting details of the historical development of EPO (Guarnieri, 2015a) have been described previously.

2. THE SEARCH FOR EPO MIMETICS

Even before EPO was brought to market, researchers were investigating the possibility of creating mimetics that had the potential to be alternative competitive products. Interestingly, the experiments were mostly interpreted to mean that the creation of a viable mimetic was unlikely or impossible. Carbohydrates (Jacobs et al., 1985; Lin et al., 1985) make up approximately 40% of the molecular weight of EPO. Even partial interference with this posttranslational modification (Imai et al., 1990) can result in an EPO that is 500-fold (Delorme et al., 1992; Wasley et al., 1991; Yamaguchi et al., 1991) less active in vivo. The importance of these findings motivated Amgen (Egrie & Browne, 2001) to engineer two additional N-linked glycosylation sites, creating an artificial sugar-laden EPO and a new second-generation FDA-approved product with threefold longer serum half-life called Aranesp. The vital importance of posttranslational modification seems to suggest that endogenous natural EPO requires glycosylation to be fully active leading to the conclusion that making a truly functional mimetic is likely impossible.

The Robert Wood Johnson (RWJ) group produced an especially impressive body of work that included the discovery of a 20 amino acid (EMP1) partial agonist EPO (Wrighton et al., 1996) mimetic, the high-resolution (Livnah et al., 1996) cocrystal structure of the EMP1 dimer with the extracellular part of the receptor (EPOR), and the demonstration that EMP1-PEG-EMP1 is a full (Johnson et al., 1997) agonist. This last result provided conclusive proof that EPO was not the only molecule that can uniquely fully activate the EPOR and by extension proves that posttranslational glycosylation is not actually required. It is important to realize that EMP1-PEG-EMP1 molecule is about the size of 60 amino acids, which led the RWJ group to conclude that this was the minimum size for fully activating the receptor. This size requirement is further supported by their mutagenesis experiments showing that EPO contact with F93 (Middleton et al., 1996) on both receptors is essential for receptor activation—the F93 pair is separated by over 20A. Merck (Qureshi et al., 1999) and Scripps (Goldberg et al., 2002) created large complicated organic molecules that are weak agonists demonstrating that nonpeptides could activate EPOR, but confirming the RWJ conclusion that only something the size of ~60 residues could be a full agonist. Additional interesting details of these developments have been described (Guarnieri, 2015a) previously.

3. COMPARING THE AMGEN AND RWJ WORK ON EPO

Amgen (Syed et al., 1998) states that EPO induces EPOR into the unique fully active signal transducing conformational state that can be seen in the one special receptor orientation manifested in their EPO–EPOR cocrystal structure. The RWJ EMP1-EMP1-EPOR does not have this unique receptor arrangement, which Amgen explains is the reason why the EMP1 EPO mimetic peptide is only a partial agonist. The homodimer arms that bind the ligand are 120 degrees in the Amgen structure and 180 degrees in the RWJ structure. Since Amgen stresses these issues so strongly and claims that it is the essence of the on-state, it is important to clearly understand what they mean, so it is summarized in Figs. 1 and 2. The crucial F93 pair is shown in Fig. 3. When F93 is mutated EPO and EMP1 binding are both negated,

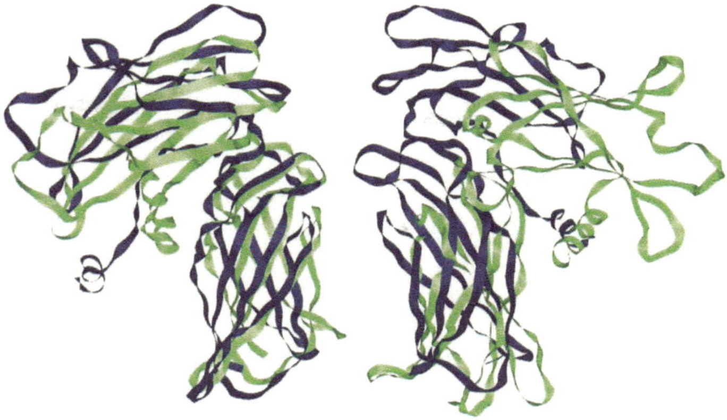

Fig. 1 The RWJ EPOR structure is shown in *purple ribbons* with the partial agonist EMP1 peptides removed. The Amgen EPOR structure is shown in *green ribbons*. Residues 202–204 on the *left side* were used to align the structures—no residues were used from the *right side*. The *bottom* domain inserts into the membrane and the *top* domain binds the ligand.

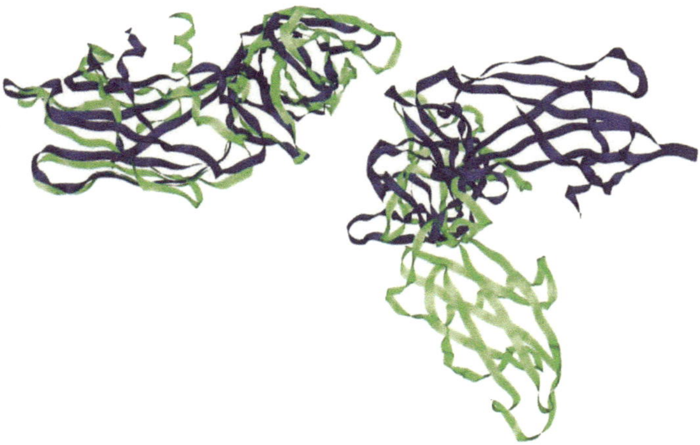

Fig. 2 This is just Fig. 1 rotated down 90 degrees. The membrane insertion domains are pointed directly back into the page and the EPO-binding domain is pointed toward the reader. The critical point is that the *green Amgen structure* has a 120 degree angle between the homodimer EPO-binding domains. The *purple RWJ structure* has a 180 degree angle between the EPO-binding domains. Amgen highlights this result as the critical difference between their fully activated state and the RWJ partially active state.

which leads to the natural conclusion that this is an absolutely essential ligand–protein interaction that must be preserved. The interatomic distance between the F93 pair is >20A. This finding definitively rules out the possibility that any small molecule (defined as MW < 600) can activate EPOR.

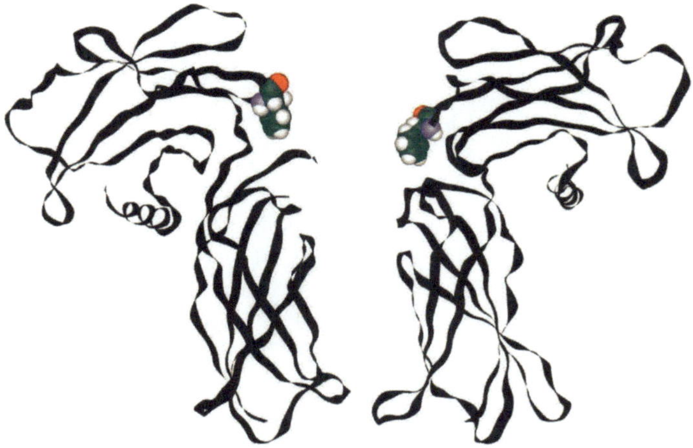

Fig. 3 This is the Amgen EPOR structure in *black ribbons*. The F93 pair is shown in space fill. The F93 pair is separated by >20A, so if interacting with both is essential for activating the receptor then it is impossible to make a small molecule agonist. The F93 pair is also separated by >20A in the RWJ EPOR structure.

4. ACTIVATION REQUIRES INDUCED ASYMMETRY NOT SPECIFIC INTERACTIONS

In 1992 Watowich et al. (1992) showed that the R130C mutation caused a cross-linked constitutively active EPOR–EPOR homodimer. Fig. 4 shows that the R130 pair is separated by over 27A in the EPO–EPOR structure. The Amgen group (Syed et al., 1998) explains this away by stating, "Signaling in this case might be the result of a subset of receptor molecules dimerizing intracellularly." An alternative simpler explanation is that the on-state requires inducing more intimate contact between the two halves of the dimer. Figs. 5 and 6 show a comparative interreceptor distance analysis between the Amgen and RWJ structure. There are several unique intimate interactions at the center of the RWJ receptor conformation, while the two receptors in the Amgen conformation essentially are noninteracting.

I believe that a more cogent hypothesis that explains all of the data is—receptor activation requires stabilizing an intrinsic asymmetry between the homodimers that only exists when the two halves come together to form an intimate interaction responsible for transducing the signal. This hypothesis actually leads to a specific conclusion that is experimentally testable with a computer simulation. The explanation is a bit complicated, so I will try to make it as simple as possible. Amgen assumes that the RWJ EMP1–EPOR structure is completely twofold symmetric, because it appears to look that

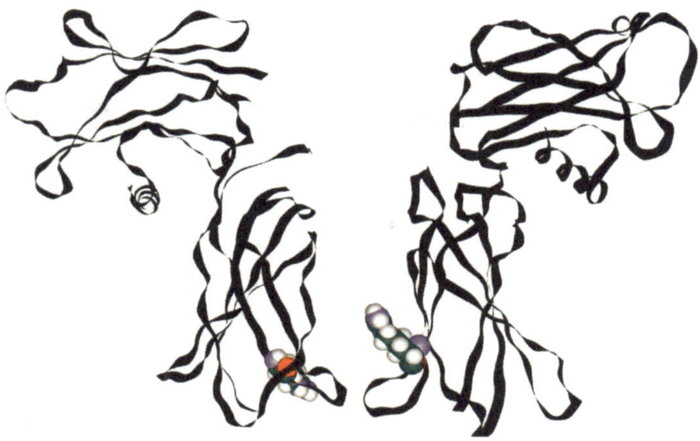

Fig. 4 This is the Amgen EPOR structure in *black ribbons*. The R130 pair shown in space fill is separated by >25A. The R130C mutation results in a disulfide bridge and constitutive activation of EPOR. This result calls into question the Amgen claim that their structure represents a unique active state.

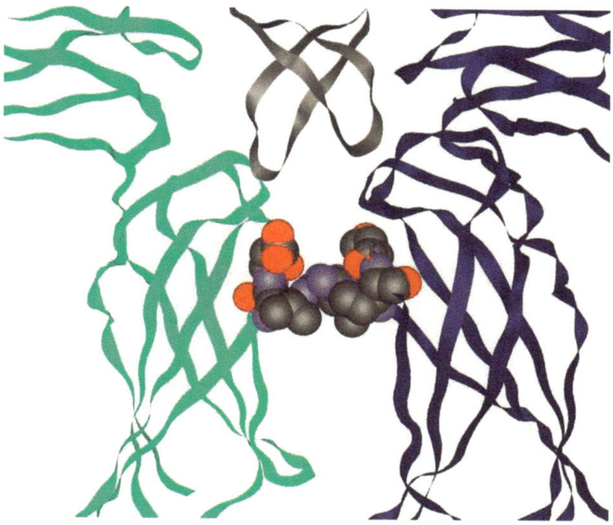

Fig. 5 This is the RWJ structure with the EPOR dimer shown in *light and dark blue ribbons* and the EMP1 peptide ligand dimer in *gray ribbons*. R178 from one receptor is in VDW contact with L175 of the other receptor forming a homodimer interacting quartet at the *center* of the receptor. The E176 from both receptors packs against this quartet.

way and I suspect that even the RWJ folks would not dispute this. This leads Amgen to conclude that the RWJ structure represents a state that can never be fully activated, because it apparently has symmetric ligand binding. In the last line of their paper (Syed et al., 1998) they state, "Asymmetrical

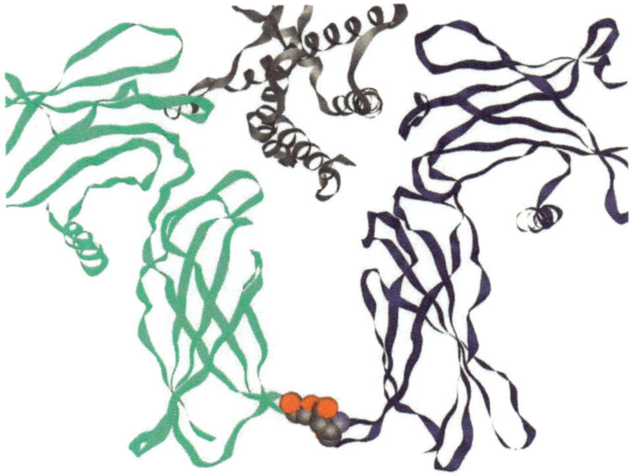

Fig. 6 This is the Amgen structure with the EPOR dimer shown in *light and dark blue ribbons* and EPO in *gray ribbons*. The EPOR homodimers are separated by >8A everywhere except at the bottom, where D133 is in VDW contact with S135.

molecules might be required to achieve these orientations most efficiently." By these orientations, they mean the EPOR orientation induced by the binding of EPO not EMP1. In any case, they are explicitly stating that the RWJ structure is symmetric and that is the reason why it is only partially active instead of fully active. I hypothesize that Amgen is correct that asymmetry is an essential component of receptor activation, but I also hypothesize that Amgen is wrong about the RWJ structure—it has subtle intrinsic asymmetry at the point of intimate contact and this is an on-state of the receptor. The fundamental question is—can such an asymmetry be demonstrated.

5. SIMULATED ANNEALING OF CHEMICAL POTENTIAL

How do small organic molecules, typically referred to as fragments, bind to proteins? The formally correct answer is that they will bind with a Boltzmann distribution, $\exp[-E/RT]$ with $E=$ energy, $R=$ universal gas constant, and $T=$ temperature. Imagine that there is also an exchange of mass. A typical example is when a concentrated salt solution is put into contact with a dilute salt solution through a water-permeable membrane. The high chemical potential of the concentrated solution induces water to come across the membrane in an attempt to make the ion concentrations of both solutions the same. In this case, the Boltzmann expression must be augmented with this chemical potential (u) difference so the expression becomes $\exp[-E/RT + u/RT]$. Monte Carlo simulations using these exponential expressions are fairly standard in the computational chemistry community.

Imagine that a protein is placed in a simulation cell and put into contact with a bath of organic fragments or water through a membrane. Further imagine that the chemical potential of the protein is set to a very high value relative to the bath. These conditions will result in fragments rapidly crossing the membrane and entirely filling the protein simulation cell including every buried crevice within the protein that is large enough to accommodate a fragment. This will happen even if the protein–fragment interaction energy is highly repulsive if the chemical potential is set high enough. Now imagine that the chemical potential is slowly and gradually reversed. As the chemical potential of the bath becomes larger and larger, fragments from the protein simulation cell cross the membrane and return to the bath. What is very interesting and highly relevant is the fact that every protein has localized high affinity sites for any given fragment and this fragment continues to hang on and bind to these sites even as the rest of the protein is vacated. There is a subset of especially important localized sites that have a high affinity for a broad diversity of fragments and a low affinity for water. It turns out that this is a defining characteristic of a small molecule binding site.

The process of setting the chemical potential very high and then gradually lowering it to find all of the high affinity binding sites on a protein is called (Guarnieri, 2004; Guarnieri & Mezei, 1996) simulated annealing of chemical potential (SACP) (Bradbury, Stump, Guarnieri, & Berk, 2011; Clark, Guarnieri, Shkurko, & Wiseman, 2006; Guarnieri, 2015b; Kulp, Blumenthal, Wang, Bryan, & Guarnieri, 2012; Kulp, Kulp, Pompliano, & Guarnieri, 2011; Moffett et al., 2011; Vallee et al., 2014). SACP simulations using furan, methylacetate, and methanol as fragments were carried out on the RWJ structure with the EMP1 peptide removed. Fig. 7 shows the RWJ structure at high chemical potential in contact with a bath of furan. When the chemical potential is dramatically lowered, all of the fragments vacate the protein simulation cell with only high affinity clusters remaining that represent a Boltzmann distribution of binding fragments (Fig. 8).

Recall, that the purpose of conducting these simulation was to find evidence for the proposed asymmetry in the RWJ structure. When we focus in where the homodimers interact, we see something quite remarkable—the fragment distributions have asymmetry (Fig. 9). Even better, the asymmetry presents as a stereocenter on one side that is nonexistent on the other side. Chirality is one of the most fundamental asymmetric motifs in chemistry and

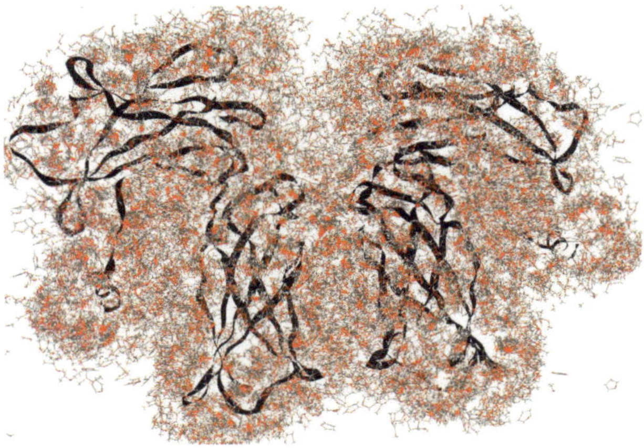

Fig. 7 The RWJ structure is in *black ribbons*. At high chemical potential fragments bind and completely cover every part of the protein.

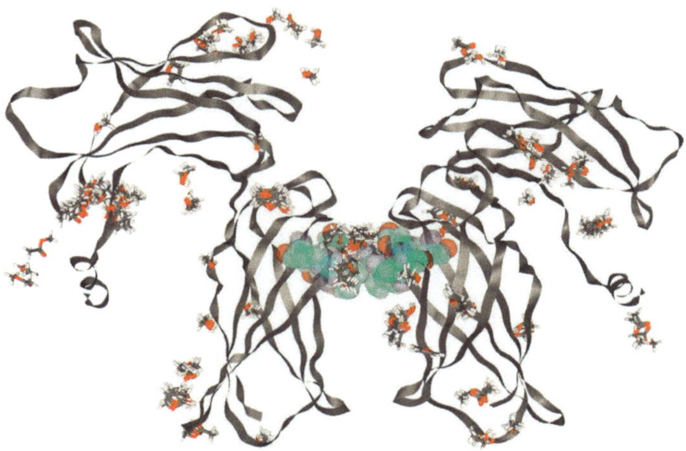

Fig. 8 The RWJ structure is in *gray ribbons* with interreceptor contact residues R178, L175, and E176 highlighted. The methanol, furan, and methylacetate high affinity binding distributions predicted from the SACP computer simulations at low chemical potential. Something pretty remarkable happens at the EPOR–EPOR contact residues—the distributions are asymmetric.

biology. I interpreted these simulation results as providing compelling evidence supporting my hypothesis that the RWJ structure is the on-state. Additionally, I interpreted these fragment patterns as the core of a very small molecule EPOR agonist. The chiral center of the molecule is in essence the mirror image asymmetry that exists in the receptor, so together they cancel out the asymmetry and form an on-state that transduces the signal.

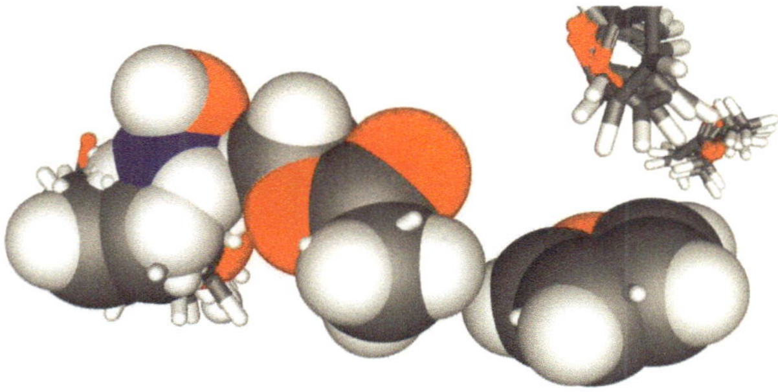

Fig. 9 Zooming in to the highlighted site in Fig. 8 where the receptors meet. The fragments to the far *left* and *right* are furan that the simulation finds symmetrically placed between the two receptors, which are linked by multiple methylacetates that contact both receptors. The *blue atom* is the carbon of a methanol fragment that merges into a carbon of one methylacetate linker. This just adds a hydroxyl group at this position, which forms a stereocenter.

The SACP simulations are extremely compelling, because the only inputs into the algorithm are the protein structure, fragments to be run, and force field. There is no human intervention whatsoever, so how can the asymmetry predicted by the simulation occur unless it exists in the structure. An important question is—what is the connection between the receptor and the asymmetry in the fragment distributions. Because they are binding to each other in a complementary fashion, they are mirror images of each other in some sense. When they are put together in a united package, the asymmetry of one should cancel out the asymmetry of the other. This means that together, the ligand and receptor form a new type of transient symmetric complex. In essence, the ligand has to stabilize the asymmetry of the activated receptor in order to indeed turn it on. Soon as the ligand is displaced, then the unstable asymmetric receptor immediately reverts back to a symmetric off-state. Alternatively, as in the Amgen crystal, shutdown mechanisms are immediately activated when EPO binds and the receptor dimer is pushed apart into an uncommunicating stable symmetric postsignaling off-state that releases the ligand.

6. EXPERIMENTAL CONFIRMATION OF THE THEORIES AND COMPUTATION PREDICTIONS

Fig. 10 shows the first molecule that I designed from the fragment distributions and a summary of the proliferation assays. The simulations show the symmetric binding of the furan fragments across the two receptors. Methyl acetate has a translational distribution across the two receptors that has subtle asymmetry, because the ketone moiety is somewhat biased to one side. The

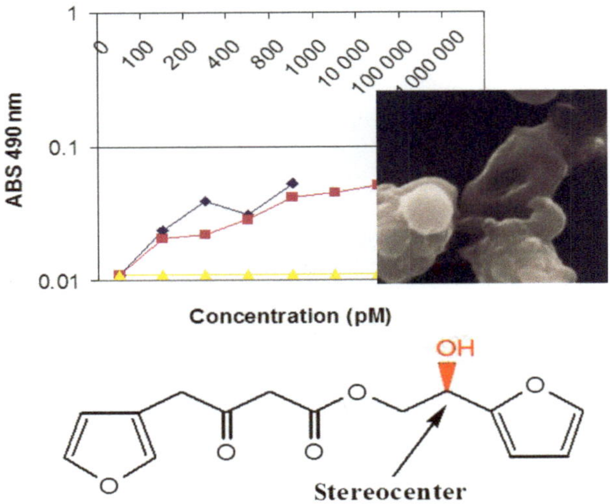

Fig. 10 A summary of the experimental results. The first compound that I designed from the fragment distributions is shown. The *top graph* is a standard cell proliferation assay using human hematopoietic stem cells showing that the compound is a subnanomolar agonist. The inset is a scanning electron micrograph taken after 4 days showing that the compound has induced differentiation into a red blood cell. These results were repeated many times.

methanol binding is striking. It exists only on one side of the molecule in the receptor binding pocket and not the other. The way it merges into the ester distribution creates a stereocenter on one side of the molecule and not the other. This is a critical asymmetric element and what I was hypothesizing would come out of the simulation. This molecule was synthesized and it displaced radiolabeled EPO in competition binding experiment with a nanomolar IC_{50}. The compound was then tested in standard assays and shown to proliferate human hematopoietic stem cells with subnanomolar potency. Finally, after 4 days scanning electron microscopy revealed the existence of a fully formed red blood cell. All of the experimental details for doing the competition binding with radiolabeled EPO and the proliferation assays using human hematopoietic stem cells have been previously described and the interested reader is referred to this chapter (Guarnieri, 2015a).

7. CONCLUSIONS

High-resolution protein crystal structures provide singularly important valuable information to the research community. They are snapshots of the protein interactions responsible for driving vital biological processes and Nobel prizes have been awarded for creating crystal structures of a K-channel (MacKinnon, 2004) and a GPCR (Hausch & Holsboer, 2012). We must remember, however, that any structure is just one snapshot of a dynamic biological process. A deeper understanding of the biology may be gained by putting the crystal structure into the larger context provided by the information from other types of experiments. Using EPO as an example, the most natural easiest interpretation of the EPO–EPOR cocrystal is that it must be the active state of the receptor. If we consider, however, that EPO binding almost immediately triggers EPOR shutdown, then we have to ask is this really the on-state or a postsignal transduction off-state. We also have to consider that the process of making high-resolution crystals intrinsically requires that the protein be in a highly stable motion-free state. EPO binding activates EPOR and simultaneously drives the receptor into a state that prepares it for ubiquitin degradation (Bulut, Sulahian, Ma, Chi, & Huang, 2011; Kang et al., 2011; Walrafen et al., 2005) even while it is still in the membrane. Rapidly turning off EPOR signaling is so important that recent research indicates that microRNA (Alural et al., 2014; Ferracin et al., 2013; Gonsalves et al., 2015; Rivkin et al., 2016; Sun et al., 2015; Zhou et al., 2016) processes are involved creating redundant shutdown mechanisms. Additionally, activated EPOR upregulates (Ren, Komatsu, Shimizu, Okada, & Miura, 1994) PLC-γ, and it is now known that PLC-γ downregulates (Kang et al., 2011) EPOR—so this is another shutdown loop. These considerations indicate that the transient EPO-bound receptor active state is unlikely to be stable and thus the EPO–EPOR cocrystal probably is a snapshot of an off-state. Simple investigation of the RWJ structure shows that the two homodimeric receptors make an intimate contact right at the center, while in the Amgen structure all contact is absent, with the homodimers being separated by >8A except for one receptor–receptor contact near the intracellular side. The receptor must be predominantly in the off-state in order to prevent unnecessary and uncontrolled growth, so it seems most likely that the two halves are kept apart in the inactive state. Agonist binding causes intimate interaction between the homodimers, signal transduction, and immediate transitioning to a separated off-state. Because the RWJ EPO mimetic peptides are partial agonists, the receptor does not have the normal activity that triggers the shutdown mechanisms so EPOR becomes trapped in a partial on-state. These considerations formed the basis of my hypotheses that, (1) the RWJ structure is an on-state, (2) the Amgen structure is an off-state, (3) the RWJ structure contains intrinsic subtle asymmetry at the point where the receptors meet, (4) a small molecule that binds and stabilizes this state with a complimentary mirror image asymmetry will be an agonist, and (5) removing the asymmetric component from the small molecule will transform it into an antagonist. A key concept intertwined with these hypotheses is that symmetry is intrinsically more stable than asymmetry and thus an off-state must be more symmetric than an on-state. Just to be clear, there has to be more than one off-state, the state prior to ligand binding is an inactive receptor, while the postsignaling off-state is an inactive receptor that triggers shutdown mechanisms. Interestingly, relative rotation of receptor domains has been shown to be an activation mechanism (Standfuss et al., 2011) in GPCRs. This concept provides further evidence that the RWJ structure is in an on-state and the Amgen structure is in an off-state. When the homodimers are positioned so that one maximally eclipses the other (Fig. 11), it is clear that a plane of symmetry exists that cuts the two dimers in half in the EPOR from Amgen. In the RWJ EPOR this plane of symmetry does not exist, because one dimer is rotated relative to the other much like the description of GPCR activation.

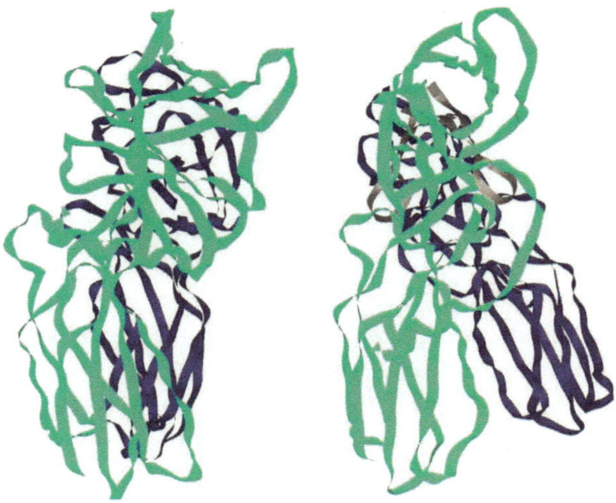

Fig. 11 The Amgen structure is on the *left* and the RWJ structure is on the *right*. The structures were rotated so that one receptor is pointing back and one is pointing forward. The object is to see how much one homodimer will eclipse the other homodimer as a measure of global symmetry. In the Amgen structure, there is an almost perfect plane of symmetry perpendicular to the page and slicing both receptors in half. The RWJ structure does not have this plane of symmetry and thus is intrinsically more asymmetric. The relative rotation of helices out of the plane has been described as part of the activation of GPCRs.

The key to pursuing this research beyond just theory and computer simulation was finding evidence from the simulations to support my hypothesis that the RWJ structure has intrinsic asymmetry at the point where the receptors meet. The simulations simply have no mechanism for spuriously creating such precise data at one point in the receptor, so when I analyzed the results and found the striking fragment data indicating a stereocenter on one side of the molecule and not the other, I decided to commit experimental resources to this project. The experiments were carried out when I was the head of biophysics at a small company called the Sarnoff Corporation. Anyone in a comparable position to mine knows that resources are extremely limited at small companies and there is only one chance to succeed or fail, so this is a major commitment that would have destroyed my fragment-based drug discovery efforts if it failed. I decided to go ahead with the project, because it seemed to me that it was impossible that the simulation could so specifically conform to my hypotheses and analysis, and I really believed that I was on the right track.

8. ACKNOWLEDGMENTS

The agonist presented in this work obviously cannot be in vivo active and thus could never be a drug. The ester bond in the middle of the structure is unstable and the molecule will fall apart when subjected to any stability studies. This work was meant to provide evidence that a very small EPO mimetic could be created based on asymmetry considerations and that the interactions deemed essential by RWJ and Amgen were in fact not really important. In order to move this forward as a drug discovery program, the project had to be restarted, but with the view of creating stable small molecules that conformed to the asymmetry principles. The one additional principle that I have not yet mentioned is that the molecule must also possess rotational degrees of freedom, because it has to ride with the receptor from the inactive to the active state. If the molecule is too rigid, it will get stuck somewhere in transit and prevent receptor activation. The company that I founded to use my fragment-based drug discovery technology and pursue the EPO project, Locus Pharmaceuticals, never understood these principles and the EPO project died. Current work on prolyl hydroxylase has apparently made the search for EPO mimetics unnecessary and my work obsolete. Prolyl hydroxylase operates on the hypoxia-inducible factor (HIF) by adding a hydroxyl group to HIF, which targets it for degradation. Thus, inhibiting prolyl hydroxylase will block HIF degradation extending its lifetime and triggering erythropoiesis. The biotechnology company's Fibrogen (Provenzano et al., 2016) and Akebia (Pergola, Spinowitz, Hartman, Maroni, & Haase, 2016) have positive clinical data showing that small molecule inhibitors of prolyl hydroxylase are effective at reversing anemia in renoprival patients. HIF is upregulated in response to low blood oxygen levels. So in 2016, we have returned to the early 20th century prevailing concept that the amount of oxygen in the blood is the force driving the creation of new red blood cells.

REFERENCES

Alural, B., Duran, G. A., Tufekci, K. U., Allmer, J., Onkal, Z., Tunali, D., et al. (2014). EPO mediates neurotrophic, neuroprotective, anti-oxidant, and anti-apoptotic effects via downregulation of miR-451 and miR-885-5p in SH-SY5Y neuron-like cells. *Frontiers in Immunology*, *5*, 475.

Bradbury, M. W., Stump, D., Guarnieri, F., & Berk, P. D. (2011). Molecular modeling and functional confirmation of a predicted fatty acid binding site of mitochondrial aspartate aminotransferase. *Journal of Molecular Biology*, *412*(3), 412–422.

Browne, J. K., Cohen, A. M., Egrie, J. C., Lai, P. H., Lin, F. K., Strickland, T., et al. (1986). Erythropoietin: Gene cloning, protein structure, and biological properties. *Cold Spring Harbor Symposia on Quantitative Biology*, *51*(Pt. 1), 693–702.

Bulut, G. B., Sulahian, R., Ma, Y., Chi, N. W., & Huang, L. J. (2011). Ubiquitination regulates the internalization, endolysosomal sorting, and signaling of the erythropoietin receptor. *The Journal of Biological Chemistry*, *286*(8), 6449–6457.

Carnot, P., & Deflandre, C. (1906). The hemopoietic activity of serum during the regeneration of blood. *Comptes Rendus de l'Académie des Sciences*, *1434*, 384–387.

Clark, M., Guarnieri, F., Shkurko, I., & Wiseman, J. (2006). Grand canonical Monte Carlo simulation of ligand-protein binding. *Journal of Chemical Information and Modeling*, *46*(1), 231–242.

Delorme, E., Lorenzini, T., Giffin, J., Martin, F., Jacobsen, F., Boone, T., et al. (1992). Role of glycosylation on the secretion and biological activity of erythropoietin. *Biochemistry*, *31*(41), 9871–9876.

Dukes, P. P., Egrie, J. C., Strickland, T. W., Browne, J. K., & Lin, F. K. (1986). Megakaryocyte colony stimulating activity of recombinant human and monkey erythropoietin. *Progress in Clinical and Biological Research*, *215*, 105–109.

Egrie, J. C., & Browne, J. K. (2001). Development and characterization of novel erythropoiesis stimulating protein (NESP). *British Journal of Cancer*, *84*(Suppl. 1), 3–10.

Egrie, J. C., Browne, J., Lai, P., & Lin, F. K. (1985). Characterization of recombinant monkey and human erythropoietin. *Progress in Clinical and Biological Research*, *191*, 339–350.

Egrie, J. C., Strickland, T. W., Lane, J., Aoki, K., Cohen, A. M., Smalling, R., et al. (1986). Characterization and biological effects of recombinant human erythropoietin. *Immunobiology*, *172*(3–5), 213–224.

Erslev, A. (1953). Humoral regulation of red cell production. *Blood*, *8*(4), 349–357.

Eschbach, J. W., Egrie, J. C., Downing, M. R., Browne, J. K., & Adamson, J. W. (1987). Correction of the anemia of end-stage renal disease with recombinant human erythropoietin. Results of a combined phase I and II clinical trial. *The New England Journal of Medicine*, *316*(2), 73–78.

Ferracin, M., Bassi, C., Pedriali, M., Pagotto, S., D'Abundo, L., Zagatti, B., et al. (2013). miR-125b targets erythropoietin and its receptor and their expression correlates with metastatic potential and ERBB2/HER2 expression. *Molecular Cancer*, *12*(1), 130.

Goldberg, J., Jin, Q., Ambroise, Y., Satoh, S., Desharnais, J., Capps, K., et al. (2002). Erythropoietin mimetics derived from solution phase combinatorial libraries. *Journal of the American Chemical Society*, *124*(4), 544–555.

Goldwasser, E., & Kung, C. K. (1971). Purification of erythropoietin. *Proceedings of the National Academy of Sciences of the United States of America*, *68*(4), 697–698.

Gonsalves, C. S., Li, C., Mpollo, M. S., Pullarkat, V., Malik, P., Tahara, S. M., et al. (2015). Erythropoietin-mediated expression of placenta growth factor is regulated via activation of hypoxia-inducible factor-1alpha and post-transcriptionally by miR-214 in sickle cell disease. *The Biochemical Journal*, *468*(3), 409–423.

Guarnieri, F., Computational protein probing to identify binding sites. US Patent No. 6,735,530, 2004. Issued May 11, 2004.

Guarnieri, F. (2015a). Designing a small molecule erythropoietin mimetic. *Methods in Molecular Biology*, *1289*, 185–210.

Guarnieri, F. (2015b). Designing an orally available nontoxic p38 inhibitor with a fragment-based strategy. *Methods in Molecular Biology*, *1289*, 211–226.

Guarnieri, F., & Mezei, M. (1996). Simulated annealing of chemical potential: A general procedure for locating bound waters. Application to the study of the differential hydration propensities of the major and minor grooves of DNA. *Journal of the American Chemical Society*, *118*, 8493–8494.

Hausch, F., & Holsboer, F. (2012). The seven pillars of molecular pharmacology: GPCR research honored with Nobel prize for chemistry. *Angewandte Chemie (International Ed. in English)*, *51*(49), 12172–12175.

Imai, N., Kawamura, A., Higuchi, M., Oh-eda, M., Orita, T., Kawaguchi, T., et al. (1990). Physicochemical and biological comparison of recombinant human erythropoietin with human urinary erythropoietin. *Journal of Biochemistry*, *107*(3), 352–359.

Jacobs, K., Shoemaker, C., Rudersdorf, R., Neill, S. D., Kaufman, R. J., Mufson, A., et al. (1985). Isolation and characterization of genomic and cDNA clones of human erythropoietin. *Nature*, *313*(6005), 806–810.

Jacobson, L. O., Goldwasser, E., Fried, W., & Plzak, L. (1957). Role of the kidney in erythropoiesis. *Nature*, *179*(4560), 633–634.

Johnson, D. L., Farrell, F. X., Barbone, F. P., McMahon, F. J., Tullai, J., Kroon, D., et al. (1997). Amino-terminal dimerization of an erythropoietin mimetic peptide results in increased erythropoietic activity. *Chemistry & Biology*, *4*(12), 939–950.

Kang, J. K., Chang, C. H., Nam, H. J., Kim, S. K., Ahn, K. J., Seok, H., et al. (2011). Down-regulation of erythropoietin receptor by overexpression of phospholipase C-gamma 1 is critical for decrease on focal adhesion in transformed cells. *Cellular Oncology (Dordrecht)*, *34*(1), 11–21.

Kulp, J. L., 3rd, Blumenthal, S. N., Wang, Q., Bryan, R. L., & Guarnieri, F. (2012). A fragment-based approach to the SAMPL3 challenge. *Journal of Computer-Aided Molecular Design*, *26*(5), 583–594.

Kulp, J. L., 3rd, Kulp, J. L., Jr., Pompliano, D. L., & Guarnieri, F. (2011). Diverse fragment clustering and water exclusion identify protein hot spots. *Journal of the American Chemical Society*, *133*(28), 10740–10743.

Lai, P. H., Everett, R., Wang, F. F., Arakawa, T., & Goldwasser, E. (1986). Structural characterization of human erythropoietin. *The Journal of Biological Chemistry*, *261*(7), 3116–3121.

Lin, F.K., *Production of Erythropoietin.* Patent # 1985002610, 1984 Filing Date.

Lin, F. K., Lin, C. H., Lai, P. H., Browne, J. K., Egrie, J. C., Smalling, R., et al. (1986). Monkey erythropoietin gene: Cloning, expression and comparison with the human erythropoietin gene. *Gene*, *44*(2–3), 201–209.

Lin, F. K., Suggs, S., Lin, C. H., Browne, J. K., Smalling, R., Egrie, J. C., et al. (1985). Cloning and expression of the human erythropoietin gene. *Proceedings of the National Academy of Sciences of the United States of America*, *82*(22), 7580–7584.

Livnah, O., Stura, E. A., Johnson, D. L., Middleton, S. A., Mulcahy, L. S., Wrighton, N. C., et al. (1996). Functional mimicry of a protein hormone by a peptide agonist: The EPO receptor complex at 2.8 Å. *Science*, *273*(5274), 464–471.

Locatelli, F., & Vecchio, L. D. (2001). Darbepoetin alfa.Amgen. *Current Opinion in Investigational Drugs*, *2*(8), 1097–1104.

MacKinnon, R. (2004). Nobel lecture. Potassium channels and the atomic basis of selective ion conduction. *Bioscience Reports*, *24*(2), 75–100.

Middleton, S. A., Johnson, D. L., Jin, R., McMahon, F. J., Collins, A., Tullai, J., et al. (1996). Identification of a critical ligand binding determinant of the human erythropoietin receptor. Evidence for common ligand binding motifs in the cytokine receptor family. *The Journal of Biological Chemistry*, *271*(24), 14045–14054.

Miyake, T., Kung, C. K., & Goldwasser, E. (1977). Purification of human erythropoietin. *The Journal of Biological Chemistry*, *252*(15), 5558–5564.

Moffett, K., Konteatis, Z., Nguyen, D., Shetty, R., Ludington, J., Fujimoto, T., et al. (2011). Discovery of a novel class of non-ATP site DFG-out state p38 inhibitors utilizing computationally assisted virtual fragment-based drug design (vFBDD). *Bioorganic & Medicinal Chemistry Letters*, *21*(23), 7155–7165.

No-Authors-Listed. (1972). Erythropoietin. *British Medical Journal*, *1*(5795), 263.

Pavlou, A. K., & Reichert, J. M. (2004). Recombinant protein therapeutics—Success rates, market trends and values to 2010. *Nature Biotechnology*, *22*(12), 1513–1519.

Pergola, P. E., Spinowitz, B. S., Hartman, C. S., Maroni, B. J., & Haase, V. H. (2016). Vadadustat, a novel oral HIF stabilizer, provides effective anemia treatment in nondialysis-dependent chronic kidney disease. *Kidney International*, *90*, 1115–1122.

Plzak, L. F., Fried, W., Jacobson, L. O., & Bethard, W. F. (1955). Demonstration of stimulation of erythropoiesis by plasma from anemic rats using Fe59. *The Journal of Laboratory and Clinical Medicine*, *46*(5), 671–678.

Provenzano, R., Besarab, A., Sun, C. H., Diamond, S. A., Durham, J. H., Cangiano, J. L., et al. (2016). Oral hypoxia-inducible factor prolyl hydroxylase inhibitor Roxadustat (FG-4592) for the treatment of anemia in patients with CKD. *Clinical Journal of the American Society of Nephrology*, *11*(6), 982–991.

Qureshi, S. A., Kim, R. M., Konteatis, Z., Biazzo, D. E., Motamedi, H., Rodrigues, R., et al. (1999). Mimicry of erythropoietin by a nonpeptide molecule. *Proceedings of the National Academy of Sciences of the United States of America*, *96*(21), 12156–12161.

Radtke, H. W., Claussner, A., Erbes, P. M., Scheuermann, E. H., Schoeppe, W., & Koch, K. M. (1979). Serum erythropoietin concentration in chronic renal failure: Relationship to degree of anemia and excretory renal function. *Blood*, *54*(4), 877–884.

Reissmann, K. R. (1950). Studies on the mechanism of Erythropoietic stimulation in parabiotic rats during hypoxia. *Blood*, *5*, 372–380.

Ren, H. Y., Komatsu, N., Shimizu, R., Okada, K., & Miura, Y. (1994). Erythropoietin induces tyrosine phosphorylation and activation of phospholipase C-gamma 1 in a human erythropoietin-dependent cell line. *The Journal of Biological Chemistry*, *269*(30), 19633–19638.

Rivkin, M., Simerzin, A., Zorde-Khvalevsky, E., Chai, C., Yuval, J. B., Rosenberg, N., et al. (2016). Inflammation-induced expression and secretion of microRNA 122 leads to reduced blood levels of kidney-derived erythropoietin and anemia. *Gastroenterology*, *151*, 999–1010.

Standfuss, J., Edwards, P. C., D'Antona, A., Fransen, M., Xie, G., Oprian, D. D., et al. (2011). The structural basis of agonist-induced activation in constitutively active rhodopsin. *Nature*, *471*(7340), 656–660.

Sun, Z., Wang, Y., Han, X., Zhao, X., Peng, Y., Li, Y., et al. (2015). miR-150 inhibits terminal erythroid proliferation and differentiation. *Oncotarget*, *6*(40), 43033–43047.

Syed, R. S., Reid, S. W., Li, C., Cheetham, J. C., Aoki, K. H., Liu, B., et al. (1998). Efficiency of signalling through cytokine receptors depends critically on receptor orientation. *Nature*, *395*(6701), 511–516.

Vallee, M., Vitiello, S., Bellocchio, L., Hebert-Chatelain, E., Monlezun, S., Martin-Garcia, E., et al. (2014). Pregnenolone can protect the brain from cannabis intoxication. *Science*, *343*(6166), 94–98.

Walrafen, P., Verdier, F., Kadri, Z., Chretien, S., Lacombe, C., & Mayeux, P. (2005). Both proteasomes and lysosomes degrade the activated erythropoietin receptor. *Blood*, *105*(2), 600–608.

Ward, H. P., Kurnick, J. E., & Pisarczyk, M. J. (1971). Serum level of erythropoietin in anemias associated with chronic infection, malignancy, and primary hematopoietic disease. *The Journal of Clinical Investigation*, *50*(2), 332–335.

Wasley, L. C., Timony, G., Murtha, P., Stoudemire, J., Dorner, A. J., Caro, J., et al. (1991). The importance of N- and O-linked oligosaccharides for the biosynthesis and in vitro and in vivo biologic activities of erythropoietin. *Blood*, 77(12), 2624–2632.

Watowich, S. S., Yoshimura, A., Longmore, G. D., Hilton, D. J., Yoshimura, Y., & Lodish, H. F. (1992). Homodimerization and constitutive activation of the erythropoietin receptor. *Proceedings of the National Academy of Sciences of the United States of America*, *89*(6), 2140–2144.

Wrighton, N. C., Farrell, F. X., Chang, R., Kashyap, A. K., Barbone, F. P., Mulcahy, L. S., et al. (1996). Small peptides as potent mimetics of the protein hormone erythropoietin. *Science*, *273*(5274), 458–464.

Yamaguchi, K., Akai, K., Kawanishi, G., Ueda, M., Masuda, S., & Sasaki, R. (1991). Effects of site-directed removal of N-glycosylation sites in human erythropoietin on its production and biological properties. *The Journal of Biological Chemistry*, *266*(30), 20434–20439.

Zhou, Y., Fang, L., Yu, Y., Niu, J., Jiang, L., Cao, H., et al. (2016). Erythropoietin protects the tubular basement membrane by promoting the bone marrow to release extracellular vesicles containing tPA-targeting miR-144. *American Journal of Physiology Renal Physiology*, *310*(1), F27–F40.

CHAPTER THREE

Endogenous Erythropoietin

Steven J. Korzeniewski[1], Athina Pappas
Wayne State University School of Medicine, Detroit, United States
[1]Corresponding author: e-mail address: skorzeni@med.wayne.edu

Contents

Abstract

We sought to briefly describe current models of endogenous erythropoietin (EPO) pleiotropic properties to make four points clear. *First*, endogenous EPO regulates erythroid cell apoptosis so that red blood cell production is balanced against the number of cells destroyed in order to maintain optimal tissue oxygen levels (i.e., consistent with provision of homeostatic functional signaling information). *Second*, preclinical and clinical studies alike provide additional evidence of other (i.e., extraerythropoietic) immune-related and growth/trophic properties. *Third*, EPO might also be increased as an antiinflammatory response to other proinflammatory cytokines, and not because it is an inflammatory protein, in which case, it would make the association between EPO and these other proteins an *epiphenomenon*. *Fourth*, on the other hand, EPO might also act as a tissue protector or it could reflect immaturity/vulnerability of the brain or of the systems responsible for protecting it. Each of these scenarios is plausible, and all are probably true in certain circumstances.

1. INTRODUCTION

In spite of the fascinating history of impressive accomplishments developing *exogenous* or recombinant human erythropoietin (rhEPO) for therapeutic purposes, only recently have clinical studies begun to examine

Vitamins and Hormones, Volume 105
ISSN 0083-6729
http://dx.doi.org/10.1016/bs.vh.2017.03.003
© 2017 Elsevier Inc.
All rights reserved.

information provided by circulating *endogenous* erythropoietin (EPO) about human diseases. Consequently, our understanding of human EPO is perhaps best described as nascent, albeit an appreciation now growing in fertile soils afforded by over a century of hypothesis-driven research and more recently by rapid technological advances.

This chapter aims to briefly describe how different types of information about endogenous EPO might relate to perturbations of *biological systems or physiologic processes* putatively involved in the pathogenesis of multiple human diseases, though our writing surely reflects our primary interests in perinatal brain injury. We use a systems epidemiology approach (Dammann, Gray, Gressens, Wolkenhauer, & Leviton, 2014) and we integrate clinical and preclinical evidence within a translational epidemiology framework (Khoury, Gwinn, & Ioannidis, 2010) together to describe putative pleiotropic actions or properties of endogenous EPO (Fig. 1).

1.1 A Brief History

James W. Fisher's review, *Landmarks in the Development of Erythropoietin* (Fisher, 2010), explains that the idea of a humoral substance controlling red blood cell production dates to the first years of the 20th century

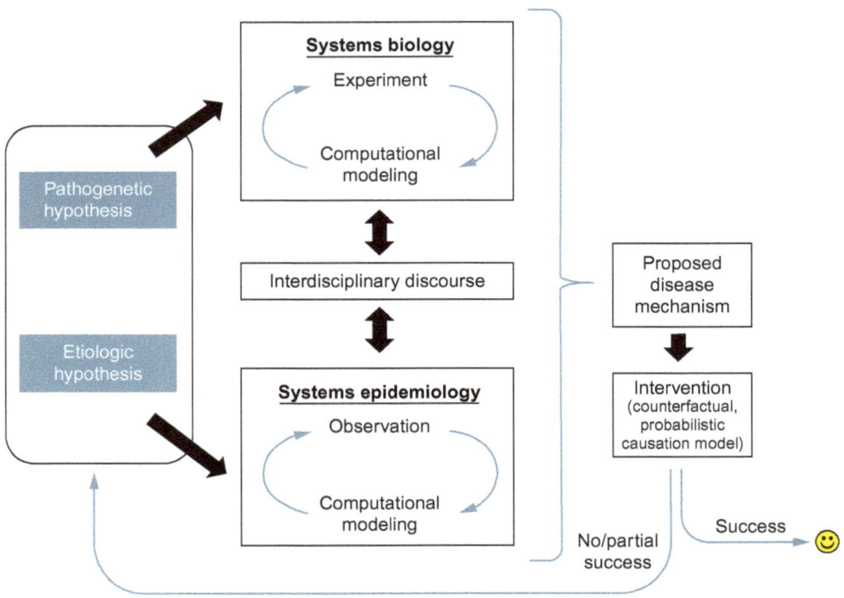

Fig. 1 Systems epidemiology framework (Dammann et al., 2014).

(Carnot & Deflandre, 1906). It followed observations that people living at high altitude had more viscous blood than those living at lower altitudes, but the concept was proposed by investigators working with animals who demonstrated that serum from a bleeding or hypoxic rabbit increased production of immature red blood cells when injected into a normal or healthy control rabbit. The name "Erythropoietin," coined later in 1948 by (Bonsdorff & Jalavisto, 1948), was derived from the Greek "erythro" meaning red and "poiein" meaning to make, or to indicate a stimulatory effect on growth or multiplication, in this case of red blood cells or erythrocytes.

The first quantitative assay for EPO was developed in 1955 (Plzak, Fried, Jacobson, & Bethard, 1955) a bioassay followed in 1961 (Cotes & Bangham, 1961) and a sensitive and specific radioimmunoassay followed with the availability of purified human urinary EPO (Garcia, Sherwood, & Goldwasser, 1979; Miyake, Kung, & Goldwasser, 1977). Around the same time during the 1960s–70s, investigators provided direct evidence that the kidney produced EPO (Fisher & Birdwell, 1961; Kuratows, Michalak, & Lewartowicz, 1961), then the liver was identified as the primary production site in the fetus (Zanjani, Poster, Burlington, Mann, & Wasserman, 1977). But it was not until the 1980s that exogenous or recombinant EPO was reported to be an effective treatment in anemic renal disease patients (Eschbach, Egrie, Downing, Browne, & Adamson, 1987; Fisher, 2010).

More nascent is the concept that EPO protects the brain against adversity. It was just over two decades ago in the 1990s that EPO receptors were first identified in the brain (Nagao, Masuda, Abe, Ueda, & Sasaki, 1992), the capacity for astrocytes to synthesize EPO was reported (Masuda et al., 1994), as was the first in vivo evidence that EPO has the capacity to protect the brain (Morishita, Masuda, Nagao, Yasuda, & Sasaki, 1997). The latter was provided by investigators who found that exposing cultured hippocampal and cerebral cortical neurons exposed to EPO exhibited less of the glutamate-induced cell death seen in other neurons not exposed to EPO.

As recently as 2003 came the first evidence that it also has the capacity to protect the myocardium against ischemia–perfusion-related injury (Calvillo et al., 2003). Even more recent is evidence that early administration of exogenous EPO increases the risk of retinopathy of prematurity, *an inflammatory disease* (Hellstrom, Smith, & Dammann, 2013) characterized by abnormal vasculogenesis.

2. WHAT IS ENDOGENOUS HUMAN EPO, WHAT DOES IT DO, AND HOW DO WE KNOW?

EPO is a glycoprotein produced primarily by the kidney in adults (Fisher & Birdwell, 1961; Kuratows et al., 1961) and by the liver in fetuses (Zanjani et al., 1977). About half its molecular weight is comprised of sugar moieties that protect it from degradation, and the mature circulating protein contains 165 amino acids (Fisher, 2010). Historically, EPO is described as regulating or initiating production of circulating red blood cells in concert with its receptor to maintain adequate tissue oxygenation.

But, stressed tissues can also produce EPO locally (Brines & Cerami, 2008); its receptors appear upregulated in response to insults like trauma, hypoxia, and/or infection, and both animal and human studies provide evidence that circulating concentrations are correlated with that of multiple inflammation-related proteins (Brines & Cerami, 2006). Hence, preclinical and clinical studies alike provide evidence that, in addition to actions on erythropoiesis, endogenous EPO has other (i.e., *extraerythropoietic*) immune-related properties or functions as well as growth/trophic (Brines & Cerami, 2006; Sasaki, 2003; Sasaki, Masuda, & Nagao, 2001) and neuropoietic functions. In fact, neuroregenerative effects in insects that lack hematopoiesis suggest a primordial phylogenetic role in neuroplasticity (Ostrowski, Ehrenreich, & Heinrich, 2011).

2.1 Regulation and Actions on Erythropoiesis

An oxygen deficit (hypoxia) initiates processes that involve EPO and its receptors to increase production of red blood cells (Maxwell, 2003). Two so-called hypoxia-inducible growth factors have been postulated to control EPO gene expression (Semenza & Wang, 1992; Wang & Semenza, 1993a, 1993b) but each seems to have unique targets (Hu, Wang, Chodosh, Keith, & Simon, 2003); only HIF-2 knockout mice are anemic (Gruber et al., 2007), and these mice also have reduced renal and hepatic EPO production (Rankin et al., 2007; Scortegagna et al., 2005). Signal transduction pathways have also been postulated to play some role in EPO messenger RNA production, though evidence provided in prior studies is somewhat inconsistent (Fisher, 2010).

The first report that prostaglandins increased EPO production was published in 1973 (Paulo, Wilkerson, Roh, George, & Fisher, 1973) and subsequent studies provided consistent evidence over the following decade.

2.2.2 Clinical Studies

EPO does not cross the human placenta, nor is it stored, so fetal plasma and amniotic fluid concentrations are indicative of fetal synthesis and elimination (Widness, Schmidt, & Sawyer, 1995). Moreover, nonerythroid cells in the CNS of mid-trimester human fetuses express functional EPO receptors suggesting an important functional role in the fetal brain (Li, Juul, Morris-Wiman, Calhoun, & Christensen, 1996). The half-life of EPO ranges from 2 to 4h in newborns and 4–8h in adult humans, but it is less clear during fetal development. Clinical studies provide evidence that amniotic fluid EPO concentrations increase exponentially during presumed fetal hypoxia, for instance, in the setting of preeclampsia, diabetes, and Rh-immunized pregnancies (see Teramo & Widness, 2009). Investigators studying hypertensive disorders of pregnancy or type-1 diabetes report that amniotic fluid EPO thresholds of 36–37 mU/mL can identify about half or more newborns who develop low umbilical artery pH (<7.2) and/or low pO_2 (<12 mmHg) with false-positive rates ranging from about 10% to 20% (Teramo, Kari, Eronen, Markkanen, & Hiilesmaa, 2004; Teramo, Hiilesmaa, Schwartz, Clemons, & Widness, 2004).

Amniotic fluid EPO concentrations also correlate with umbilical cord or fetal plasma levels, with fetal plasma concentrations averaging more than twice that found in the corresponding amniotic fluid. Both umbilical cord and amniotic fluid endogenous EPO concentrations correlate inversely with umbilical artery pH, base excess, and pO_2, and directly with pCO_2 and lactate (Buescher, Hertwig, Wolf, & Dudenhausen, 1998; Rollins, Maxwell, Afrasiabi, Halliday, & Lappin, 1993; Shannon, Davis, Kitzmiller, Fulcher, & Koenig, 1986; Stangenberg et al., 1993; Teramo et al., 1987; Teramo, Kari, et al., 2004). It is not clear how EPO accesses the amniotic cavity but since it and its receptors and also EPO mRNA are each present in human fetal membranes (i.e., the amnion) after about 20 weeks of gestation, it is possible that the placenta produces EPO found in amniotic fluid, perhaps by way of mechanisms initiated to protect the fetal brain (Biron-Shental et al., 2007; Conrad, Benyo, Westerhausen-Larsen, & Miles, 1996; Kim, Bogic, Cheung, & Brace, 2001; Ogawa et al., 2003). Alternatively, amniotic fluid EPO may arise from fetal urine after the 16th week of gestation when the fetal kidneys become functional; increased urinary EPO levels are reported in the first 72h of life in critically ill human neonates in association with brain and acute kidney injury (Li et al., 2013). Moreover, elevated endogenous EPO levels are detected in the CNS and serum of neonates with encephalopathy due to presumed hypoxia–ischemia, and serum levels are

4.6 times higher than CSF levels (Zhu et al., 2009). A correlation between the severity of presumed hypoxia and the magnitude of change in EPO levels also has been demonstrated in adults (Eckardt et al., 1989).

2.3 Extraerythropoietic Immune-Related Properties

Preclinical and clinical evidence additionally supports that endogenous human EPO concentrations vary in accordance with other inflammation-related proteins, perhaps acting as an acute phase reactant (like C-reactive protein). Serum EPO concentrations increase sharply when mice are exposed to lipopolysaccharide (Ramadori, Ahmad, & Ramadori, 2010). Patients with inflammation-initiating illnesses, like sepsis or septic shock and others who suffered different acute inflammation-related conditions, each had higher circulating concentrations of EPO than their peers who were not affected by those conditions (Abel, Spannbrucker, Fandrey, & Jelkmann, 1996; Ferrucci et al., 2005). The increased risk of inflammation-related disorders, like retinopathy of prematurity (Hellstrom et al., 2013) with early administration of exogenous EPO further buttresses the line of evidence linking endogenous circulating concentrations with other inflammation-related proteins whose concentrations vary in response to infection or injury.

2.3.1 Preclinical Studies

Animal models of experimental autoimmune encephalomyelitis and of experimental stroke provide compelling evidence that EPO has antiinflammatory properties. Administration of EPO daily from day 3 after immunization with myelin basic protein delayed the onset of autoimmune encephalitis and decreased its severity, whereas immunohistochemical analysis of the spinal cord demonstrated that EPO markedly diminished inflammation and glial activation/proliferation (Agnello et al., 2002). EPO supplementation delayed an increase of TNF levels and markedly reduced those of IL-6, suggesting that it might act as a protective cytokine in inflammatory pathologies.

Next, other investigators provided consistent evidence showing that EPO administration markedly reduced astrocyte activation and the recruitment of leukocytes and microglia into an experimental infarction produced in rats by middle cerebral artery occlusion (Villa et al., 2003). Administration of exogenous EPO reduced production of ischemia-induced proinflammatory cytokines TNF, IL- 6, and monocyte chemoattractant protein 1 concentration by more than half, suggesting, along with other results, that EPO attenuates ischemia-induced inflammation by reducing neuronal death rather than by affecting EPO receptor expressing inflammatory cells. More recent studies provide additional evidence that postnatal EPO administration prevents lipopolysaccharide-induced white matter injury in rats (Kumral et al., 2007). *But these are experimental animal studies involving exogenous EPO.*

2.3.2 The Extremely Low Gestational Age Newborn Study

Compelled by evidence that EPO has antiinflammatory properties in the kidney (Lee et al., 2005) and in muscle (Contaldo et al., 2011) as well as its apparent growth/trophic properties, Wells Logan and the extremely low gestational age newborn (ELGAN) study investigators recently examined information, provided by circulating levels of endogenous EPO, about perinatal brain damage. This large cohort study of children born before 28 weeks of gestation reported that increased or highest quartile endogenous EPO concentrations correlated or cooccurred with persistent or intermittently increased or highest quartile concentrations of multiple inflammation-related proteins during the first 2 postnatal weeks (Logan et al., 2014).

In a subsequent analysis of the same cohort (Korzeniewski et al., 2015), the ELGAN study investigators demonstrated that highest quartile EPO concentrations were associated with elevated risks of very low mental development and psychomotor index scores and microcephaly, but *importantly* not with any form of cerebral palsy when compared to children who had neither EPO nor other inflammation-related protein concentrations in the highest quartile for gestational age. By contrast, children who had highest quartile concentrations for gestational age for *both* endogenous EPO and other inflammation-related proteins were at higher risk of neurodevelopmental impairment, ventriculomegaly, microcephaly, and hemiparetic cerebral palsy. In short, the combination of highest quartile EPO with highest quartile concentration of other inflammation-related proteins appeared more strongly associated with brain damage and neurodevelopmental dysfunction than either condition alone. Only one prior study was identified that had associated increased circulating endogenous EPO concentrations with perinatal brain damage but this 1988 study associated increased cord blood concentrations with cerebral palsy diagnosed at age two (Ruth, Autti-Ramo, Granstrom, Korkman, & Raivio, 1988).

The ELGAN study investigators also reported an association between a sustained increase of circulating IL-6 receptor (IL-6R) concentrations with appreciably lower frequencies of ventriculomegaly and quadriparesis. Though this might have been a chance finding, it is also possible that IL-6R might protect against some forms of brain damage since IL-6 signaling via its membrane receptor IL-6R can have antiinflammatory and tissue regenerating effects (Grivennikov et al., 2009). The ELGAN study investigators had previously reported a lower likelihood of motor impairment (O'Shea et al., 2012) and quadriparesis (Kuban et al., 2014) in ELGANS with elevated IL-6R.

A third report by Mari Holm and the ELGAN study investigators additionally found that relatively increased endogenous EPO blood concentrations in extremely preterm newborns during the first 2 weeks of postnatal life conveyed information about increased risks of inflammatory diseases of the bowel, lung, and retina (Holm et al., 2016). Interestingly, while hypoxia is recognized as a main contributor to EPO production, this latter study found no association between increased circulating concentrations and indicators of presumed hypoxia including blood gas abnormalities during the first few postnatal days.

3. SUMMARY AND CONCLUSION

We sought to briefly describe current models of endogenous EPO pleiotropic effects to make four points clear. *First*, endogenous EPO regulates erythroid cell apoptosis so that red blood cell production is balanced against the number of cells destroyed in order to maintain optimal tissue oxygen levels. *Second*, preclinical and clinical studies alike provide additional evidence of other (i.e., extraerythropoietic) immune-related and growth/trophic properties. *Third*, by contrast, EPO might be increased as an antiinflammatory response to other proinflammatory cytokines, and not because it is an inflammatory protein, in which case, it would make the association between EPO and these other proteins an *epiphenomenon*. *Fourth*, on the other hand, EPO might also act as a tissue protector (Rathod & Salahudeen, 2011) or it could reflect immaturity/vulnerability of the brain or of the systems responsible for protecting it (Leviton, Blair, Dammann, & Allred, 2005). Indeed, each of these scenarios is plausible, and all might be true in certain circumstances (Holm et al., 2016; Korzeniewski et al., 2015; Logan et al., 2014).

ACKNOWLEDGMENTS

This work was supported in part by the Wayne State University Perinatal Research Initiative, and we additionally acknowledge the influences of Alan Leviton and the ELGAN study investigators on our thinking about endogenous EPO.

REFERENCES

Abel, J., Spannbrucker, N., Fandrey, J., & Jelkmann, W. (1996). Serum erythropoietin levels in patients with sepsis and septic shock. *European Journal of Haematology*, *57*(5), 359–363. Epub 1996/11/01. PubMed PMID: 9003476.

Agnello, D., Bigini, P., Villa, P., Mennini, T., Cerami, A., Brines, M. L., et al. (2002). Erythropoietin exerts an anti-inflammatory effect on the CNS in a model of experimental autoimmune encephalomyelitis. *Brain Research*, *952*(1), 128–134. PubMed PMID: 12363412.

Alvarez Arroyo, M. V., Castilla, M. A., Gonzalez Pacheco, F. R., Tan, D., Riesco, A., Casado, S., et al. (1998). Role of vascular endothelial growth factor on erythropoietin-related endothelial cell proliferation. *Journal of the American Society of Nephrology*, *9*(11), 1998–2004. Epub 1998/11/10. PubMed PMID: 9808085.

Ashley, R. A., Dubuque, S. H., Dvorak, B., Woodward, S. S., Williams, S. K., & Kling, P. J. (2002). Erythropoietin stimulates vasculogenesis in neonatal rat mesenteric microvascular endothelial cells. *Pediatric Research*, *51*(4), 472–478. http://dx.doi.org/10.1203/00006450-200204000-00012. PubMed PMID: 11919332.

Biron-Shental, T., Schaiff, W. T., Ratajczak, C. K., Bildirici, I., Nelson, D. M., & Sadovsky, Y. (2007). Hypoxia regulates the expression of fatty acid-binding proteins in primary term human trophoblasts. *American Journal of Obstetrics and Gynecology*, *197*(5), 516. e1–5516. e6. http://dx.doi.org/10.1016/j.ajog.2007.03.066. PubMed PMID: 17826730. PubMed Central PMCID: PMC2151846.

Bonsdorff, E., & Jalavisto, E. (1948). A humoral mechanism in anoxic erythocytosis. *Acta Physiologica Scandinavica*, *16*(2–3), 150–170. http://dx.doi.org/10.1111/j.1748-1716.1948.tb00535.x. PubMed PMID: WOS:A1948XW04200006.

Brines, M., & Cerami, A. (2006). Discovering erythropoietin's extra-hematopoietic functions: Biology and clinical promise. *Kidney International*, *70*(2), 246–250. Epub 2006/06/02. http://dx.doi.org/10.1038/sj.ki.5001546. PubMed PMID: 16738535.

Brines, M., & Cerami, A. (2008). Erythropoietin-mediated tissue protection: Reducing collateral damage from the primary injury response. *Journal of Internal Medicine*, *264*(5), 405–432. Epub 2008/11/20. http://dx.doi.org/10.1111/j.1365-2796.2008.02024.x. PubMed PMID: 19017170.

Buescher, U., Hertwig, K., Wolf, C., & Dudenhausen, J. W. (1998). Erythropoietin in amniotic fluid as a marker of chronic fetal hypoxia. *International Journal of Gynaecology and Obstetrics*, *60*(3), 257–263. PubMed PMID: 9544710.

Calvillo, L., Latini, R., Kajstura, J., Leri, A., Anversa, P., Ghezzi, P., et al. (2003). Recombinant human erythropoietin protects the myocardium from ischemia-reperfusion injury and promotes beneficial remodeling. *Proceedings of the National Academy of Sciences of the United States of America*, *100*(8), 4802–4806. http://dx.doi.org/10.1073/pnas.0630444100. PubMed PMID: WOS:000182306100083.

Carlini, R. G., Reyes, A. A., & Rothstein, M. (1995). Recombinant human erythropoietin stimulates angiogenesis in vitro. *Kidney International*, *47*(3), 740–745. http://dx.doi.org/10.1038/ki.1995.113. PubMed PMID: 7752572.

Carlini, R. G., Alonzo, E. J., Dominguez, J., Blanca, I., Weisinger, J. R., Rothstein, M., et al. (1999). Effect of recombinant human erythropoietin on endothelial cell apoptosis. *Kidney International*, *55*(2), 546–553. Epub 1999/02/13. http://dx.doi.org/10.1046/j.1523-1755.1999.00266.x. PubMed PMID: 9987078.

Carnot, P., & Deflandre, C. (1906). The hemopoietic activity of serum during the regeneration of blood. *Comptes Rendus Hebdomadaires Des Seances De L Academie Des Sciences*, *143*, 384–386. PubMed PMID: WOS: 000200949400154.

Chang, Y. S., Mu, D., Wendland, M., Sheldon, R. A., Vexler, Z. S., McQuillen, P. S., et al. (2005). Erythropoietin improves functional and histological outcome in neonatal stroke.

Pediatric Research, 58(1), 106–111. Epub 2005/05/10. http://dx.doi.org/10.1203/01.PDR.0000163616.89767.69. PubMed PMID: 15879287.

Chong, Z. Z., Kang, J. Q., & Maiese, K. (2003). Erythropoietin fosters both intrinsic and extrinsic neuronal protection through modulation of microglia, Akt1, Bad, and caspase-mediated pathways. *British Journal of Pharmacology, 138*(6), 1107–1118. Epub 2003/04/10. http://dx.doi.org/10.1038/sj.bjp.0705161. PubMed PMID: 12684267 PubMed Central PMCID: PMC1573758.

Conrad, K. P., Benyo, D. F., Westerhausen-Larsen, A., & Miles, T. M. (1996). Expression of erythropoietin by the human placenta. *The FASEB Journal, 10*(7), 760–768. PubMed PMID: 8635693.

Contaldo, C., Lindenblatt, N., Elsherbiny, A., Hogger, D. C., Borozadi, M. K., Vetter, S. T., et al. (2011). Erythropoietin requires endothelial nitric oxide synthase to counteract TNF-[alpha]-induced microcirculatory dysfunction in murine striated muscle. *Shock, 35*(3), 315–321. http://dx.doi.org/10.1097/SHK.0b013e3181fd0700. PubMed PMID: 20926979.

Cotes, P. M., & Bangham, D. R. (1961). Bio-assay of erythropoietin in mice made polycythaemic by exposure to air at a reduced pressure. *Nature, 191*(479), 1065–1067. http://dx.doi.org/10.1038/1911065a0. PubMed PMID: WOS: A19618742B00238.

Dame, C., Bartmann, P., Wolber, E., Fahnenstich, H., Hofmann, D., & Fandrey, J. (2000). Erythropoietin gene expression in different areas of the developing human central nervous system. *Brain Research Developmental Brain Research, 125*(1–2), 69–74. Epub 2001/01/13. PubMed PMID: 11154762.

Dammann, O., Gray, P., Gressens, P., Wolkenhauer, O., & Leviton, A. (2014). Systems epidemiology: What's in a name? *Online Journal of Public Health Informatics, 6*(3), e198Epub 2015/01/20. http://dx.doi.org/10.5210/ojphi.v6i3.5571. PubMed PMID: 25598870. PubMed Central PMCIDPMCPmc4292535.

Demers, E. J., McPherson, R. J., & Juul, S. E. (2005). Erythropoietin protects dopaminergic neurons and improves neurobehavioral outcomes in juvenile rats after neonatal hypoxia-ischemia. *Pediatric Research, 58*(2), 297–301. Epub 2005/08/02. http://dx.doi.org/10.1203/01.PDR.0000169971.64558.5A. PubMed PMID: 16055937.

Eckardt, K. U., Boutellier, U., Kurtz, A., Schopen, M., Koller, E. A., & Bauer, C. (1989). Rate of erythropoietin formation in humans in response to acute hypobaric hypoxia. *Journal of Applied Physiology (Bethesda, Md: 1985), 66*(4), 1785–1788. Epub 1989/04/01. PubMed PMID: 2732171.

Eschbach, J. W., Egrie, J. C., Downing, M. R., Browne, J. K., & Adamson, J. W. (1987). Correction of the anemia of end-stage renal-disease with recombinant-human-erythropoietin—Results of a combined phase-I and phase-II clinical-trial. *New England Journal of Medicine, 316*(2), 73–78. http://dx.doi.org/10.1056/nejm198701083160203. PubMed PMID: WOS: A1987F482900003.

Fandrey, J., & Jelkmann, W. E. (1991). Interleukin-1 and tumor necrosis factor-alpha inhibit erythropoietin production in vitro. *Annals of the New York Academy of Sciences, 628*, 250–255. Epub 1991/01/01. PubMed PMID: 1712553.

Ferrucci, L., Guralnik, J. M., Woodman, R. C., Bandinelli, S., Lauretani, F., Corsi, A. M., et al. (2005). Proinflammatory state and circulating erythropoietin in persons with and without anemia. *The American Journal of Medicine, 118*(11), 1288. Epub 2005/11/08. http://dx.doi.org/10.1016/j.amjmed.2005.06.039. PubMed PMID: 16271918.

Fisher, J. W. (2010). Landmark advances in the development of erythropoietin. *Experimental Biology and Medicine (Maywood, NJ), 235*(12), 1398–1411. Epub 2010/12/04. http://dx.doi.org/10.1258/ebm.2010.010137. PubMed PMID: 21127338.

Fisher, J. W., & Birdwell, B. J. (1961). Production of an erythropoietic factor by in situ perfused kidney. *Acta Haematologica, 26*(4), 224–232. PubMed PMID: WOS: A19611516A00002.

Garcia, J. F., Sherwood, J., & Goldwasser, E. (1979). Radioimmunoassay of erythropoietin. *Blood Cells*, *5*(3), 405–419. PubMed PMID: WOS: A1979HL31700005.

Grasso, G., Buemi, M., Alafaci, C., Sfacteria, A., Passalacqua, M., Sturiale, A., et al. (2002). Beneficial effects of systemic administration of recombinant human erythropoietin in rabbits subjected to subarachnoid hemorrhage. *Proceedings of the National Academy of Sciences of the United States of America*, *99*(8), 5627–5631. Epub 2002/04/12. http://dx.doi.org/10.1073/pnas.082097299. PubMed PMID: 11943864 PubMed Central PMCID: PMC122821.

Grivennikov, S., Karin, E., Terzic, J., Mucida, D., Yu, G. Y., Vallabhapurapu, S., et al. (2009). IL-6 and Stat3 are required for survival of intestinal epithelial cells and development of colitis-associated cancer. *Cancer Cell*, *15*(2), 103–113. http://dx.doi.org/10.1016/j.ccr.2009.01.001. PubMed PMID: 19185845. PubMed Central PMCID: PMC2667107.

Gruber, M., Hu, C.-J., Johnson, R. S., Brown, E. J., Keith, B., & Simon, M. C. (2007). Acute postnatal ablation of Hif-2 alpha results in anemia. *Proceedings of the National Academy of Sciences of the United States of America*, *104*(7), 2301–2306. http://dx.doi.org/10.1073/pnas.0608382104. PubMed PMID: WOS: 000244438500048.

Haller, H., Christel, C., Dannenberg, L., Thiele, P., Lindschau, C., & Luft, F. C. (1996). Signal transduction of erythropoietin in endothelial cells. *Kidney International*, *50*(2), 481–488. Epub 1996/08/01. PubMed PMID: 8840276.

Heeschen, C., Aicher, A., Lehmann, R., Fichtlscherer, S., Vasa, M., Urbich, C., et al. (2003). Erythropoietin is a potent physiologic stimulus for endothelial progenitor cell mobilization. *Blood*, *102*(4), 1340–1346. Epub 2003/04/19. http://dx.doi.org/10.1182/blood-2003-01-0223. PubMed PMID: 12702503.

Hellstrom, A., Smith, L. E., & Dammann, O. (2013). Retinopathy of prematurity. *Lancet*, *382*(9902), 1445–1457. Epub 2013/06/21. http://dx.doi.org/10.1016/s0140-6736(13)60178-6. PubMed PMID: 23782686. PubMed Central PMCID: PMCPmc4389630.

Holm, M., Skranes, J., Dammann, O., Fichorova, R. N., Allred, E. N., & Leviton, A. (2016). Systemic endogenous erythropoietin and associated disorders in extremely preterm newborns. *Archives of Disease in Childhood. Fetal and Neonatal Edition*, *101*(5), F458–F463. Epub 2015/10/22. http://dx.doi.org/10.1136/archdischild-2015-309127. PubMed PMID: 27173415 PubMed Central PMCIDPMCPmc4590239.

Hu, C. J., Wang, L. Y., Chodosh, L. A., Keith, B., & Simon, M. C. (2003). Differential roles of hypoxia-inducible factor 1alpha (HIF-1alpha) and HIF-2alpha in hypoxic gene regulation. *Molecular and Cellular Biology*, *23*(24), 9361–9374. Epub 2003/12/04. PubMed PMID: 14645546. PubMed Central PMCID: PMCPmc309606.

Iwai, M., Cao, G., Yin, W., Stetler, R. A., Liu, J., & Chen, J. (2007a). Erythropoietin promotes neuronal replacement through revascularization and neurogenesis after neonatal hypoxia/ischemia in rats. *Stroke: A Journal of Cerebral Circulation*, *38*(10), 2795–2803. Epub 2007/08/19. http://dx.doi.org/10.1161/STROKEAHA.107.483008. PubMed PMID: 17702962.

Iwai, M., Liu, J. L., Yin, W., Cao, G. D., Signore, A. P., & Chen, J. (2007b). Erythropoietin enhances angiogenesis and neuronal replacement in cerebral cortex after neonatal hypoxia-ischemia. *Stroke*, *38*(2), 557–558. PubMed PMID: WOS: 000244122600534.

Iwai, M., Stetler, R. A., Xing, J., Hu, X., Gao, Y., Zhang, W., et al. (2010). Enhanced oligodendrogenesis and recovery of neurological function by erythropoietin after neonatal hypoxic/ischemic brain injury. *Stroke*, *41*(5), 1032–1037. http://dx.doi.org/10.1161/STROKEAHA.109.570325. PubMed PMID: 20360553 PubMed Central PMCID: PMC2919308.

Jelkmann, W. (1998). Proinflammatory cytokines lowering erythropoietin production. *Journal of Interferon and Cytokine Research*, *18*(8), 555–559. http://dx.doi.org/10.1089/jir.1998.18.555. PubMed PMID: WOS: 000075511900002.

Jia, Y., Suzuki, N., Yamamoto, M., Gassmann, M., & Noguchi, C. T. (2012). Endogenous erythropoietin signaling facilitates skeletal muscle repair and recovery following pharmacologically induced damage. *The FASEB Journal*, *26*(7), 2847–2858. Epub 2012/04/12. http://dx.doi.org/10.1096/fj.11-196618. PubMed PMID: 22490927. PubMed Central PMCID: PMCPmc3382092.

Juul, S. (2012). Neuroprotective role of erythropoietin in neonates. *The Journal of Maternal-Fetal & Neonatal Medicine*, *25*(Suppl. 4), 105–107. Epub 2012/09/14. http://dx.doi.org/10.3109/14767058.2012.715025. PubMed PMID: 22958035.

Kastner, A., Grube, S., El-Kordi, A., Stepniak, B., Friedrichs, H., Sargin, D., et al. (2012). Common variants of the genes encoding erythropoietin and its receptor modulate cognitive performance in schizophrenia. *Molecular Medicine*, *18*, 1029–1040. Epub 2012/06/07. http://dx.doi.org/10.2119/molmed.2012.00190. PubMed PMID: 22669473. PubMed Central PMCID PMC3459483.

Kato, S., Aoyama, M., Kakita, H., Hida, H., Kato, I., Ito, T., et al. (2011). Endogenous erythropoietin from astrocyte protects the oligodendrocyte precursor cell against hypoxic and reoxygenation injury. *Journal of Neuroscience Research*, *89*(10), 1566–1574. Epub 2011/08/13. http://dx.doi.org/10.1002/jnr.22702. PubMed PMID: 21833990.

Kertesz, N., Wu, J., Chen, T. H., Sucov, H. M., & Wu, H. (2004). The role of erythropoietin in regulating angiogenesis. *Developmental Biology*, *276*(1), 101–110. http://dx.doi.org/10.1016/j.ydbio.2004.08.025. PubMed PMID: 15531367.

Khoury, M. J., Gwinn, M., & Ioannidis, J. P. (2010). The emergence of translational epidemiology: From scientific discovery to population health impact. *American Journal of Epidemiology*, *172*(5), 517–524. Epub 2010/08/07. http://dx.doi.org/10.1093/aje/kwq211. PubMed PMID: 20688899. PubMed Central PMCID: PMC2927741.

Kim, M. J., Bogic, L., Cheung, C. Y., & Brace, R. A. (2001). Placental expression of erythropoietin mRNA, protein and receptor in sheep. *Placenta*, *22*(5), 484–489. http://dx.doi.org/10.1053/plac.2001.0681. PubMed PMID: 11373159.

Kim, S. N., Moon, J. H., Kim, J. G., Chae, Y. S., Cho, Y. Y., Lee, S. J., et al. (2009). Mobilization effects of G-CSF, GM-CSF, and darbepoetin-alpha for allogeneic peripheral blood stem cell transplantation. *Journal of Clinical Apheresis*, *24*(5), 173–179. http://dx.doi.org/10.1002/jca.20207. PubMed PMID: 19753648.

Korzeniewski, S. J., Allred, E., Logan, J. W., Fichorova, R. N., Engelke, S., Kuban, K. C., et al. (2015). Elevated endogenous erythropoietin concentrations are associated with increased risk of brain damage in extremely preterm neonates. *PLoS One*, *10*(3), e0115083. http://dx.doi.org/10.1371/journal.pone.0115083. PubMed PMID: 25793991. PubMed Central PMCID: PMC4368546.

Kuban, K. C., O'Shea, T. M., Allred, E. N., Paneth, N., Hirtz, D., Fichorova, R. N., et al. (2014). Systemic inflammation and cerebral palsy risk in extremely preterm infants. *Journal of Child Neurology*, *29*(12), 1692–1698. [Epub ahead of print, March 18]. Epub Mar 18. http://dx.doi.org/10.1177/0883073813513335. PubMed PMID: 24646503.

Kumral, A., Ozer, E., Yilmaz, O., Akhisaroglu, M., Gokmen, N., Duman, N., et al. (2003). Neuroprotective effect of erythropoietin on hypoxic-ischemic brain injury in neonatal rats. *Biology of the Neonate*, *83*(3), 224–228. Epub 2003/03/28. http://dx.doi.org/10.1159/000068926. PubMed PMID: 12660442.

Kumral, A., Uysal, N., Tugyan, K., Sonmez, A., Yilmaz, O., Gokmen, N., et al. (2004). Erythropoietin improves long-term spatial memory deficits and brain injury following neonatal hypoxia-ischemia in rats. *Behavioural Brain Research*, *153*(1), 77–86. Epub 2004/06/29. http://dx.doi.org/10.1016/j.bbr.2003.11.002. PubMed PMID: 15219709.

Kumral, A., Baskin, H., Yesilirmak, D. C., Ergur, B. U., Aykan, S., Genc, S., et al. (2007). Erythropoietin attenuates lipopolysaccharide-induced white matter injury in the neonatal rat brain. *Neonatology*, *92*(4), 269–278. http://dx.doi.org/10.1159/000105493. PubMed PMID: 17627093.

Kuratows, Z., Michalak, E., & Lewartowicz, B. (1961). Studies on production of erythropoietin by isolated perfused organs. *Blood, 18*(5), 527–534. PubMed PMID: WOS: A19617352A00006.
Lee, S. H., Li, C., Lim, S. W., Ahn, K. O., Choi, B. S., Kim, Y. S., et al. (2005). Attenuation of interstitial inflammation and fibrosis by recombinant human erythropoietin in chronic cyclosporine nephropathy. *American Journal of Nephrology, 25*(1), 64–76. http://dx.doi.org/10.1159/000084275. PubMed PMID: 15746540.
Leviton, A., Blair, E., Dammann, O., & Allred, E. (2005). The wealth of information conveyed by gestational age. *The Journal of Pediatrics, 146*(1), 123–127. http://dx.doi.org/10.1016/j.jpeds.2004.09.028. PubMed PMID: 15644836.
Li, Y., Juul, S. E., Morris-Wiman, J. A., Calhoun, D. A., & Christensen, R. D. (1996). Erythropoietin receptors are expressed in the central nervous system of mid-trimester human fetuses. *Pediatric Research, 40*(3), 376–380. Epub 1996/09/01. http://dx.doi.org/10.1203/00006450-199609000-00002. PubMed PMID: 8865271.
Li, Y., Xiao, Z., Yan, J., Li, X., Wang, Q., Zhu, H., et al. (2013). Urine erythropoietin level is associated with kidney and brain injury in critically ill neonates. *Neonatology, 104*(2), 87–94. Epub 2013/07/03. http://dx.doi.org/10.1159/000350546. PubMed PMID: 23817218.
Logan, J. W., Allred, E. N., Fichorova, R. N., Engelke, S., Dammann, O., & Leviton, A. (2014). Endogenous erythropoietin varies significantly with inflammation-related proteins in extremely premature newborns. *Cytokine, 69*(1), 22–28. Epub 2014/10/07. http://dx.doi.org/10.1016/j.cyto.2014.04.009. doi: 10.1111/ejn.12747. PubMed PMID: 25022958. PubMed Central PMCID: PMCPmc4285695.
Marti, H. H., Bernaudin, M., Petit, E., & Bauer, C. (2000). Neuroprotection and angiogenesis: Dual role of erythropoietin in brain ischemia. *News in Physiological Sciences: An International Journal of Physiology Produced Jointly by the International Union of Physiological Sciences and the American Physiological Society, 15*, 225–229. PubMed PMID: 11390915.
Masuda, S., Okano, M., Yamagishi, K., Nagao, M., Ueda, M., & Sasaki, R. (1994). A novel site of erythropoietin production. Oxygen-dependent production in cultured rat astrocytes. *The Journal of Biological Chemistry, 269*(30), 19488–19493. PubMed PMID: 8034718.
Matsushita, H., Johnston, M. V., Lange, M. S., & Wilson, M. A. (2003). Protective effect of erythropoietin in neonatal hypoxic ischemia in mice. *Neuroreport, 14*(13), 1757–1761. http://dx.doi.org/10.1097/01.wnr.0000086250.76504.71. PubMed PMID: WOS: 000186170400020.
Maxwell, P. (2003). HIF-1: An oxygen response system with special relevance to the kidney. *Journal of the American Society of Nephrology: JASN, 14*(11), 2712–2722. Epub 2003/10/22. PubMed PMID: 14569080.
McClure, M. M., Threlkeld, S. W., & Fitch, R. H. (2006). The effects of erythropoietin on auditory processing following neonatal hypoxic-ischemic injury. *Brain Research, 1087*(1), 190–195. Epub 2006/04/29. http://dx.doi.org/10.1016/j.brainres.2006.03.016. PubMed PMID: 16643862.
Miyake, T., Kung, C. K. H., & Goldwasser, E. (1977). Purification of human erythropoietin. *Journal of Biological Chemistry, 252*(15), 5558–5564. PubMed PMID: WOS: A1977DS11600062.
Morishita, E., Masuda, S., Nagao, M., Yasuda, Y., & Sasaki, R. (1997). Erythropoietin receptor is expressed in rat hippocampal and cerebral cortical neurons, and erythropoietin prevents in vitro glutamate-induced neuronal death. *Neuroscience, 76*(1), 105–116.
Nagao, M., Masuda, S., Abe, S., Ueda, M., & Sasaki, R. (1992). Production and ligand-binding characteristics of the soluble form of murine erythropoietin receptor. *Biochemical and Biophysical Research Communications, 188*(2), 888–897. PubMed PMID: 1445329.
Ogawa, A., Terada, S., Sakuragawa, N., Masuda, S., Nagao, M., & Miki, M. (2003). Progesterone, but not 17beta-estradiol, up-regulates erythropoietin (EPO) production in

human amniotic epithelial cells. *Journal of Bioscience and Bioengineering, 96*(5), 448–453. http://dx.doi.org/10.1016/S1389-1723(03)70130-3. PubMed PMID: 16233554.

O'Shea, T. M., Allred, E. N., Kuban, K. C., Dammann, O., Paneth, N., Fichorova, R., et al. (2012). Elevated concentrations of inflammation-related proteins in postnatal blood predict severe developmental delay at 2 years of age in extremely preterm infants. *The Journal of Pediatrics, 160*(3), 395–401.e4. Epub 2011/10/18. http://dx.doi.org/10.1016/j.jpeds.2011.08.069. PubMed PMID: 22000304. PubMed Central PMCID: PMCPmc3279610.

Ostrowski, D., Ehrenreich, H., & Heinrich, R. (2011). Erythropoietin promotes survival and regeneration of insect neurons in vivo and in vitro. *Neuroscience, 188*, 95–108. Epub 2011/05/24. http://dx.doi.org/10.1016/j.neuroscience.2011.05.018. PubMed PMID: 21600963.

Paulo, L. G., Wilkerson, R. D., Roh, B. L., George, W. J., & Fisher, J. W. (1973). Effects of prostaglandin-E1 on erythropoietin production. *Proceedings of the Society for Experimental Biology and Medicine, 142*(3), 771–775. PubMed PMID: WOS: A1973P157600010.

Plzak, L. F., Fried, W., Jacobson, L. O., & Bethard, W. F. (1955). Demonstration of stimulation of erythropoiesis by plasma from anemic rats using FE59. *Journal of Laboratory and Clinical Medicine, 46*(5), 671–678. PubMed PMID: WOS: A1955WD60800002.

Ramadori, P., Ahmad, G., & Ramadori, G. (2010). Cellular and molecular mechanisms regulating the hepatic erythropoietin expression during acute-phase response: A role for IL-6. *Laboratory Investigation: A Journal of Technical Methods and Pathology, 90*(9), 1306–1324. Epub 2010/05/12. http://dx.doi.org/10.1038/labinvest.2010.85. PubMed PMID: 20458283.

Rankin, E. B., Biju, M. P., Liu, Q., Unger, T. L., Rha, J., Johnson, R. S., et al. (2007). Hypoxia-inducible factor-2 (HIF-2) regulates hepatic erythropoietin in vivo. *The Journal of Clinical Investigation, 117*(4), 1068–1077. http://dx.doi.org/10.1172/jci30117. PubMed PMID: WOS: 000245451700029.

Rathod, D. B., & Salahudeen, A. K. (2011). Nonerythropoietic properties of erythropoietin: Implication for tissue protection. *Journal of Investigative Medicine, 59*(7), 1083–1085. Epub 2011/10/21. http://dx.doi.org/10.231/JIM.0b013e31822cf86e. PubMed PMID: 22011619.

Rollins, M. D., Maxwell, A. P., Afrasiabi, M., Halliday, H. L., & Lappin, T. R. (1993). Cord blood erythropoietin, pH, PaO2 and haematocrit following caesarean section before labour. *Biology of the Neonate, 63*(3), 147–152. PubMed PMID: 8324093.

Ruth, V., Autti-Ramo, I., Granstrom, M. L., Korkman, M., & Raivio, K. O. (1988). Prediction of perinatal brain damage by cord plasma vasopressin, erythropoietin, and hypoxanthine values. *The Journal of Pediatrics, 113*(5), 880–885. Epub 1988/11/01. PubMed PMID: 3183848.

Sargin, D., El-Kordi, A., Agarwal, A., Muller, M., Wojcik, S. M., Hassouna, I., et al. (2011). Expression of constitutively active erythropoietin receptor in pyramidal neurons of cortex and hippocampus boosts higher cognitive functions in mice. *BMC Biology, 9*, 27. Epub 2011/04/30. http://dx.doi.org/10.1186/1741-7007-9-27. PubMed PMID: 21527022. PubMed Central PMCID: PMC3120735.

Sasaki, R. (2003). Pleiotropic functions of erythropoietin. *Internal Medicine, 42*(2), 142–149. PubMed PMID: 12636232.

Sasaki, R., Masuda, S., & Nagao, M. (2001). Pleiotropic functions and tissue-specific expression of erythropoietin. *News in Physiological Sciences: An International Journal of Physiology Produced Jointly by the International Union of Physiological Sciences and the American Physiological Society, 16*, 110–113. PubMed PMID: 11443227.

Scortegagna, M., Ding, K., Zhang, Q., Oktay, Y., Bennett, M. J., Bennett, M., et al. (2005). HIF-2alpha regulates murine hematopoietic development in an erythropoietin-dependent manner. *Blood, 105*(8), 3133–3140. Epub 2005/01/01. http://dx.doi.org/10.1182/blood-2004-05-1695. PubMed PMID: 15626745.

Semenza, G. L., & Wang, G. L. (1992). A nuclear factor induced by hypoxia via denovo protein-synthesis binds to the human erythropoietin gene enhancer at a site required for transcriptional activation. *Molecular and Cellular Biology*, *12*(12), 5447–5454. PubMed PMID: WOS: A1992JZ25800018.

Shannon, K., Davis, J. C., Kitzmiller, J. L., Fulcher, S. A., & Koenig, H. M. (1986). Erythropoiesis in infants of diabetic mothers. *Pediatric Research*, *20*(2), 161–165. http://dx.doi.org/10.1203/00006450-198602000-00014. PubMed PMID: 3945525.

Shingo, T., Sorokan, S. T., Shimazaki, T., & Weiss, S. (2001). Erythropoietin regulates the in vitro and in vivo production of neuronal progenitors by mammalian forebrain neural stem cells. *The Journal of Neuroscience: The Official Journal of the Society for Neuroscience*, *21*(24), 9733–9743. Epub 2001/12/12. PubMed PMID: 11739582.

Siren, A. L., Fratelli, M., Brines, M., Goemans, C., Casagrande, S., Lewczuk, P., et al. (2001). Erythropoietin prevents neuronal apoptosis after cerebral ischemia and metabolic stress. *Proceedings of the National Academy of Sciences of the United States of America*, *98*(7), 4044–4049. Epub 2001/03/22. http://dx.doi.org/10.1073/pnas.051606598. PubMed PMID: 11259643. PubMed Central PMCID: PMC31176.

Solaroglu, I., Solaroglu, A., Kaptanoglu, E., Dede, S., Haberal, A., Beskonakli, E., et al. (2003). Erythropoietin prevents ischemia-reperfusion from inducing oxidative damage in fetal rat brain. *Child's Nervous System: ChNS: Official Journal of the International Society for Pediatric Neurosurgery*, *19*(1), 19–22. Epub 2003/01/24. http://dx.doi.org/10.1007/s00381-002-0680-2. PubMed PMID: 12541081.

Spandou, E., Papadopoulou, Z., Soubasi, V., Karkavelas, G., Simeonidou, C., Pazaiti, A., et al. (2005). Erythropoietin prevents long-term sensorimotor deficits and brain injury following neonatal hypoxia-ischemia in rats. *Brain Research*, *1045*(1–2), 22–30. Epub 2005/05/25. http://dx.doi.org/10.1016/j.brainres.2005.03.013. PubMed PMID: 15910759.

Spandou, E., Soubasi, V., Papoutsopoulou, S., Karkavelas, G., Simeonidou, C., Kaiki-Astara, A., et al. (2004). Erythropoietin prevents hypoxia/ischemia-induced DNA fragmentation in an experimental model of perinatal asphyxia. *Neuroscience Letters*, *366*(1), 24–28. Epub 2004/07/22. http://dx.doi.org/10.1016/j.neulet.2004.05.032. PubMed PMID: 15265583.

Stangenberg, M., Legarth, J., Cao, H. L., Lingman, G., Persson, B., Rahman, F., et al. (1993). Erythropoietin concentrations in amniotic fluid and umbilical venous blood from Rh-immunized pregnancies. *Journal of Perinatal Medicine*, *21*(3), 225–234. PubMed PMID: 8229614.

Taniguchi, N., Nakamura, T., Sawada, T., Matsubara, K., Furukawa, K., Hadase, M., et al. (2010). Erythropoietin prevention trial of coronary restenosis and cardiac remodeling after ST-elevated acute myocardial infarction (EPOC-AMI): A pilot, randomized, placebo-controlled study. *Circulation Journal: Official Journal of the Japanese Circulation Society*, *74*(11), 2365–2371. PubMed PMID: 20834185.

Teramo, K. A., & Widness, J. A. (2009). Increased fetal plasma and amniotic fluid erythropoietin concentrations: Markers of intrauterine hypoxia. *Neonatology*, *95*(2), 105–116. Epub 2008/09/09. http://dx.doi.org/10.1159/000153094. PubMed PMID: 18776724. PubMed Central PMCID: PMCPmc2863306.

Teramo KA, Widness JA, Clemons GK, Voutilainen P, McKinlay S, Schwartz R. Amniotic fluid erythropoietin correlates with umbilical plasma erythropoietin in normal and abnormal pregnancy. Obstetrics and Gynecology 1987; 69(5):710–716. PubMed PMID: 3574798.

Teramo, K., Kari, M. A., Eronen, M., Markkanen, H., & Hiilesmaa, V. (2004a). High amniotic fluid erythropoietin levels are associated with an increased frequency of fetal and neonatal morbidity in type 1 diabetic pregnancies. *Diabetologia*, *47*(10), 1695–1703. http://dx.doi.org/10.1007/s00125-004-1515-3. PubMed PMID: 15502930.

Teramo, K. A., Hiilesmaa, V. K., Schwartz, R., Clemons, G. K., & Widness, J. A. (2004b). Amniotic fluid and cord plasma erythropoietin levels in pregnancies complicated by preeclampsia, pregnancy-induced hypertension and chronic hypertension. *Journal of Perinatal Medicine*, *32*(3), 240–247. http://dx.doi.org/10.1515/JPM.2004.045. PubMed PMID: 15188798.

Villa, P., Bigini, P., Mennini, T., Agnello, D., Laragione, T., Cagnotto, A., et al. (2003). Erythropoietin selectively attenuates cytokine production and inflammation in cerebral ischemia by targeting neuronal apoptosis. *The Journal of Experimental Medicine*, *198*(6), 971–975. http://dx.doi.org/10.1084/jem.20021067. PubMed PMID: 12975460. PubMed Central PMCID: PMC2194205.

Wang, G. L., & Semenza, G. L. (1993a). Characterization of hypoxia-inducible factor-I and regulation of DNA-binding activity by hypoxia. *Journal of Biological Chemistry*, *268*(29), 21513–21518. PubMed PMID: WOS: A1993MC80900016.

Wang, G. L., & Semenza, G. L. (1993b). General involvement of hypoxia-inducible factor-I in transcriptional response to hypoxia. *Proceedings of the National Academy of Sciences of the United States of America*, *90*(9), 4304–4308. http://dx.doi.org/10.1073/pnas.90.9.4304. PubMed PMID: WOS: A1993LA43000109.

Wang, X., Zhu, C., Gerwien, J. G., Schrattenholz, A., Sandberg, M., Leist, M., et al. (2004). The nonerythropoietic asialoerythropoietin protects against neonatal hypoxia-ischemia as potently as erythropoietin. *Journal of Neurochemistry*, *91*(4), 900–910. Epub 2004/11/05. http://dx.doi.org/10.1111/j.1471-4159.2004.02769.x. PubMed PMID: 15525344.

Wen, T. C., Sadamoto, Y., Tanaka, J., Zhu, P. X., Nakata, K., Ma, Y. J., et al. (2002). Erythropoietin protects neurons against chemical hypoxia and cerebral ischemic injury by up-regulating Bcl-xL expression. *Journal of Neuroscience Research*, *67*(6), 795–803. Epub 2002/03/14. PubMed PMID: 11891794.

Widness, J. A., Schmidt, R. L., & Sawyer, S. T. (1995). Erythropoietin transplacental passage—Review of animal studies. *Journal of Perinatal Medicine*, *23*(1–2), 61–70. Epub 1995/01/01. PubMed PMID: 7658322.

Yu, X., Shacka, J. J., Eells, J. B., Suarez-Quian, C., Przygodzki, R. M., Beleslin-Cokic, B., et al. (2002). Erythropoietin receptor signalling is required for normal brain development. *Development*, *129*(2), 505–516. Epub 2002/01/25. PubMed PMID: 11807041.

Zanjani, E. D., Poster, J., Burlington, H., Mann, L. I., & Wasserman, L. R. (1977). Liver as the primary site of erythropoietin formation in the fetus. *The Journal of Laboratory and Clinical Medicine*, *89*(3), 640–644. Epub 1977/03/01. PubMed PMID: 839121.

Zhu, C., Kang, W., Xu, F., Cheng, X., Zhang, Z., Jia, L., et al. (2009). Erythropoietin improved neurologic outcomes in newborns with hypoxic-ischemic encephalopathy. *Pediatrics*, *124*(2), e218–e226. Epub 2009/08/05. http://dx.doi.org/10.1542/peds.2008-3553. PubMed PMID: 19651565.

CHAPTER FOUR

Erythropoietin Regulation by Angiotensin II

Yong-Chul Kim, Ognoon Mungunsukh, Regina M. Day[1]
Uniformed Services University of the Health Sciences, Bethesda, MD, United States
[1]Corresponding author: e-mail address: regina.day@usuhs.edu

Contents

Abstract

The renin–angiotensin system (RAS) is a key regulator of blood pressure and blood volume homeostasis. The RAS is primarily comprised of the precursor protein angiotensinogen and the two proteases, renin and angiotensin-converting enzyme (ACE). Angiotensin I (Ang I) is derived from angiotensinogen by renin, but appears to have no biological activity. In contrast, angiotensin II (Ang II) that has a variety of biological functions in the cells is converted from Ang I through removal of two-C-terminal residues by ACE. The physiological effects of Ang II are due to Ang II signaling through specific receptor binding, resulting in muscle contraction leading to increased blood pressure and volume. To modulate RAS, three classes of drugs have been developed: (1) renin inhibitors to prevent angiotensinogen conversion to Ang I, (2) ACE inhibitors, to prevent Ang I processing to Ang II and (3) angiotensin receptor blockers, to inhibit Ang II signaling through its receptor. Studies using the RAS inhibitors and Ang II demonstrated that RAS signaling mediates actions of Ang II in the regulation of proliferation and differentiation of specific hematopoietic cell types, especially in the red blood cell lineage. Accumulating evidence indicates that RAS regulates EPO, an essential mediator of red cell production, for human anemia and erythropoiesis in vivo and in vitro. The regulation of EPO expression by Ang II may be responsible for maintaining red blood

Vitamins and Hormones, Volume 105
ISSN 0083-6729
http://dx.doi.org/10.1016/bs.vh.2017.02.001
2017 Published by Elsevier Inc.

cell homeostasis. This review highlights the biological roles of RAS for blood cell and EPO homeostasis through Ang II signaling. The molecular mechanism for Ang II-induced EPO production of the cell or tissue type-specific expression is discussed.

1. INTRODUCTION

The renin–angiotensin system (RAS), an endocrine system of the peripheral blood, has been established as a key regulator of blood pressure and blood volume homeostasis (Reid, 1992; Stroth & Unger, 1999; Young & Pan, 1984). RAS is primarily comprised of the precursor protein angiotensinogen and the two proteases, renin and angiotensin-converting enzyme (ACE), that are required for its maturation to the active form, angiotensin II (Ang II). Human angiotensinogen, a 452 amino acid glycoprotein, is secreted constitutively by the liver into the plasma, but must be activated by the two proteases to be processed to its active form (Deschepper, 1994; Lu, Cassis, Kooi, & Daugherty, 2016; Lynch & Peach, 1991). Angiotensinogen is cleaved to a 10 amino acid peptide, angiotensin I (Ang I; Asp-Arg-Val-Tyr-Ile-His-Pro-Phe-His-Leu), by renin, a protease produced primarily in afferent arterioles of the kidney by juxtaglomerular cells (Gonzalez & Prieto, 2015; Lu et al., 2016). Ang I is then further processed by the removal of the two carboxy terminal residues to produce Ang II by ACE, found primarily in endothelium of the lung (Bernstein, 1998).

The effects of Ang II on the cardiovascular system are produced both directly and indirectly. Ang II effects on vascular tone were demonstrated to be due to Ang II signaling through specific receptor binding on vascular smooth muscle cells (VSMCs), resulting in muscle contraction leading to increased blood pressure (Matsubara, 1998; Pueyo & Michel, 1997). Ang II also indirectly affects blood volume, blood pressure, and mineral balance through the regulation of the aldosterone hormone release from the *zona glomerulosa* of the adrenal cortex (Bollag, 2014; Rojas-Vega & Gamba, 2016). Aldosterone regulates sodium and water retention and potassium excretion in the kidney, thus providing an additional mechanism for the regulation of blood pressure (Bollag, 2014; Rojas-Vega & Gamba, 2016).

Medications for the blockade of the RAS pathway were introduced more than 20 years ago, and these drugs have become first line treatments for primary hypertension. At least three classes of drugs have been developed: renin inhibitors to prevent angiotensin to Ang I conversion,

ACE inhibitors to prevent Ang I processing to Ang II by the inhibition of ACE, and angiotensin receptor blockers, or ARBs, to inhibit Ang II signaling through its receptors (Aronow, 2016; Castrop, 2015; Taddei, 2015). These drug classes prevent Ang II-induced vascular smooth muscle contraction and also inhibit Ang II-induced intrarenal reduction of sodium excretion (Navar, Harrison-Bernard, Imig, Cervenka, & Mitchell, 2000). At this time, at least 10 ACE inhibitors (benazepril, captopril, enalapril, fosinopril, lisinopril, moexipril, perindopril, quinapril, ramipril, and trandolapril) and at least seven ARBs (candesartan, eprosartan, irbesartan, losartan, olmesartan, telmisartan, and valsartan) are currently available in the United States. The reduction of blood pressure by these drugs in hypertensive patients reduces cardiovascular morbidity and mortality and prevents damage to other organ systems, which is often superior to other drugs that target alternative blood pressure mechanisms (Aronow, 2016; Carson et al., 2001; Dezsi, 2014; Gradman, 1999; Negro, 2008; Xie et al., 2015; Zhao, Wang, & Wang, 2015).

2. RAS IN THE REGULATION OF BLOOD CELL HOMEOSTASIS IN VIVO

Soon after the introduction of ACE inhibitors as therapeutics, it was discovered that ACE inhibitors caused adverse effects related to the reduction of blood cell levels in patients. Uncommonly, agranulocytosis, neutropenia, aplastic anemia, or pancytopenia were observed, but mild to moderate anemia was found to be more common. Studies of angiotensin also revealed that along with the regulation of blood pressure, modulation of erythropoiesis was observed. A series of studies in animal models as well as in humans revealed a complex role for the regulation of blood cell homeostasis by RAS.

2.1 Role of ACE Inhibitors and ARBs in the Induction of Agranulocytosis, Neutropenia, Aplastic Anemia, and Pancytopenia

As early as 1980, agranulocytosis and neutropenia were reported in patients administered captopril, usually occurring within the first 3 months of therapy (Amann et al., 1980; Frank, 1989; Gleason et al., 1993; Moussa, Schiano, Spatoliatore, & Salman, 1992; van Brummelen, Willemze, Tan, & Thompson, 1980; Young, Lin, Ko, Hwang, & Ding, 1993). It was later found that other nonsulfhydryl ACE inhibitors could also result in agranulocytosis or neutropenia, although with lower frequency compared with captopril (Elis, Lishner, Lang, & Ravid, 1991; Frank, 1989; Hamidou et al., 1996; Horowitz, Molnar, Levy, & Pollack, 2005). An overall rate for the development of neutropenia (defined as two consecutive neutrophil counts less than 1000/mm^3) was found to be ~0.02% for captopril in patients without comorbidities, and 7.2% of patients with renal impairment and collagen vascular disease (Irvin & Viau, 1986). Increased adverse effects in patients with renal disease were hypothesized to be due to lack of renal clearance of the drugs, resulting in higher drug concentration in the plasma (Waeber, Gavras, Brunner, & Gavras, 1981).

Aplastic anemia and pancytopenia were also reported in response to captopril and other ACE inhibitors in the absence of hemolytic anemia or other immune-related mechanisms (Chisi, Wdzieczak-Bakala, Thierry, Briscoe, & Riches, 1999; Gavras, Graff, Rose, et al., 1981; Harrison, Laidlaw, & Reilly, 1995; Israeli, Or, & Leitersdorf, 1985; Schratzlseer, Lipp, Riess, Antoni, & Delius, 1994). In several cases, blood cell suppression from ACE inhibitors was fatal, in part due to delayed diagnosis (el-Makri, Larabi, Kechrid, Belkahia, & Ben Ayed, 1981; Gavras et al., 1981; Harrison et al., 1995; Waeber et al., 1981). In other cases, these conditions could be reversed by discontinuation of the ACE inhibitors and by treatment with agents such as granulocyte colony-stimulating factor (Israeli et al., 1985; Schratzlseer et al., 1994).

2.2 RAS Regulation of Erythropoiesis

Animal studies of RAS elements and inhibitors have provided evidence for the role of Ang II in the regulation of erythropoiesis. Early studies of Ang II effects in rats demonstrated that intravenous drip Ang II (100 μg/kg) accelerated the incorporation of radioiron (Fe^{59}) in erythrocytes (Nakao, Shirakura, Azuma, & Maekawa, 1967). Conversely, ACE inhibitors reduced hemoglobin, hematocrit, and erythrocyte counts in rodents as well as reduced blood pressure (Barshishat-Kupper et al., 2011; Gould & Goodman, 1990; Imai, Yoshimura, Ohtki, & Hashimoto, 1985; Naeshiro, Sato, Chatani, & Sato, 1998; Wong et al., 1991). ARBs were also reported to induce dose-dependent reductions in hematocrit and red blood cell levels (Neaeshiro, Ishimura, Okada, Chantani, & Sato, 1997; Sato et al., 1996). Interestingly, erythrocyte counts, hematocrit values, and hemoglobin concentrations were reduced by similar percentages (~7%–9%) in rats treated with the ARB candesartan and in mice treated with the ACE inhibitor captopril (72 mg/kg/day), suggesting that a similar mechanism, inhibition of Ang II signaling, is likely involved (Barshishat-Kupper et al., 2011; Naeshiro et al., 1998). Bone marrow histological analysis revealed decreased erythroid cell numbers, but overall histopathology showed no evidence of hemorrhage, hemolysis, or compensatory erythropoiesis (Neaeshiro et al., 1997). In mice treated in a short-time course with the ACE inhibitor captopril, reductions in hematocrit, reticulocytes, and red blood cell populations were shown to be transient, with a significant decrease in these parameters occurring within 7–10 days, but then rebounding thereafter (Barshishat-Kupper et al., 2011).

Studies of transgenic and knockout mice allowed the investigation of RAS effects on erythropoiesis in the background of healthy animal with wild-type control littermates (Doan, Gletsu, Cole, & Bernstein, 2001). Knockout mice for elements of RAS generally exhibit anemia, including knockout mice for angiotensinogen and ACE (Cole et al., 2000; Doan et al., 2001). Also, knockout mice for ACE2, an enzyme that catalyses the conversion of angiotensin I to the peptide angiotensin-(1–9), which also have reduced levels of plasma Ang II, also had anemia, reduced hematocrit, and reduced hemoglobin (Doan et al., 2001). In these mice, administration of Ang II (0.3 mg/kg) for 2 weeks via implanted osmotic minipumps was sufficient to elevate the hematocrit to normal levels (Cole et al., 2000; Doan et al., 2001). Knockout mice for renin and Ang II likewise exhibited reduced hematocrits and red blood cell levels that could be rescued by infusion of Ang II (Kato et al., 2015).

In double transgenic mice (THM: Tsukuba hypertensive mice), coexpression of human renin and angiotensin resulted in $\sim$25% increase in hematocrit levels and reticulocytes compared with wild-type mice (Kato et al., 2005). However, when bone marrow transplantations were performed on THM mice using donor knockout mice for Ang II receptors, the hematocrit and reticulocytes remained elevated, suggesting that Ang II signaling in the bone marrow progenitor or stem cells was not responsible for increased erythrocytosis in the THM mice (Kato et al., 2005). Ang II induces signal transduction to cells by binding specific seven transmembrane G protein-coupled receptors (GPCRs): either the type 1a or type 1b (AT_1a and AT_1b) receptor, or the type 2 (AT_2) receptors, which are differentially expressed on cells of different tissues (Akazawa, Yano, Yabumoto, Kudo-Sakamoto, & Komuro, 2013; Vinturache & Smith, 2014). A series of Ang II receptor knockout experiments was performed in the THM mice to determine which of the Ang II receptors could be removed to restore erythropoiesis to normal levels. Knockout of the AT_1a receptor was sufficient to restore hematocrit levels to near wild-type levels. However, a comparison of knockout mice for the different receptors revealed that both AT_1a and AT_1b receptor subtypes must be knocked out in order to result in the same level of anemia observed in animals that were deficient for angiotensinogen (Kato et al., 2015). Together these findings suggest that Ang II signaling through the AT_1 receptors can regulate erythropoiesis, although this signaling does not involve the hematopoietic stem or progenitor cells of the bone marrow.

The role of RAS in erythropoiesis in humans was investigated using ACE inhibitors and ARBs. Unlike severe agranulocytosis, neutropenia, aplastic anemia, and pancytopenia, patient evidence suggested that mild to moderate anemia and reduced hematocrits are common, usually reversible, side effects of both ACE inhibitors and ARBs (Cheungpasitporn, Thongprayoon, Chiasakul, Korpaisarn, & Erickson, 2015; Vlahakos, Marathias, & Madias, 2010). Hypertension patients administered either ACE inhibitors or ARBs exhibit reduced hemoglobin levels, averaging ~0.3 g/dL, although the degree of anemia and/or reductions in hematocrit rarely required withdrawal of the medication (Griffing & Melby, 1982; Lindholm et al., 2003; Marathias, Agroyannis, Mavromoustakos, Matsoukas, & Vlahakos, 2004; Naito, Kawashima, Akiba, Takanashi, & Nihei, 2003; Olsen et al., 2009; Onoyama, Sanai, Motomura, & Fujishima, 1989; Vlahakos et al., 2010). ACE inhibitors and ARBs were also shown to reduce erythrocytosis and hematocrits in kidney transplant patients (Vlahakos, Canzanello, Madaio, & Madias, 1991).

Patients with diseases or mutations that affected RAS were also examined for differential effects on erythropoiesis and anemia; however, the significant comorbidities in many of these patient populations make it difficult to draw firm conclusions (Vlahakos et al., 2010). For example, patients with hypertension or patients on hemodialysis with increased renin levels displayed increased hematocrit levels (Vlahakos et al., 2010). Conversely, in patients with mutations affecting prorenin or pathways for renin biosynthesis and secretion, low renin levels were found to be correlated with anemia, but these patients also have early onset kidney failure (Zivna et al., 2009). In lung cancer patients with the *ACE* DD genotype that results in relatively high levels of ACE activity, a correlation was found with higher hematocrit levels compared with lung cancer patients with other ACE genotypes (Rigat et al., 1990; Yaren et al., 2008). These findings in human patients were suggestive of a role of RAS for modulation of erythropoiesis, but the studies are complicated by other potential effects of the diseases.

2.3 RAS Regulation of Hematopoietic Progenitors

Following the observations of blood cell deficiencies in patients receiving ACE inhibitors and ARBs, several studies were conducted to identify direct effects of Ang II on hematopoietic stem or progenitor cell proliferation or maturation. Chisi et al. (1999) investigated the ability of the captopril (10^{-6} M) to prevent proliferation in hematopoietic stem and progenitor cells in long-term bone marrow cultures and demonstrated a reduction in nonadherent granulocyte-macrophage colony-forming cells (GM-CFCs) as well as a transient reduction of high proliferative potential colony-forming cells (HPP-CFCs) in culture (Chisi et al., 1999). Captopril was also found to reduce "primitive" hematopoietic cell proliferation following radiation exposure in vivo (Chisi et al., 2000). The mechanism of the inhibition by captopril was thought to be at least in part due to the inhibition of ACE activity for the proteolytic degradation of the tetrapeptide acetyl-Ser-Asp-Lys-Pro (AcSDKP), a potent inhibitor of hematopoietic cell proliferation (Chisi et al., 1999).

Conversely, in a separate set of studies, it was found that Ang II (100 μg/mL in culture) could increase the proliferation of hematopoietic progenitor cells isolated from murine bone marrow or from human cord blood in a dose-dependent manner (Rodgers, Xiong, Steer, & diZerega, 2000). This effect was blocked by the ARB losartan, indicating that specific Ang II receptors were required (Rodgers et al., 2000). Analysis of cell types affected by Ang II revealed ~5-fold increase in the formation of colony-forming unit-granulocyte/macrophage (CFU-GM) and ~20-fold increase in erythroid progenitor populations (Rodgers et al., 2000). A more modest increase was also observed in granulocyte, erythrocyte, monocyte/macrophage colonies (CFU-GEMM) but no change was observed in blast-forming units-erythroid (BFU-E) (Rodgers et al., 2000). In a separate study of Ang II effects on erythroid progenitors, it was found that $CD34^+$ hematopoietic progenitors cultured with 1–100 nM Ang II augmented numbers of BFU-Es (Mrug, Stopka, Julian, Prchal, & Prchal, 1997). Our laboratory also demonstrated that in mice, captopril caused transient delays in the cell cycle of short-term hematopoietic progenitor cell (ST-HSC), inducing a pause in the G_0/G_1 phase of the cycle (Davis et al., 2010). Ang II type 1a receptor expression was identified on human T cells, B cells, $CD34^+CD38^+$ cells, and $CD34^+CD38^-$ cells, providing evidence of the Ang II-specific receptor on mature hematopoietic cells and progenitors (Mrug et al., 1997; Rodgers et al., 2000).

Expression of ACE (referred to as CD143) was identified as a marker for primitive lymphohematopoietic cells derived from human pluripotent stem cells (Jokubaitis et al., 2008; Zambidis et al., 2008). Increasing evidence points toward a "local" bone marrow RAS, in which all components of the RAS system are provided locally for the generation of Ang II, which is believed to play a role in normal hematopoiesis as well as some forms of myeloid leukemia (Beyazit et al., 2007; Haznedaroglu & Ozturk, 2003; Kato et al., 2015; Uz et al., 2014).

3. RAS REGULATION OF ERYTHROPOIETIN IN VIVO

Although a number of studies indicated that Ang II could directly affect hematopoiesis through receptors on hematopoietic progenitors, accumulating evidence also indicated that Ang II acted indirectly on hematopoiesis via regulation of gene expression of erythropoietin (EPO), a potent kidney glycopeptide hormone that is an essential growth factor regulating erythropoiesis (Dunn, Lo, & Donnelly, 2007; Elliott, Pham, & Macdougall, 2008; Freudenthaler, Lucht, Schenk, Brink, & Gleiter, 2000; Freudenthaler, Schreeb, Korner, & Gleiter, 1999; Vlahakos et al., 2010).

In early studies of the effects of Ang II on blood flow and blood pressure, administration of Ang II protein (0.2 and 0.4 μg/kg/min) induced renal arterial constriction and increased plasma levels of EPO within 2 h in dogs (Fisher, Samuels, & Langston, 1967). In a separate study in rats, intravenous administration of Ang II (100 μg/kg) induced EPO expression (Nakao et al., 1967). Also in rats, EPO concentrations in the plasma was positively correlated with renin, angiotensinogen, Ang I, and Ang II (Gould, Goodman, DeWolf, Onesti, & Swartz, 1980). Consistent with these observations, activation of the RAS in double transgenic THM mice enhanced serum EPO levels and kidney EPO mRNA expression (Kato et al., 2005). EPO expression was reduced in THM/AT_1R-knockout mice, suggesting that Ang II signaling through an AT_1 receptor was required for regulation of EPO (Kato et al., 2005).

Our laboratory demonstrated that administration of the ACE inhibitor captopril to wild-type C57BL/6 mice transiently reduced plasma EPO levels to below 70% within 2 days (Barshishat-Kupper et al., 2011). This was reflected by a reduction of ~2-fold in the mRNA levels of EPO in the kidney (Barshishat-Kupper et al., 2011). The levels of EPO returned to baseline or above within 10 days; the rebound effect of EPO expression was demonstrated to be due to the activation of HIF factors in the kidney (Barshishat-Kupper et al., 2011).

Studies using healthy human volunteers as well as human patients demonstrated a link between RAS activity and EPO levels in the plasma. ACE inhibitors enalapril (5 mg/daily) and captopril (6.25 mg/twice daily) were demonstrated to decrease plasma levels of EPO by ~20% within 28 days in healthy volunteers (Pratt et al., 1992). In this experiment, the drugs were washed out for 28 days, resulting in a return of EPO to baseline levels. The drugs were then readministered and washed out in a repeat of the experiment, resulting in a second reduction of EPO followed by a return of EPO levels to baseline levels with drug removal (Pratt et al., 1992).

In complementary studies, administration of Ang II (i.v., 1–3 μg/min for 6 h) in healthy human volunteers increased serum EPO concentration in a dose-dependent manner; serum EPO levels increased by as much as ~3.5-fold (Freudenthaler et al., 2000, 1999). The increase of plasma EPO by Ang II was reduced or blocked by ARBs losartan or valsartan, indicating the requirement of an AT_1 receptor for the action of Ang II on EPO expression (Freudenthaler et al., 2000; Gossmann et al., 2001). These studies performed with healthy individuals with the appropriate control groups provided strong evidence for the regulation of EPO by Ang II.

4. MOLECULAR MECHANISMS OF ANG II REGULATION OF EPO EXPRESSION

Initially, the regulation of EPO gene expression by Ang II was believed to be at least in part in response to activation of angiotensin type 1 receptors of the vasculature and/or altered renal hemodynamics (Benohr, Harsch, Proksch, & Gleiter, 2004; Gossmann et al., 2001). Although the hemodynamics of the kidney may influence EPO expression, more recent evidence indicates that Ang II-specific signaling in human renal cells can drive EPO gene expression.

As stated earlier, the biological actions of Ang II are mediated primarily by two specific receptors, AT_1R and AT_2R receptors of the GPCR family (Mogi, Iwai, & Horiuchi, 2007). The expression of AT_1R or AT_2R is cell type- and tissue type-specific and induce different signaling pathways resulting in a variety of biological outcomes (Kim & Day, 2012; Lee, Mungunsukh, Tutino, Marquez, & Day, 2010; Mogi et al., 2007).

Our laboratory demonstrated that treatment of murine kidney slices in situ with Ang II (100 μm) increased EPO mRNA within 2 h, with a maximal increase within 4 h. The increased EPO protein was detected within 1 h of treatment, suggesting that nontranscriptional mechanisms may also be involved in Ang II regulation (Kim & Day, 2012). In order to define the signal transduction pathway required for Ang II-induced EPO, we examined EPO regulation in human renal 786-O cells. Ang II-induced transcriptional activation of EPO was inhibited by the AT_1R inhibitor telmisartan but not by the AT_2R inhibitor PD 123319. AT_1R signals through a variety of downstream molecules, including G proteins, Src homology region 2 domain-containing phosphatases 1 and 2 (SHP1/2), and p44/p42 mitogen-activated protein kinase (ERK) (Day, Rafty, Chesterman, & Khachigian, 1999; Godeny et al., 2007; Lee et al., 2010; Olson, Shamhart, Naugle, & Meszaros, 2008). Inhibitor studies indicated that the MEK1/2 inhibitor U0126 blocked Ang II-induced EPO, but not the SHP1/2 inhibitor NSC-87877. Consistent with this observation, phosphorylation of ERK1/2 (at Thr202/Tyr204) was induced within 15 min of Ang II treatment.

Both heterotrimeric G proteins and small monomeric G protein p21/Ras may be activated by GPCRs (Dong & Wu, 2013; Marty & Ye, 2010). Ang II induced direct association between AT_1R and p21/Ras for MAPK activation in neurons (Yang, Lu, Yu, & Raizada, 1996). Our study investigating the role of p21/Ras in EPO regulation demonstrated that Ang II-induced phosphorylation of ERK1/2 when wild-type p21/Ras was overexpressed in 786-O cells. ERK1/2 phosphorylation was reduced >40% in cells expressing of p21/Ras-S17N dominant-negative mutant. Consistent with this, EPO protein was increased by Ang II in cells expressing wild-type p21/Ras, but EPO was reduced in cells expressing p21/Ras-S17N. The endogenous and exogenous interaction between AT_1R and p21/Ras was confirmed by coimmunoprecipitation of p21/Ras using an antibody against the C-terminal of AT_1R (306–359). The association was increased by Ang II treatment within 5 min, but was eradicated by telmisartan. Furthermore, the binding of GFP-p21/Ras to HA-tagged AT_1R was reduced by AT_1R truncations in the cytoplasmic C-terminal domain (>60%), suggesting that the C-terminus of AT_1R contains a majority of the binding requirements for p21/Ras interactions. A variety of adaptor proteins have been demonstrated to transduce the activation of p21/Ras by GPCRs. GRB2-associated binding protein 1 (Gab1) activates the Ras/ERK/MAPK pathway in response to a number of ligands (Cai, Nishida, Hirano, & Khavari, 2002; Montagner et al., 2005). Reciprocal coimmunoprecipitation with anti-AT_1R (C-terminal cytoplasmic 306–359 regions) or Gab1 antibodies for endogenous AT_1R and Gab1 interaction indicated that Gab1 was constitutively associated with AT_1R in 786-O cells. Unlike the AT_1R-p21/Ras interaction, Ang II and telmisartan had no effect on Gab1 association.

AT_1R mutation studies demonstrated that specific amino acids in second and seventh transmembrane domains (TM2 and TM7) are important for either the activated conformation of AT_1R, and for ERK activation and G protein/PLC/PKC signal transduction (Yee, Suzuki, Luo, & Fluharty, 2006). D74 (TM2) was crucial for G protein-dependent signaling but not for G protein-independent activation of ERK1/2 by AT_1R, and this amino acid is highly conserved among GPCRs (Aplin, Bonde, & Hansen, 2009; Bihoreau et al., 1993). Using ectopic expression of WT AT_1R, or mutant D74E (in TM2), or mutants T287V, F293L, or N295S (in TM7), we investigated regions of AT_1R required for ERK1/2 activation. Ang II increased ERK1/2 phosphorylation by ~2-fold in cells overexpressing wild-type AT_1R. However, the D74E mutant blocked Ang II-induced ERK1/2.

The ERK-dependent early growth response gene-1 (Egr-1) transcription factor mediates Ang II regulation of platelet-derived growth factor A-chain and cyclin D in VSMCs and in Chinese hamster ovary (CHO) cells, respectively (Day et al., 1999; Guillemot, Levy, Raymondjean, & Rothhut, 2001). We examined the potential role of Egr-1 in Ang II regulation of EPO gene expression. Ang II induced nuclear accumulation of Egr-1. In contrast, the Wilm's tumor 1 (WT1), another transcription factor implicated in EPO regulation, was increased ~70% in cytoplasm after Ang II treatment suggesting that this factor is not activated by Ang II. Ang II increased the expression of reporter plasmid containing the human EPO gene promoter ~2-fold. Overexpression of wild-type Egr-1 further enhanced Ang II activation of the reporter by ~1.5-fold, but overexpression of dominant-negative Egr-1 suppressed Ang II induction. Together, these studies indicate that Ang II binds to AT_1R for activation of the Ras/ERK1/2 MAPK pathway, and subsequent activation of the Egr-1 transcription factor which can regulate EPO expression through binding to the EPO promoter (Fig. 1).

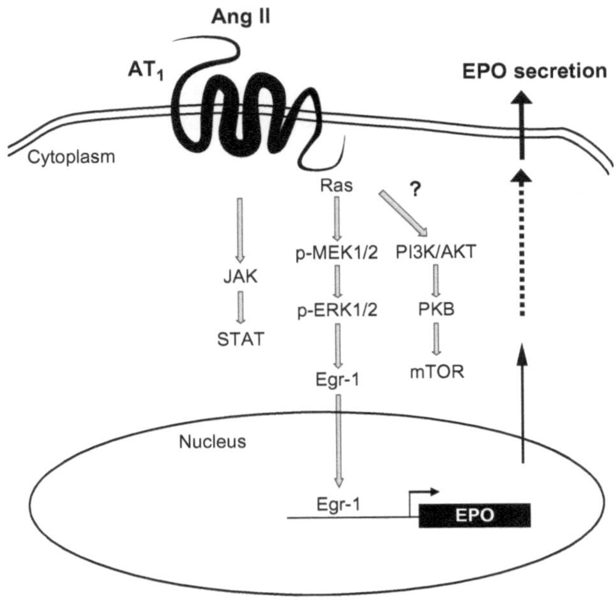

Fig. 1 Mechanism of EPO regulation by Ang II. Ang II binds to AT_1R for activation of the Ras/ERK1/2 MAPK pathway, and subsequent activation of the Egr-1 transcription factor which can regulate EPO expression through binding to the EPO promoter. The JAK/STAT and PI3K/AKT pathways are not thought to participate in this regulation.

5. BIPHASIC EFFECTS OF ACE INHIBITORS ON RADIATION-INDUCED HEMATOPOIETIC INJURY

In recognition of the effects of Ang II on hematopoiesis, EPO, Ang II, ACE inhibitors, and ARBs have been examined for prevention and mitigation of hematopoietic injuries following total body irradiation (TBI) in mice. Ang II administration was demonstrated to improve hematopoiesis following TBI. Ang II (10 or 100 μg/kg/day) increased white blood cell recovery and higher numbers of myeloid progenitor cells in the bone marrow following a sublethal total body radiation exposure in mice (Rodgers, Xiong, & diZerega, 2002). In this study, Ang II was administered by subcutaneous injection on either day 0 (the same day as irradiation) or 2 days after irradiation. The effect of administration of Ang II on blood cell recovery could be due to the direct effects of Ang II on hematopoietic cell proliferation and differentiation as well as regulation of EPO.

Initial studies with ACE inhibitors suggested that these agents might not aid in hematopoietic recovery following radiation exposure. Captopril administration for 3 days prior to radiation exposure (500 mg/L supplied in the drinking water) did not mitigate radiation-induced bone marrow injuries (Moulder, Cohen, Fish, & Hill, 1993). However, a later study using the ACE inhibitor perindopril (10–90 mg/kg twice a day) administered for 4 days beginning 48 h prior to irradiation increased survival from TBI and improved hematopoietic recovery of platelets and red blood cells as well as CFU-GM, BFU-E, and megakaryocyte colony-forming unit (CFU-MK) in the bone marrow (Charrier et al., 2004). Another study demonstrated that captopril (10–50 mg/kg) administered 1 h prior to TBI reduced the frequency of micronucleated polychromatic erythrocytes in the bone marrow (Hosseinimehr, Zakaryaee, & Froughizadeh, 2006). The protective effect of captopril could be blocked by the AT1 receptor antagonist telmisartan, but administration of AcSDKP had no effect (Charrier et al., 2004).

In order to better understand the divergent results from the ACE inhibitor studies, our laboratory investigated several time courses of captopril administration relative to radiation exposure in a murine model of TBI. Our data indicated that the ACE inhibitor captopril can sensitize or reduce hematopoietic radiation injuries depending upon the time of captopril administration relative to radiation exposure (Davis et al., 2010). Captopril (79 mg/kg in the drinking water) administered for 7 days to mice prior to a half lethal dose of radiation at 30 days ($LD_{50/30}$) sensitized the hematopoietic system, resulting in increased red blood cell depletion and reduced overall survival. In contrast, treatment with captopril initiated 4 h postirradiation for 7–30 days increased survival and improved recovery of platelets, RBC, and reticulocytes (Barshishat-Kupper et al., 2011; Davis et al., 2010).

As stated earlier, in nonirradiated mice, captopril has biphasic effects on the rate of proliferation of ST-HSC (Barshishat-Kupper et al., 2011). Captopril administration for 2 consecutive days induced transient quiescence (increased G_0) of the ST-HSC population, which correlates with both blockade of Ang II processing as well as reduction in plasma EPO levels (Barshishat-Kupper et al., 2011). However, after 7 consecutive days of administration, the ST-HSC population reentered the cell cycle, indicating increased rates of proliferation; reentry into the cell cycle correlated with the activation of HIF factors in the kidney and the reinduction of EPO expression (Barshishat-Kupper et al., 2011).

Studies of cell death and radiation have demonstrated a correlation between radiation sensitivity and specific phases of the cell cycle (Tamulevicius, Wang, & Iliakis, 2007). Improper repair of DNA double-stranded breaks (DNA DSBs) has been hypothesized to be a leading cause of cell death following radiation exposure. Repair of DNA DSB occurs primarily via two distinct mechanisms: homologous recombination repair which can occur during the S and G_2 phases of the cell cycle, and nonhomologous end joining which can occur during G_1/G_0 phase of the cell cycle (Tamulevicius et al., 2007).

Additionally, the length of time of the cell cycle determines the time during which DNA repair can occur; for example, cells with long G_1 cycles have a higher rate of survival following radiation exposure than cells with shorter G_1 cycles (Tamulevicius et al., 2007). Thus, irradiation of fast cycling cells results in less DNA repair and more cell death than irradiation of slow cycling or quiescent cells. In the case of captopril administration, when radiation exposure occurs within ~48 h of drug administration, the ST-HSC population is transiently quiescent, resulting in more DNA repair and improved cell survival. However, when radiation exposure occurs after 7 days of captopril treatment, the ST-HSC population has reentered the cell cycle, possibly due to the rebound expression of EPO, and less DNA repair and less cell survival occurs (Fig. 2).

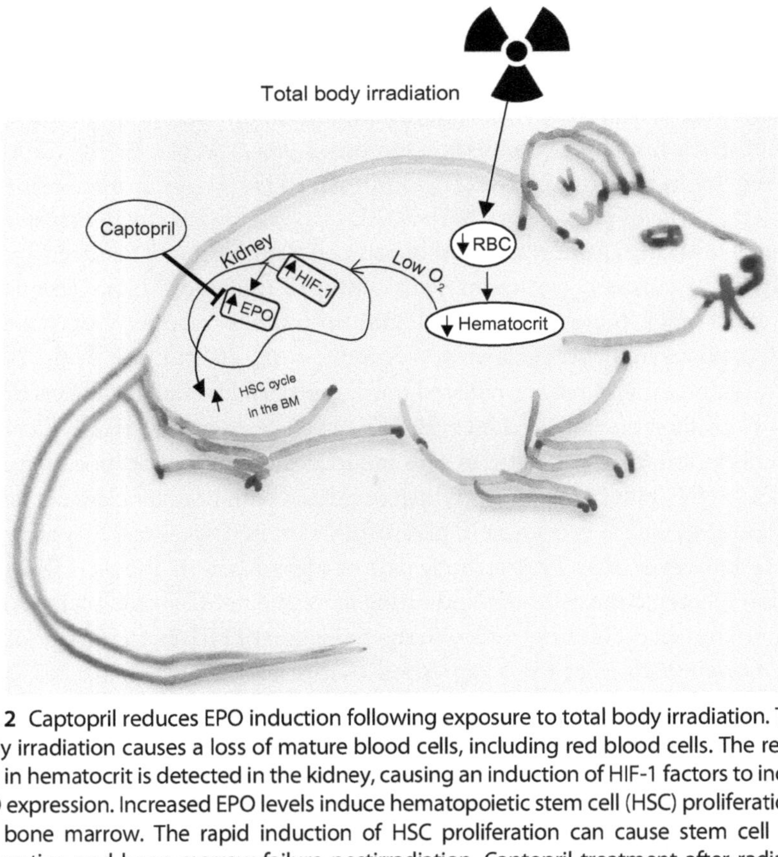

Fig. 2 Captopril reduces EPO induction following exposure to total body irradiation. Total body irradiation causes a loss of mature blood cells, including red blood cells. The reduction in hematocrit is detected in the kidney, causing an induction of HIF-1 factors to induce EPO expression. Increased EPO levels induce hematopoietic stem cell (HSC) proliferation in the bone marrow. The rapid induction of HSC proliferation can cause stem cell pool exhaustion and bone marrow failure postirradiation. Captopril treatment after radiation exposure causes a reduction of EPO levels and attenuates radiation-induced EPO. This results in a delay in signaling for HSC proliferation and prevents stem cell pool exhaustion.

6. SUMMARY AND CONCLUSIONS

The RAS is critical for blood pressure and blood volume homeostasis, and more recent evidence indicates that it also plays a critical role in hematopoietic cell homeostasis. Ang II exerts direct effects through its receptors on a variety of cells and has more far-reaching effects by its regulation of aldosterone and EPO. Independent of its actions as an endocrine system, angiotensin, renin, and ACE expression have also been demonstrated in a variety of other tissue types, along with the AT_1 and AT_2 receptors, suggesting that localized actions exist for Ang II (Phillips, Speakman, & Kimura, 1993). The overall impact of Ang II on other systems beyond the vascular and hematopoietic systems is still being elucidated.

ACKNOWLEDGMENTS

We thank Ms. Elizabeth McCart for critical reading of this manuscript. One of the authors is an employee of the US Government, and this manuscript was prepared as part of their official duties. Title 17 U.S.C. §105 provides that "Copyright protection under this title is not available for any work of the United States Government." Title 17 U.S.C. §101 defined a US Government work as a work prepared by a military service member or employees of the US Government as part of that person's official duties. The views in this article are those of the authors and do not necessarily reflect the views, official policy, or position of the Uniformed Services University of the Health Sciences, Department of the Navy, Department of Defense, or the US Federal Government.

REFERENCES

Akazawa, H., Yano, M., Yabumoto, C., Kudo-Sakamoto, Y., & Komuro, I. (2013). Angiotensin II type 1 and type 2 receptor-induced cell signaling. *Current Pharmaceutical Design*, *19*, 2988–2995.

Amann, F. W., Buhler, F. R., Conen, D., Brunner, F., Ritz, R., & Speck, B. (1980). Captoril-associated agranulocytosis. *Lancet*, *1*, 150.

Aplin, M., Bonde, M. M., & Hansen, J. L. (2009). Molecular determinants of angiotensin II type 1 receptor functional selectivity. *Journal of Molecular and Cellular Cardiology*, *46*, 15–24.

Aronow, W. S. (2016). Current treatment of hypertension in patients with coronary artery disease recommended by different guidelines. *Expert Opinion on Pharmacotherapy*, *17*, 205–215.

Barshishat-Kupper, M., Mungunsukh, O., Tipton, A. J., McCart, E. A., Panganiban, R. A., Davis, T. A., … Day, R. M. (2011). Captopril modulates hypoxia-inducible factors and erythropoietin responses in a murine model of total body irradiation. *Experimental Hematology*, *39*, 293–304.

Benohr, P., Harsch, S., Proksch, B., & Gleiter, C. H. (2004). Does angiotensin II modulate erythropoietin production in HepG2 cells? *Nephron. Experimental Nephrology*, *98*, e124–e131.

Bernstein, K. E. (1998). Examining the renin-angiotensin system one hundred years after its discovery. *Néphrologie, 19*, 391–395.

Beyazit, Y., Aksu, S., Haznedaroglu, I. C., Kekilli, M., Misirlioglu, M., Tuncer, S., ... Goker, H. (2007). Overexpression of the local bone marrow renin-angiotensin system in acute myeloid leukemia. *Journal of the National Medical Association, 99*, 57–63.

Bihoreau, C., Monnot, C., Davies, E., Teutsch, B., Bernstein, K. E., Corvol, P., & Clauser, E. (1993). Mutation of Asp74 of the rat angiotensin II receptor confers changes in antagonist affinities and abolishes G-protein coupling. *Proceedings of the National Academy of Sciences of the United States of America, 90*, 5133–5137.

Bollag, W. B. (2014). Regulation of aldosterone synthesis and secretion. *Comprehensive Physiology, 4*, 1017–1055.

Cai, T., Nishida, K., Hirano, T., & Khavari, P. A. (2002). Gab1 and SHP-2 promote Ras/MAPK regulation of epidermal growth and differentiation. *The Journal of Cell Biology, 159*, 103–112.

Carson, P., Giles, T., Higginbotham, M., Hollenberg, N., Kannel, W., & Siragy, H. M. (2001). Angiotensin receptor blockers: Evidence for preserving target organs. *Clinical Cardiology, 24*, 183–190.

Castrop, H. (2015). A role for AT1 receptor-associated proteins in blood pressure regulation. *Current Opinion in Pharmacology, 21*, 43–47.

Charrier, S., Michaud, A., Badaoui, S., Giroux, S., Ezan, E., Sainteny, F., ... Vainchenker, W. (2004). Inhibition of angiotensin I-converting enzyme induces radioprotection by preserving murine hematopoietic short-term reconstituting cells. *Blood, 104*, 978–985.

Cheungpasitporn, W., Thongprayoon, C., Chiasakul, T., Korpaisarn, S., & Erickson, S. B. (2015). Renin-angiotensin system inhibitors linked to anemia: A systematic review and meta-analysis. *QJM, 108*, 879–884.

Chisi, J. E., Briscoe, C. V., Ezan, E., Genet, R., Riches, A. C., & Wdzieczak-Bakala, J. (2000). Captopril inhibits in vitro and in vivo the proliferation of primitive haematopoietic cells induced into cell cycle by cytotoxic drug administration or irradiation but has no effect on myeloid leukaemia cell proliferation. *British Journal of Haematology, 109*, 563–570.

Chisi, J. E., Wdzieczak-Bakala, J., Thierry, J., Briscoe, C. V., & Riches, A. C. (1999). Captopril inhibits the proliferation of hematopoietic stem and progenitor cells in murine long-term bone marrow cultures. *Stem Cells, 17*, 339–344.

Cole, J., Ertoy, D., Lin, H., Sutliff, R. L., Ezan, E., Guyene, T. T., ... Bernstein, K. E. (2000). Lack of angiotensin II-facilitated erythropoiesis causes anemia in angiotensin-converting enzyme-deficient mice. *Journal of Clinical Investigation, 106*, 1391–1398.

Davis, T., Landauer, M., Mog, S., Barshishat-Kupper, M., Zins, S., Amare, M., & Day, R. (2010). Timing of captopril administration determines radiation protection or radiation sensitization in a murine model of total body irradiation. *Experimental Hematology, 38*, 270–281.

Day, F. L., Rafty, L. A., Chesterman, C. N., & Khachigian, L. M. (1999). Angiotensin II (ATII)-inducible platelet-derived growth factor a-chain gene expression is p42/44 extracellular signal-regulated kinase-1/2 and Egr-1-dependent and mediated via the ATII type 1 but not type 2 receptor. Induction by ATII antagonized by nitric oxide. *Journal of Biological Chemistry, 274*, 23726–23733.

Deschepper, C. F. (1994). Angiotensinogen: Hormonal regulation and relative importance in the generation of angiotensin II. *Kidney International, 46*, 1561–1563.

Dezsi, C. A. (2014). Differences in the clinical effects of angiotensin-converting enzyme inhibitors and angiotensin receptor blockers: A critical review of the evidence. *American Journal of Cardiovascular Drugs, 14*, 167–173.

Doan, T. N., Gletsu, N., Cole, J., & Bernstein, K. E. (2001). Genetic manipulation of the renin-angiotensin system. *Current Opinion in Nephrology and Hypertension, 10*, 483–491.

Dong, C., & Wu, G. (2013). G-protein-coupled receptor interaction with small GTPases. *Methods in Enzymology, 522*, 97–108.

Dunn, A., Lo, V., & Donnelly, S. (2007). The role of the kidney in blood volume regulation: The kidney as a regulator of the hematocrit. *The American Journal of the Medical Sciences, 334*, 65–71.

Elis, A., Lishner, M., Lang, R., & Ravid, M. (1991). Agranulocytosis associated with enalapril. *DICP, 25*, 461–462.

Elliott, S., Pham, E., & Macdougall, I. C. (2008). Erythropoietins: A common mechanism of action. *Experimental Hematology, 36*, 1573–1584.

el-Makri, A., Larabi, M. S., Kechrid, C., Belkahia, C., & Ben Ayed, H. (1981). Fatal bone-marrow suppression associated with captopril. *British Medical Journal (Clinical Research Ed.), 283*, 277–278.

Fisher, J. W., Samuels, A. I., & Langston, J. W. (1967). Effects of angiotensin and renal artery constriction on erythropoietin production. *The Journal of Pharmacology and Experimental Therapeutics, 157*, 618–625.

Frank, G. J. (1989). The safety of ACE inhibitors for the treatment of hypertension and congestive heart failure. *Cardiology, 76*(Suppl. 2), 56–67.

Freudenthaler, S. M., Lucht, I., Schenk, T., Brink, M., & Gleiter, C. H. (2000). Dose-dependent effect of angiotensin II on human erythropoietin production. *Pflügers Archiv, 439*, 838–844.

Freudenthaler, S. M., Schreeb, K., Korner, T., & Gleiter, C. H. (1999). Angiotensin II increases erythropoietin production in healthy human volunteers. *European Journal of Clinical Investigation, 29*, 816–823.

Gavras, I., Graff, L. G., Rose, B. D., McKenna, J. M., Brunner, H. R., & Gavras, H. (1981). Fatal pancytopenia associated with the use of captopril. *Annals of Internal Medicine, 94*, 58–59.

Gleason, M. M., Roloff, J. S., Cyran, S. E., Weber, H. S., Baylen, B. G., & Myers, J. L. (1993). Captopril-induced bone marrow suppression in two cardiac patients with trisomy 21. *Pediatric Cardiology, 14*, 227–229.

Godeny, M. D., Sayyah, J., VonDerLinden, D., Johns, M., Ostrov, D. A., Caldwell-Busby, J., & Sayeski, P. P. (2007). The N-terminal SH2 domain of the tyrosine phosphatase, SHP-2, is essential for Jak2-dependent signaling via the angiotensin II type AT1 receptor. *Cellular Signalling, 19*, 600–609.

Gonzalez, A. A., & Prieto, M. C. (2015). Renin and the (pro)renin receptor in the renal collecting duct: Role in the pathogenesis of hypertension. *Clinical and Experimental Pharmacology & Physiology, 42*, 14–21.

Gossmann, J., Burkhardt, R., Harder, S., Lenz, T., Sedlmeyer, A., Klinkhardt, U., … Scheuermann, E. H. (2001). Angiotensin II infusion increases plasma erythropoietin levels via an angiotensin II type 1 receptor-dependent pathway. *Kidney International, 60*, 83–86.

Gould, A. B., & Goodman, S. A. (1990). Effect of an angiotensin-converting enzyme inhibitor on blood pressure and erythropoiesis in rats. *European Journal of Pharmacology, 181*, 225–234.

Gould, A. B., Goodman, S., DeWolf, R., Onesti, G., & Swartz, C. (1980). Interrelation of the renin system and erythropoietin in rats. *The Journal of Laboratory and Clinical Medicine, 96*, 523–534.

Gradman, A. H. (1999). Long-term benefits of angiotensin II blockade: Is the consensus changing? *The American Journal of Cardiology, 84*, 16S–21S.

Griffing, G. T., & Melby, J. C. (1982). Enalapril (MK-421) and the white cell count and haematocrit. *Lancet, 1*, 1361.

Guillemot, L., Levy, A., Raymondjean, M., & Rothhut, B. (2001). Angiotensin II-induced transcriptional activation of the cyclin D1 gene is mediated by Egr-1 in CHO-AT(1A) cells. *The Journal of Biological Chemistry*, *276*, 39394–39403.

Hamidou, M., Glemarec, J., Elia-Foeillet, S., Acard, F., Chiffoleau, A., & Grolleau, J. Y. (1996). Agranulocytosis caused by perindopril. *Annales de Médecine Interne (Paris)*, *147*, 523–524.

Harrison, B. D., Laidlaw, S. T., & Reilly, J. T. (1995). Fatal aplastic anaemia associated with lisinopril. *Lancet*, *346*, 247–248.

Haznedaroglu, I. C., & Ozturk, M. A. (2003). Towards the understanding of the local hematopoietic bone marrow renin-angiotensin system. *The International Journal of Biochemistry & Cell Biology*, *35*, 867–880.

Horowitz, N., Molnar, M., Levy, Y., & Pollack, S. (2005). Ramipril-induced agranulocytosis confirmed by a lymphocyte cytotoxicity test. *The American Journal of the Medical Sciences*, *329*, 52–53.

Hosseinimehr, S. J., Zakaryaee, V., & Froughizadeh, M. (2006). Oral oxymetholone reduces mortality induced by gamma irradiation in mice through stimulation of hematopoietic cells. *Molecular and Cellular Biochemistry*, *287*, 193–199.

Imai, K., Yoshimura, S., Ohtki, T., & Hashimoto, K. (1985). One month studies on the subacute toxicity of captopril in rats. *The Journal of Toxicological Sciences*, *6*(Suppl. 2), 189–214.

Irvin, J. D., & Viau, J. M. (1986). Safety profiles of the angiotensin converting enzyme inhibitors captopril and enalapril. *The American Journal of Medicine*, *81*, 46–50.

Israeli, A., Or, R., & Leitersdorf, E. (1985). Captopril-associated transient aplastic anemia. *Acta Haematologica*, *73*, 106–107.

Jokubaitis, V. J., Sinka, L., Driessen, R., Whitty, G., Haylock, D. N., Bertoncello, I., … Simmons, P. J. (2008). Angiotensin-converting enzyme (CD143) marks hematopoietic stem cells in human embryonic, fetal, and adult hematopoietic tissues. *Blood*, *111*, 4055–4063.

Kato, H., Ishida, J., Imagawa, S., Saito, T., Suzuki, N., Matsuoka, T., … Fukamizu, A. (2005). Enhanced erythropoiesis mediated by activation of the renin-angiotensin system via angiotensin II type 1a receptor. *The FASEB Journal*, *19*, 2023–2025.

Kato, H., Ishida, J., Matsusaka, T., Ishimaru, T., Tanimoto, K., Sugiyama, F., … Fukamizu, A. (2015). Erythropoiesis and blood pressure are regulated via AT1 receptor by distinctive pathways. *PloS One*, *10*, e0129484.

Kim, Y. C., & Day, R. M. (2012). Angiotensin II regulates activation of Bim via Rb/E2F1 during apoptosis: Involvement of interaction between AMPKbeta1/2 and Cdk4. *American Journal of Physiology. Lung Cellular and Molecular Physiology*, *303*, L228–L238.

Lee, Y. H., Mungunsukh, O., Tutino, R. L., Marquez, A. P., & Day, R. M. (2010). Angiotensin-II-induced apoptosis requires regulation of nucleolin and Bcl-xL by SHP-2 in primary lung endothelial cells. *Journal of Cell Science*, *123*, 1634–1643.

Lindholm, L. H., Persson, M., Alaupovic, P., Carlberg, B., Svensson, A., & Samuelsson, O. (2003). Metabolic outcome during 1 year in newly detected hypertensives: Results of the antihypertensive treatment and lipid profile in a north of Sweden efficacy evaluation (ALPINE study). *Journal of Hypertension*, *21*, 1563–1574.

Lu, H., Cassis, L. A., Kooi, C. W., & Daugherty, A. (2016). Structure and functions of angiotensinogen. *Hypertension Research*, *39*, 492–500.

Lynch, K. R., & Peach, M. J. (1991). Molecular biology of angiotensinogen. *Hypertension*, *17*, 263–269.

Marathias, K. P., Agroyannis, B., Mavromoustakos, T., Matsoukas, J., & Vlahakos, D. V. (2004). Hematocrit-lowering effect following inactivation of renin-angiotensin system with angiotensin converting enzyme inhibitors and angiotensin receptor blockers. *Current Topics in Medicinal Chemistry*, *4*, 483–486.

Marty, C., & Ye, R. D. (2010). Heterotrimeric G protein signaling outside the realm of seven transmembrane domain receptors. *Molecular Pharmacology, 78*, 12–18.
Matsubara, H. (1998). Pathophysiological role of angiotensin II type 2 receptor in cardiovascular and renal diseases. *Circulation Research, 83*, 1182–1191.
Mogi, M., Iwai, M., & Horiuchi, M. (2007). Emerging concepts of regulation of angiotensin II receptors: New players and targets for traditional receptors. *Arteriosclerosis, Thrombosis, and Vascular Biology, 27*(12), 2532–2539. https://doi.org/10.1161/ATVBAHA.107.144154.
Montagner, A., Yart, A., Dance, M., Perret, B., Salles, J. P., & Raynal, P. (2005). A novel role for Gab1 and SHP2 in epidermal growth factor-induced Ras activation. *The Journal of Biological Chemistry, 280*, 5350–5360.
Moulder, J. E., Cohen, E. P., Fish, B. L., & Hill, P. (1993). Prophylaxis of bone marrow transplant nephropathy with captopril, an inhibitor of angiotensin-converting enzyme. *Radiation Research, 136*(3), 404–407.
Moussa, Z., Schiano, T., Spatoliatore, G., & Salman, S. (1992). Neutropenia induced by low-dose captopril. *New York State Journal of Medicine, 92*, 219–220.
Mrug, M., Stopka, T., Julian, B. A., Prchal, J. F., & Prchal, J. T. (1997). Angiotensin II stimulates proliferation of normal early erythroid progenitors. *Journal of Clinical Investigation, 100*, 2310–2314.
Naeshiro, I., Sato, K., Chatani, F., & Sato, S. (1998). Possible mechanism for the anemia induced by candesartan cilexetil (TCV-116), an angiotensin II receptor antagonist, in rats. *European Journal of Pharmacology, 354*, 179–187.
Naito, M., Kawashima, A., Akiba, T., Takanashi, M., & Nihei, H. (2003). Effects of an angiotensin II receptor antagonist and angiotensin-converting enzyme inhibitors on burst forming units-erythroid in chronic hemodialysis patients. *American Journal of Nephrology, 23*(5), 287–293.
Nakao, K., Shirakura, T., Azuma, M., & Maekawa, T. (1967). Studies on erythropoietic action of angiotensin II. *Blood, 29*, 754–760.
Navar, L. G., Harrison-Bernard, L. M., Imig, J. D., Cervenka, L., & Mitchell, K. D. (2000). Renal responses to AT1 receptor blockade. *American Journal of Hypertension, 13*(1 Pt. 2), 45S–54S.
Neaeshiro, I., Ishimura, M., Okada, J., Chantani, F., & Sato, S. (1997). Four-week oral toxicity study of TCV-116 in rats with 4- and 13-week recovery periods. *Japanese Pharmacology and Therapeutics, 25*, 25–49.
Negro, R. (2008). Endothelial effects of antihypertensive treatment: Focus on irbesartan. *Vascular Health and Risk Management, 4*(1), 89–101.
Olsen, M. H., Wachtell, K., Beevers, G., Dahlof, B., Devereux, R. B., de Faire, U., … Wedel, H. (2009). Prognostic importance of hemoglobin in hypertensive patients with electrocardiographic left ventricular hypertrophy: The losartan intervention for end point reduction in hypertension (LIFE) study. *American Heart Journal, 157*, 177–184.
Olson, E. R., Shamhart, P. E., Naugle, J. E., & Meszaros, J. G. (2008). Angiotensin II-induced extracellular signal-regulated kinase 1/2 activation is mediated by protein kinase Cdelta and intracellular calcium in adult rat cardiac fibroblasts. *Hypertension, 51*, 704–711.
Onoyama, K., Sanai, T., Motomura, K., & Fujishima, M. (1989). Worsening of anemia by angiotensin converting enzyme inhibitors and its prevention by antiestrogenic steroid in chronic hemodialysis patients. *Journal of Cardiovascular Pharmacology, 13*(Suppl. 3), S27–S30.
Phillips, M. I., Speakman, E. A., & Kimura, B. (1993). Levels of angiotensin and molecular biology of the tissue renin angiotensin systems. *Regulatory Peptides, 43*, 1–20.

Pratt, M. C., Lewis-Barned, N. J., Walker, R. J., Bailey, R. R., Shand, B. I., & Livesey, J. (1992). Effect of angiotensin converting enzyme inhibitors on erythropoietin concentrations in healthy volunteers. *British Journal of Clinical Pharmacology*, *34*, 363–365.

Pueyo, M. E., & Michel, J. B. (1997). Angiotensin II receptors in endothelial cells. *General Pharmacology*, *29*, 691–696.

Reid, I. A. (1992). Interactions between ANG II, sympathetic nervous system, and baroreceptor reflexes in regulation of blood pressure. *The American Journal of Physiology*, *262*(6 Pt. 1), E763–E778.

Rigat, B., Hubert, C., Alhenc-Gelas, F., Cambien, F., Corvol, P., & Soubrier, F. (1990). An insertion/deletion polymorphism in the angiotensin I-converting enzyme gene accounting for half the variance of serum enzyme levels. *Journal of Clinical Investigation*, *86*, 1343–1346.

Rodgers, K. E., Xiong, S., & diZerega, G. S. (2002). Accelerated recovery from irradiation injury by angiotensin peptides. *Cancer Chemotherapy and Pharmacology*, *49*, 403–411.

Rodgers, K. E., Xiong, S., Steer, R., & diZerega, G. S. (2000). Effect of angiotensin II on hematopoietic progenitor cell proliferation. *Stem Cells*, *18*, 287–294.

Rojas-Vega, L., & Gamba, G. (2016). Mini-review: Regulation of the renal NaCl cotransporter by hormones. *American Journal of Physiology. Renal Physiology*, *310*, F10–F14.

Sato, K., Nishida, N., Sasaki, H., Chatani, F., Sato, S., Shikata, M., & Mitani, M. (1996). Four-week oral toxicity study of TCV-116 in rats. *Japanese Pharmacology and Therapeutics*, *24*, s787–s804.

Schratzlseer, G., Lipp, T., Riess, G., Antoni, D., & Delius, W. (1994). Severe pancytopenia in old age after 12-month ACE inhibitor therapy. *Deutsche Medizinische Wochenschrift*, *119*, 1029–1033.

Stroth, U., & Unger, T. (1999). The renin-angiotensin system and its receptors. *Journal of Cardiovascular Pharmacology*, *33*(Suppl. 1), S21–S28. discussion S41-23.

Taddei, S. (2015). RAS inhibitors' dose-dependent efficacy: Myth or reality? *Current Medical Research and Opinion*, *31*, 1245–1256.

Tamulevicius, P., Wang, M., & Iliakis, G. (2007). Homology-directed repair is required for the development of radioresistance during S phase: Interplay between double-strand break repair and checkpoint response. *Radiation Research*, *167*, 1–11.

Uz, B., Tatonyan, S. C., Sayitoglu, M., Erbilgin, Y., Hatirnaz, O., Aksu, S., … Haznedaroglu, I. C. (2014). Local renin-angiotensin system in normal hematopoietic and multiple myeloma-related progenitor cells. *Turkish Journal of Haematology*, *31*, 136–142.

van Brummelen, P., Willemze, R., Tan, W. D., & Thompson, J. (1980). Captopril-associated agranulocytosis. *Lancet*, *1*(8160), 150.

Vinturache, A. E., & Smith, F. G. (2014). Angiotensin type 1 and type 2 receptors during ontogeny: Cardiovascular and renal effects. *Vascular Pharmacology*, *63*, 145–154.

Vlahakos, D. V., Canzanello, V. J., Madaio, M. P., & Madias, N. E. (1991). Enalapril-associated anemia in renal transplant recipients treated for hypertension. *American Journal of Kidney Diseases*, *17*, 199–205.

Vlahakos, D. V., Marathias, K. P., & Madias, N. E. (2010). The role of the renin-angiotensin system in the regulation of erythropoiesis. *American Journal of Kidney Diseases*, *56*, 558–565. http://dx.doi.org/10.1053/j.ajkd.2009.12.042.

Waeber, B., Gavras, I., Brunner, H. R., & Gavras, H. (1981). Safety and efficacy of chronic therapy with captopril in hypertensive patients: An update. *Journal of Clinical Pharmacology*, *21*, 508–516.

Wong, P. C., Barnes, T. B., Chiu, A. T., Christ, D. D., Duncia, J. V., Herblin, W. F., & Timmermans, P. B. M. W. M. (1991). Losartan (DuP 753), an oral active nonpeptide angiotensin receptor antagonist. *Cardiovascular Drug Reviews*, *9*, 317–339.

Xie, X., Liu, Y., Perkovic, V., Li, X., Ninomiya, T., Hou, W., ... Wang, H. (2015). Renin-angiotensin system inhibitors and kidney and cardiovascular outcomes in patients with CKD: A Bayesian network meta-analysis of randomized clinical trials. *American Journal of Kidney Diseases*, *67*, 728–741.

Yang, H., Lu, D., Yu, K., & Raizada, M. K. (1996). Regulation of neuromodulatory actions of angiotensin II in the brain neurons by the Ras-dependent mitogen-activated protein kinase pathway. *The Journal of Neuroscience*, *16*, 4047–4058.

Yaren, A., Oztop, I., Turgut, S., Turgut, G., Degirmencioglu, S., & Demirpence, M. (2008). Angiotensin-converting enzyme gene polymorphism is associated with anemia in non small-cell lung cancer. *Experimental Biology and Medicine (Maywood, N.J.)*, *233*(1), 32–37.

Yee, D. K., Suzuki, A., Luo, L., & Fluharty, S. J. (2006). Identification of structural determinants for G protein-independent activation of mitogen-activated protein kinases in the seventh transmembrane domain of the angiotensin II type 1 receptor. *Molecular Endocrinology*, *20*, 1924–1934.

Young, T. H., Lin, S. H., Ko, W. S., Hwang, W. S., & Ding, Y. A. (1993). Captopril-induced agranulocytosis: A case report. *Zhonghua Yi Xue Za Zhi (Taipei)*, *51*, 151–153.

Young, D. B., & Pan, Y. J. (1984). Interaction between the natriuretic effects of renal perfusion pressure and the antinatriuretic effects of angiotensin and aldosterone in control of sodium excretion. *Journal de Physiologie*, *79*, 511–517.

Zambidis, E. T., Park, T. S., Yu, W., Tam, A., Levine, M., Yuan, X., ... Peault, B. (2008). Expression of angiotensin-converting enzyme (CD143) identifies and regulates primitive hemangioblasts derived from human pluripotent stem cells. *Blood*, *112*, 3601–3614.

Zhao, D., Wang, Z. M., & Wang, L. S. (2015). Prevention of atrial fibrillation with renin-angiotensin system inhibitors on essential hypertensive patients: A meta-analysis of randomized controlled trials. *Journal of Biomedical Research*, *29*, 475–485.

Zivna, M., Hulkova, H., Matignon, M., Hodanova, K., Vylet'al, P., Kalbacova, M., ... Kmoch, S. (2009). Dominant renin gene mutations associated with early-onset hyperuricemia, anemia, and chronic kidney failure. *American Journal of Human Genetics*, *85*, 204–213.

CHAPTER FIVE

Erythropoietin Receptor Signaling and Lipid Rafts

Kathy McGraw[1], Alan List
Division of Clinical Sciences, H. Lee Moffitt Cancer Center & Research Institute, Tampa, FL, United States
[1]Corresponding author: e-mail address: kathy.mcgraw@moffitt.org

Contents

Abstract

Erythropoiesis is tightly regulated by the growth factor erythropoietin (Epo). Signal activation begins when Epo engages its cognate receptor, Epo-R, triggering receptor homodimerization, and recruitment of signaling intermediates including Jak2 that phosphorylates both the receptor cytoplasmic tail and downstream effectors including the transcription factor, STAT5. Transcription factors subsequently activate transcription of prosurvival and prodifferentiation genes responsible for red blood cell production. The fidelity of Epo-R signaling is dependent upon residence within detergent insoluble membrane lipid raft fractions. Lipid rafts are membrane microdomains that serve as signaling scaffolds composed of densely packed sphingolipids and cholesterol where receptors and intermediate signaling proteins are recruited and interact to execute stimuli. Disruption of lipid rafts is detrimental to Epo signaling, a phenomenon that may be utilized to design novel therapeutics for conditions in which Epo signaling is deficient. Here, we review the Epo signaling cascade, particularly, as it relates to localization and dependence on lipid rafts, and discuss considerations for novel therapeutic design.

Vitamins and Hormones, Volume 105
ISSN 0083-6729
http://dx.doi.org/10.1016/bs.vh.2017.02.002
© 2017 Elsevier Inc.
All rights reserved.

1. INTRODUCTION

Erythropoietin (Epo) is the primary cytokine responsible for erythropoiesis, stimulating both the proliferation and differentiation of erythroid progenitors into mature red blood cells (RBCs). Basal Epo levels are required to maintain steady state levels of RBCs in the blood, however, during times of blood loss or hypoxic stress, expression and secretion of the cytokine by renal interstitial fibroblasts is quickly upregulated in response to hypoxic sensors, thereby expanding the pool of erythroid progenitors and differentiation of mature erythrocytes. As such, tight regulation of the pathway is critical. Biological response to the cytokine is initiated upon engagement of Epo with its cognate receptor, Epo-R. This interaction induces conformational changes and receptor homodimerization that triggers downstream signaling events resulting in the activation of transcription factors that promote erythroid expansion. Recombinant human Epo (rhEpo) has widespread therapeutic use and is standard of care for a number of anemias including that arising from chronic kidney disease, chemotherapy, or infection, and a group of disorders known as the myelodysplastic syndromes (MDSs). In MDS, rhEpo treatment rescues diminished Epo signaling restoring erythropoiesis and correcting anemia resulting in transfusion independence (Hoefsloot, van Amelsvoort, Broeders, et al., 1997; Mantovani, Lentini, Hentschel, et al., 2000). Thorough investigation of Epo-R signaling is a key to understanding the pathobiology of disorders such as MDS, as well as identifying novel therapeutic strategies.

Lipid rafts are small (20–100 nm) plasma membrane microdomains composed of densely packed cholesterol and sphingolipids that are free to "float" laterally throughout the phospholipid bilayer of the plasma membrane. The role of lipid rafts in immune signaling has been well studied, particularly in B- and T-cell receptor (TCR) signal activation and in the formation of immunological synapses. These lipid microdomains act as docking stations whereby receptors and signal intermediates interact with one another. Protein recruitment and raft-associated spatial proximity facilitates downstream signal cascade activation (Kabouridis & Jury, 2008; McGraw, Fuhler, Johnson, et al., 2012). Here, we will explore the role of lipid rafts in erythroid precursors, and particularly, the relationship between lipid rafts and Epo-R signal fidelity. We will conclude that intact lipid raft microdomains are imperative to maximize Epo-R signal fidelity and that this phenomenon may be exploited to uncover novel therapeutic strategies to treat conditions with aberrant signaling.

2. EPO AND EPO RECEPTOR

Epo is the primary regulator of erythropoiesis and is critical for tissue oxygenation under both steady-state conditions and after acute blood loss or hypoxia. The *Epo* gene was first cloned in 1985 after purification by Miyake et al. who isolated the cytokine from the urine of patients with aplastic anemia. Identification of the amino acid sequence quickly led to identification of the nucleotide sequence followed by full characterization of the gene (Lacombe & Mayeux, 1998; Miyake, Kung, & Goldwasser, 1977). The *Epo* gene is located on chromosome 7q22, and transcriptional regulation is controlled by the transcription factors hypoxia-inducible factor (HIF)-1-alpha, signal transducer and activator of transcription 5 (STAT5), STAT3, and GATA-binding protein 1 (GATA-1) (Lacombe & Mayeux, 1998; Mulcahy, 2001; Zhao, Kitidis, Fleming, Lodish, & Ghaffari, 2006). These factors are critical to a cell's ability to react to changes in physiologic needs for circulating erythrocytes. The Epo protein is a 34-kDa circulating glycoprotein that is produced predominately in the kidney and embryologically in the liver (Ebert & Bunn, 1999; Fisher, 2003; Lacombe & Mayeux, 1998). rhEpo protein has great clinical utility and is used to treat a number of conditions including anemia caused by chronic renal failure, chronic inflammatory disorders, chemotherapy, and MDS. Although rhEpo administration alleviates anemia by promoting the production of RBCs, rapid correction is associated with thromboembolism and chronic use with iron deficiency (Singbartl, 1994).

The Epo-R is the primary receptor that binds to and transmits the activation signal of the Epo cytokine. The Epo-R is a classical Type-I cytokine receptor with an extracellular domain that interacts with the microenvironment, a transmembrane region that spans the phospholipid bilayer, and a cytoplasmic tail that contains eight tyrosine phosphorylation sites that serve as docking locations for signaling adapters. Specific to the Type-I family, these cytokine receptors contain no intrinsic kinase activity and therefore rely upon the recruitment of kinases and adapter molecules to potentiate the receptor signal (D'Andrea & Zon, 1990; Lacombe & Mayeux, 1998). Epo-R maturation begins in the endoplasmic reticulum, after which it is transported to the Golgi apparatus. Only a small proportion of receptors will leave the Golgi, mature via heavy glycosylation, and be transported to and expressed on the plasma membrane. Expression of mature Epo-R on the plasma membrane is relatively low under stimulus-free conditions with less than 100 receptors predicted to be expressed at any given instance (D'Andrea & Zon, 1990; Lacombe & Mayeux, 1998). Importantly, transport of the receptor through its maturation process is dependent on its interaction with Janus Kinase 2 (Jak2) (Huang, Constantinescu, & Lodish, 2001). Jak2 is not only important for the surface expression of the receptor; it is also the primary effector kinase of the Epo/Epo-R signaling axis. Ligand engagement induces receptor conformational changes and homodimerization that results in the autophosphorylation of Jak2. Jak2, in turn, phosphorylates the receptor cytoplasmic tail to recruit a number of adaptor intermediates leading to transcriptional activation of genes involved in promoting the proliferation and differentiation of erythrocytes. The Epo-R signaling cascade and activation steps will be described in further detail in the next section. Epo-R signal attenuation occurs by negative-feedback inhibition via several mechanisms including suppressor of cytokine signaling (SOCS) proteins, negative-regulating phosphatases, receptor internalization and ubiquitination, and proteasomal degradation. The delicate balance between receptor maturation and receptor turnover is highly regulated and must occur rapidly as the body responds to external stimuli. As such, defects in this regulation, or in the receptor itself, will have damaging effects leading to disease pathogenesis and fluctuations in total red cell mass.

There are a number of primary bone marrow conditions that can be responsive to treatment with rhEpo such as MDS. MDS patients have diminished STAT5 transcription factor activity following Epo stimulation, indicating impairment in Epo-R signal response. A small proportion of patients do harbor single site Epo-R mutations; however, the functional consequence of these is unknown and does not appear to be detrimental to downstream signaling (Lewinski, Floru, Cohen, & Mittelmann, 1994). In myeloproliferative neoplasms (MPNs), such as polycythemia vera (PV), there is hypersensitivity to Epo despite lower levels of circulating cytokine, owing to somatic mutations involving JH2 pseudokinase autoinhibitory domain of *JAK2*, resulting in constitutive activation (McMullin & Percy, 1999). Similarly, in both erythroleukemia, a type of acute myeloid leukemia with pancytopenia, and Diamond Blackman Anemia, an inherited bone marrow failure syndrome with profound anemia, the Epo-R remains intact (McMullin & Percy, 1999). Alternatively, in Familial Erythrocytosis, a condition characterized by erythrocytosis, inherited mutations in the *EpoR* gene introduce a premature stop codon in the amino acid sequence resulting in halted translation. Interestingly, the truncated receptor is hypersensitive to circulating cytokine resulting in increased Jak2 and STAT5 activation (Watowich, Xie, Klingmuller, et al., 1999). This leads to sustained proliferation and increased RBC mass causing dizziness, excess bleeding, and increased risk for thrombus formation along with other symptoms in affected individuals. The lack of universal alterations in Epo-R or diminished circulating cytokine in these and other disorders, despite clear differences in Epo signaling, suggest that other factors involved in the signaling axis must be explored.

3. CONDITIONS WITH ABERRANT EPO SIGNALING

With the exception of Familial Erythrocytosis, aberrant Epo signaling found in a number of disorders cannot be attributed to dysfunctional Epo-R. Instead, aberrancies exist somewhere downstream of the receptor. MPNs are hematopoietic disorders characterized by increased production of particular cell lineages. The most common MPNs include PV characterized by increased red cell mass, platelets, and often leukocytes, primary myelofibrosis (PMF) characterized by a marked increase in bone marrow fibrosis, and essential thrombocytopenia (ET) characterized by sustained elevation in platelets. In PV patients, there are lower levels of circulating cytokine, and erythroid progenitor cells are able to proliferate independent of Epo. This is attributed to hypersensitivity to Epo, and other growth factors (McLornan, Percy, & McMullin, 2006). Just over a decade ago, a mutation in *Jak2* that changes valine to phenylalanine at amino acid 617 (V617F) was identified. Since, this mutation has been identified in 95% of PV patients and 50% of PMF and ET patients. In those cases that do not harbor the V617F mutation, other mutations in either *Jak2* or Jak2-regulating genes have been identified. These data demonstrate that aberrant Jak2 signaling downstream of the Epo-R underlies aberrant signaling. The V617F mutation causes loss of Jak2 autoinhibition resulting in constitutively active kinase activity signaling resulting in over transcribed erythroid promoting genes (Griner, McGraw, Johnson, List, & Reuther, 2013).

The MDSs are hematopoietic stem cell malignancies characterized by ineffective hematopoiesis. MDS patients display bone marrow failure, dysplastic bone marrow features, cytopenias, and an increased propensity for transformation to acute myeloid leukemia. MDS primarily affects older individuals (greater than 60 years). Survival depends on a number of factors and ranges from many years to only months. Prognostication is determined by bone barrow blast percentage, number and severity of cytopenias, cytogenetic abnormalities, and presence of somatic gene mutations. Anemia, thrombocytopenia, and neutropenia are common in MDS and the pathogenesis of the disease in unclear. Recent evidence suggests that dysregulation of innate immune signaling and chronic inflammation of the bone marrow microenvironment drive disease pathogenesis. Predisposing factors to MDS are also not clear; however, individuals with therapy-related toxicity such as that from chemotherapy or radiation are at increased risk for the MDS development. Clonal hematopoiesis also appears to contribute to MDS risk. Clonal hematopoiesis is the process of somatic gene mutation acquisition and clonal expansion. As a population ages, so does the frequency of acquiring somatic mutations in genes common to MDS. Clonal hematopoiesis is associated with increased risk for hematological malignancies including MDS (Genovese, Kahler, Handsaker, et al., 2014; Jaiswal, Fontanillas, Flannick, et al., 2014). Although our knowledge of the factors that contribute to MDS pathogenesis is increasing, we do not know what causes the deficient Epo-R signal. Therefore, it is imperative to interrogate the Epo-R signaling axis to determine its role in MDS pathogenesis.

4. EPO RECEPTOR SIGNALING

The Epo-R signaling cascade extends to a number of well-characterized enzymatic pathways that result in transcriptional activation of proliferation and antiapoptotic genes. These pathways are initiated when the Epo ligand engages the receptor, triggering conformational changes and receptor homodimerization. Dimerization, in turn, directs activation of Jak2 followed by signaling intermediates leading to increased activity of transcription factors including STAT3, STAT5, and GATA-1. As the Epo-R has no intrinsic kinase activity, it relies on these intermediates for downstream phosphorylation events. The most well-studied Epo-R signaling pathway is the Jak2/STAT5 axis. Epo-R homodimerization causes autophosphorylation of Jak2 kinase. Jak2, the key effecter of Epo signaling, is physically associated with Epo-R throughout receptor maturation and its transport to the plasma membrane. Jak2 phosphorylates tyrosine residues on the cytoplasmic tail of Epo-R that serve as docking sites for other signaling intermediates to bind to the receptor via their src-homology 2 (SH2) domains (Lacombe & Mayeux, 1998). In addition, Jak2 phosphorylates a number of key effectors. The most well-known substrate for Jak2 is STAT5. Jak2 phosphorylates STAT5, which is necessary for dimerization and translocation to the nucleus. In the nucleus, STAT5 binds to DNA and transcriptionally activates proliferation, antiapoptotic, and erythroid differentiation genes. The phosphatidyl-3 kinase (PI3K)/Akt pathway is also activated by Jak2 phosphorylation of Tyr^{479} on the Epo-R cytoplasmic tail as well as through interaction with adapter proteins IRS2 and Grb-2. The serine threonine kinase, Akt, downstream of PI3K, phosphorylates GATA-1 which promotes transcription of genes involved in erythroid differentiation and transcription of *EpoR* and *GATA-1*, thereby creating a positive-feedback loop (Zhao et al., 2006). The mitogen-activated protein kinase (MAPK) pathway is also activated upon Jak2 phosphorylation of Shc that interacts with Grb2. This activates Ras which in turn activates Raf-1, and then the MAPK protein family (Miura, Miura, Ihle, & Aoki, 1994). Another key kinase potentiating Epo-R signaling is Lyn kinase. Lyn kinase not only phosphorylates tyrosines on the cytoplasmic receptor tail but also phosphorylates STAT5 and regulates MAPK, further augmenting signal activity (Lacombe & Mayeux, 1998). In addition, Lyn kinase acts on negative-regulating phosphatases that dampen Jak2, STAT5, and MAPK activity, further increasing signal output (Slavova-Azmanova, Kucera, Louw, et al., 2014). Interestingly, Lyn kinase also negatively regulates the Epo-R signaling pathway via activation of the phosphatase SHP-1, which diminishes activity of Jak/STAT and MAPK (Slavova-Azmanova et al., 2014).

Negative regulation of Epo-R signaling occurs via several mechanisms. Primary regulation is accomplished via negative-feedback loops and SOCS family proteins. Eight proteins make up the SOCS family and each member contains an SH2 domain that can bind to phosphorylated tyrosines, and a SOCS box domain that directs target substrate proteins for degradation (Jegalian & Wu, 2002). These proteins suppress and extinguish cytokine signaling by several methods including inactivation of Jak2 and STAT5, displacing intermediates from receptor docking sites, and by regulating protein levels via ubiquitin tags targeting them for proteasomal degradation (Krebs & Hilton, 2001). Phosphatases also play a major role in negative regulation of cytokine signaling pathways including that of Epo/Epo-R. Hematopoietic cell phosphatase (HCP) binds to and dephosphorylates tyrosine residues on the Epo-R tail to inactivate signal effectors and kinases (Yi, Zhang, Miura, & Ihle, 1995). Additionally, CD45 phosphatase, and Src Homology-containing phosphatase-1, SHP-1, inactivate Jak2 by dephosphorylation. Last, Epo-R signaling is regulated by quick turnover of the receptor after ligand binding. Engagement with its cognate cytokine induces rapid internalization of the receptor via endocytosis in clathrin coated pits (Sulahian, Cleaver, & Huang, 2009). Once internalized, the receptor is ubiquitinated by E3 ligases including β-TRCP and ring finger protein 41 (RNF41) (Jing, Infante, Nachtman, & Jurecic, 2008; Meyer, Deau, Forejtnikova, et al., 2007). E3 ubiquitin ligases work by transferring a 76 amino acid peptide, ubiquitin, from the E2 subunit of the ligase to the protein substrate. The ubiquitin monomers form chains that target the protein for degradation by the proteasome. A summary of Epo-R signal activation and attenuation are provided in Fig. 1.

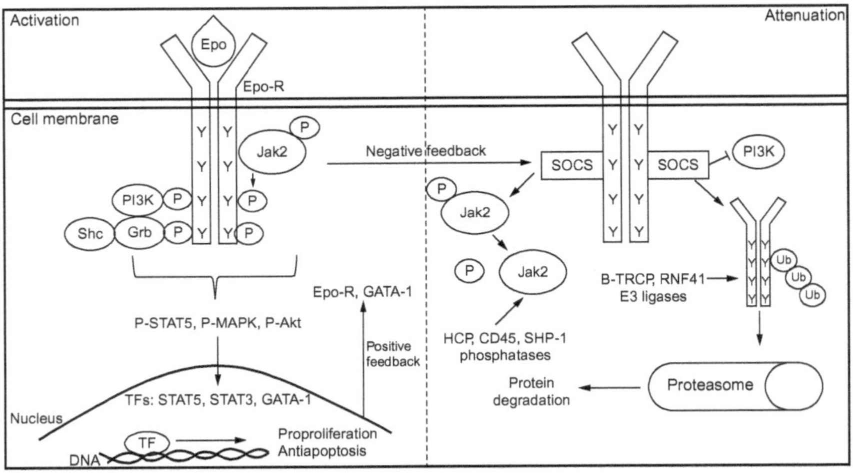

Fig. 1 Epo-R signaling pathways. Epo-R signaling is a delicate balance between activation signals (*left*) and attenuation signals (*right*). Activation is initiated when Epo binds Epo-R which then dimerizes and causes the autophosphorylation of Jak2. Jak2 in turn phosphorylates tyrosine residues on the cytoplasmic tail of the receptor. This allows binding of other signal effectors via their SH2 domains such as PI3K, Shc, and Grb. Several pathways are collectively activated including the Jak/STAT, PI3K/Akt, and MAPK pathways. These result in the activation and nuclear translocation of transcription factors (TFs) including STAT5, STAT3, and GATA-1. These TFs induce the transcription of proproliferation and antiapoptotic signals. Furthermore, they provide positive feedback as levels of Epo-R and GATA-1 transcripts are also elevated. Alternatively, negative-feedback loops are also activated. Epo stimulation induces the expression of SOCS proteins which not only inactivate signal effectors but also bind to the receptor displacing adapter molecules. SOCS proteins also target the receptor for proteasomal degradation along with the E3 ubiquitin ligases. Addition of ubiquitin monomers by these proteins creates long polyubiquitin tails that direct the receptor to the proteasome for degradation. Phosphatases also play a major role in negative regulation of the Epo-R signaling. The proteins diminish signaling by removing activating phosphate groups.

5. LIPID RAFTS

Lipid or membrane rafts are integral components of the plasma membrane whose existence was confirmed just prior to the turn of the millennium. Lipid rafts are approximately 10–200 nm in size and constitute different proportions of the plasma membrane in different cell types. The plasma membrane consists of a bilayer of phospholipids with a hydrophilic head responsible for the interactions with the microenvironment or the cytoplasm, and a hydrophobic transmembrane tail responsible for interacting with the tail of the opposing phospholipid. This phospholipid bilayer creates a highly organized, although fluidic network that is the foundation of the cell's structural integrity. Intricate compartmentalization of the plasma membrane is vital to regulate protein transport, to house receptors that detect environmental stimuli, and to facilitate engulfment of foreign matter such as bacteria or cellular debris for destruction. The plasma membrane also helps coordinate signal activation, signal inhibition, and protein turnover. The lipid raft microdomains "float" laterally throughout the plasma membrane and are indispensable for a variety of cellular processes. Lipid rafts contain high levels of very densely packed cholesterol and sphingolipids defining them as distinct entities within the plasma membrane. Sphingolipids contain amino alcohol backbones whose hydrophilic tails interact with the sterol, cholesterol, molecules forming tightly contained platforms by which they may coordinate cellular process. These sphingolipid, cholesterol-rich microdomains are relatively resistant to nonionic detergents at low temperatures that contributed to their isolation and identification. For this reason, they are sometimes referred to as detergent-resistant membranes. An example of lipid raft structure is provided in Fig. 2. It remains unclear how much of the cellular membrane is composed of lipid rafts, although we know that this differs from cell to cell. It is clear, however, that lipid rafts are involved in a number of cellular process including apoptosis, multidrug resistance, endocytosis, and perhaps most importantly cellular signaling. Lipid rafts are formed by trafficking subunits from the ER and Golgi via the actin cytoskeleton. The actin skeleton is an intricate network serving as the road structure for protein transport. The traffic signals that coordinate this complex highway are the GTPases, molecular switches that direct traffic via their fluent on/off activation state.

Coalescence of rafts is dependent upon the contribution of several proteins including dedicator of cytokinesis-2 (DOCK2), a protein involved in

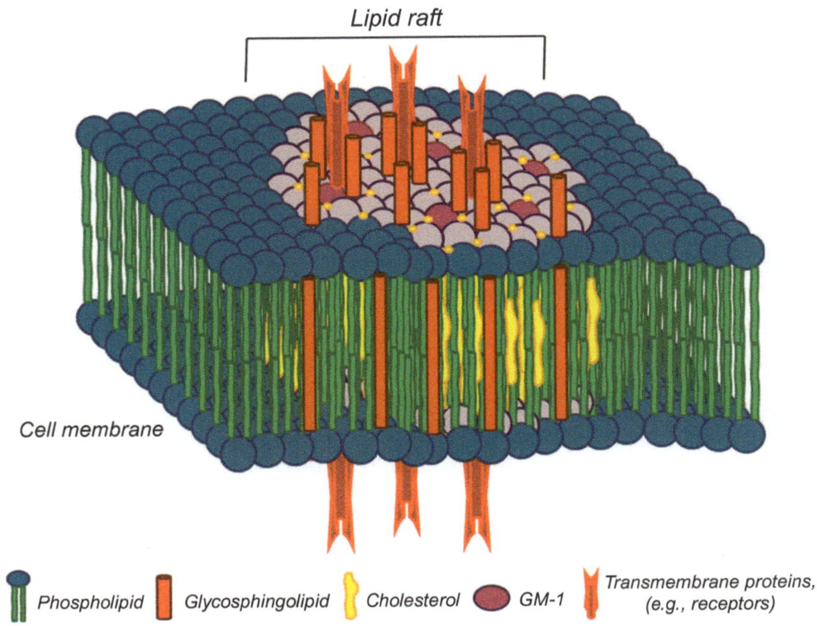

Fig. 2 Lipid raft composition. Lipid rafts are composed of tightly packed sphingolipids and cholesterol-forming platforms that contain GPI anchored proteins, cytokine receptors, kinases, and GTPases that mediate cell signaling. *From McGraw, K. (2013). Interrogation of EpoR fidelity in myelodysplastic syndrome hematopoiesis and stabilization by the immunomodulatory agent, lenalidomide (Doctoral Dissertation). Retrieved from Proquest Dissertations and Theses. (Accession Order No. 3588361).*

hematopoietic cell actin cytoskeletal rearrangement, Arp 2/3, a protein complex that facilitates actin polymerization, Wiskott-Aldrich syndrome protein (WASp) that activates ARP2/3, and the family of proteins known as the Rac GTPases and their regulators. GTPases cycle between inactive GDP-bound states and active GTP-bound states. These proteins are required for the attachment of raft fractions to the membrane via isoprenylation. The most well-studied GTPases are the Rho-like family including Rho-A, Rho-B, and Rho-C; the Rac GTPases including Rac1, Rac2, and Rac3; and Cdc42-like GTPases. Each of these has known roles in regulating the cytoskeleton and is required for protein transport, raft coalescence and aggregation, and recruitment of signaling intermediates to receptors. The most well-studied pathway of raft aggregation is the Rho GTPases and the Rho kinase (ROCK). ROCK phosphorylates and activates LIMK which then deactivates the actin-depolymerizing protein coflin via an inhibitory phosphorylation. ROCK also inhibits the myosin light chain (MLC) phosphatase, activating MLC and stimulating actin polymerization. Furthermore, mDIA is a well-studied GTPase effector that promotes actin polymerization by attaching actin monomers to branched actin filaments creating the cytoskeletal network. A summary of these proteins and their roles in actin polymerization and raft aggregation is provided in Fig. 3.

GTPases have long been known to have critical functions in hematopoiesis. They are involved in every step of hematopoietic differentiation and are critical for hematopoietic stem cell survival and proliferation. They are also required for erythropoiesis, as loss of Rac1 and Rac2 impairs erythropoiesis. These GTPases are critical for erythroid cytoskeletal organization owing to the cell's structural integrity. The dedicator of cytokinesis 4 (DOCK4), a guanine exchange factor that regulates the activity of GTPases, is critical for erythrocyte structural integrity. The *DOCK4* gene is expressed on

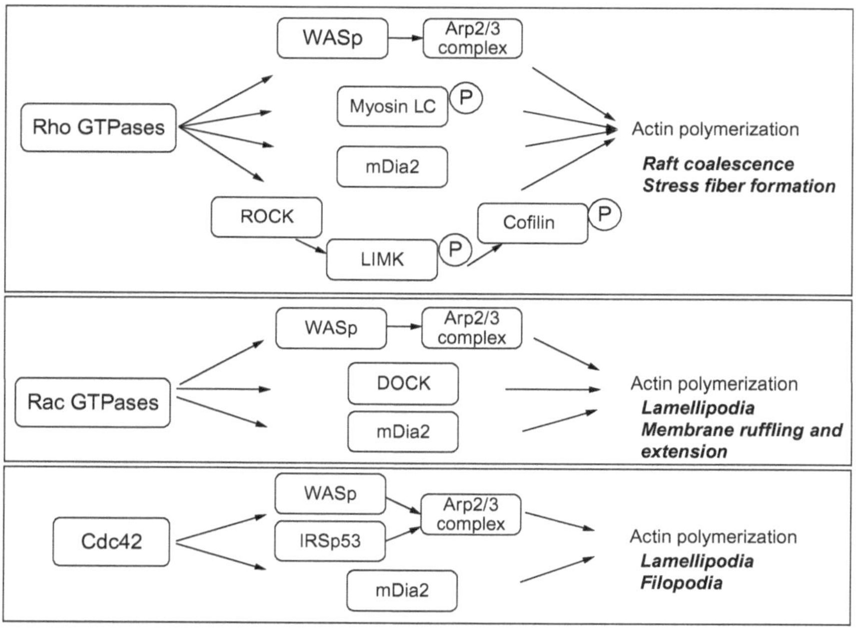

Fig. 3 Rho GTPase activation of cytoskeletal reorganization. The Rho GTPases regulate actin cytoskeletal reorganization through a number of different pathways shown here. *From McGraw, K. (2013). Interrogation of EpoR fidelity in myelodysplastic syndrome hematopoiesis and stabilization by the immunomodulatory agent, lenalidomide (Doctoral Dissertation). Retrieved from Proquest Dissertations and Theses. (Accession Order No. 3588361).*

chromosome 7, notably, a chromosomal region that is often deleted in MDS. Reduced DOCK4 expression in MDS patients with chromosome 7 deletions results in erythroid dysplasia due to disruption of the actin cytoskeleton (Sundaravel, Duggan, Bhagat, et al., 2015). This leads to erythroid frailty resulting in greater susceptibility to cell death and anemia in response to stress. These data highlight the critical role of not only the GTPases but also the actin cytoskeleton on erythropoiesis.

Residence of a variety of proteins found within lipid raft suggests their role in a number of cellular processes. Lipid rafts are highly involved in neuronal signal development, as well as signal transduction and axon directing. They are responsible for the regulation of the amyloid beta (Aβ) peptide. The Aβ peptides create plaques that affect neuronal communication and are prevalent in Alzheimer's disease patients. The Fas/CD95 death receptor complex also resides in lipid rafts suggesting a role of these microdomains in apoptosis. Multidrug resistant proteins including MRP1 and PgP are localized to rafts and function in drug extrusion and drug resistance. Rafts are also the site of viral entry playing a major role in infection. Perhaps the most well-studied role of lipid rafts is in immune signaling. Both T- and B-cells utilize lipid rafts as the site of stimulus-triggered cell signaling. B-cells, key effectors in adaptive immunity, have receptors (BCR) that interact with CD20 and CD19/21. Engagement of these receptors leads to the activation of an extensive network of signaling intermediates, all of which are recruited to the receptor within the raft fractions. This results in activation of transcription factors responsible for carrying out the necessary signal whether it is cell differentiation, protein translation to coordinate immune response, or negative-feedback signals to halt activation of the receptor. T-cells, similarly key to adaptive immunity, include several subsets each with particular function. Some of these include cytotoxic T-cells (CD8+) which are responsible for killing infected cells, helper T-cells (CD4+) which facilitate a number of responses based on the different cytokines they secrete, and suppressor T-cells that function to regulate T-cell activation. Lipid rafts have perhaps been most extensively studied in T-cells. When stimulated, TCRs aggregate in coalesced lipid raft platforms. This localization initiates the recruitment of signaling intermediates Zap-70 and PLCγ1. Of note, the negative regulatory phosphatase, CD45, is displaced from lipid raft fractions upon signal activation to further augment the signal. Lipid rafts are also largely aggregated at the immune synapse, the site of cell-to-cell contact between immune effectors such as T- and natural killer-cells. The well-studied role of lipid rafts in immune cells and receptor signaling, as well as the role of the cytoskeleton and GTPases on erythropoiesis, prompted the investigation of the role of lipid rafts in erythropoiesis, specifically as it relates to Epo signaling.

6. LIPID RAFTS AND EPO SIGNALING

Until recently the role of lipid rafts in Epo signaling had not been investigated. It was known, however, that rafts were abundant in RBC membranes (Mikhalyov & Samsonov, 2011; Murphy, Samuel, Harrison, et al., 2004). Most work on rafts in erythrocytes centered around infection by the malaria parasite, *Plasmodium falciparum*, which is facilitated by lipid rafts (Murphy, Fernandez-Pol, Chung, et al., 2007). There is also evidence to suggest that lipid rafts may have a role in erythrocyte enucleation; although reports are limited (Kalfa & Zheng, 2014). The role of lipid rafts in erythropoiesis, and Epo signaling was not investigated until recently by our laboratory. In order to determine what, if any, role of rafts existed in Epo/Epo-R signaling, we first investigated the EPO-responsive erythroleukemia cell line, UT-7. To confirm the presence of lipid rafts in these cells, we exploited the detergent-resistant properties of the cholesterol-rich fractions. Lipid rafts are undisturbed by traditional cell lysis with most nonionic detergents, and as such, one is able to remove these fractions from the cell lysates by gradient centrifugation. Purification of the rafts can be confirmed by detecting raft constituents, the most common of which is the ganglioside, GM-1, which can be detected in isolated raft fractions by immunoblotting. Indeed, GM-1 was demonstrable by immunoblotting after gradient centrifugation isolation in UT-7 cells in lower density fractions containing the lipid rafts. Direct visualization of lipid rafts can be performed by several methods including electron and fluorescence microcopy. In our study, we used fluorescent microscopy to visualize GM-1, the raft surrogate, in UT-7 cells. Unstimulated cells had a diffuse staining pattern and membrane rafts were present, although not prominently distinct on the membrane. However, after stimulation with rhEpo, there was a striking upregulation of both the number and size of rafts demonstrating rapid coalescence into clearly punctate, distinct fractions in the membrane. This was observed in both cultured UT-7 cells and erythroid progenitors isolated from healthy bone marrow. These findings are illustrated in Fig. 4. We next investigated whether the Epo-R was localized in the raft microdomains. To do so, we first utilized confocal microscopy and 3D rendering to demonstrate that upon Epo stimulation, there was colocalization of the receptor with raft

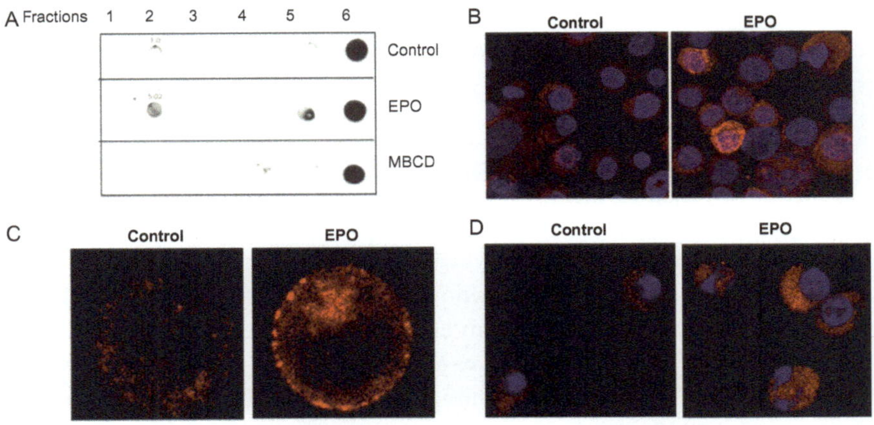

Fig. 4 Detection of lipid rafts in UT-7 and primary erythroid progenitors. (A) Dot blot detection of GM-1 in UT7 cell lysates in nonraft (fractions 5, 6) and raft fractions (fraction 2) in controls, and after Epo or MβCD (raft disrupter) treatment. (B) Immunofluorescence of UT7 cells showing an increase in raft (*red*) accumulation after Epo exposure. (C) Immunofluorescence of UT7 cells before and after Epo stimulation showing increased raft aggregates (*red*) in the plasma membrane. (D) Immunofluorescence of primary erythroid progenitors showing an increase in cellular membrane raft fluorescence intensity (*red*). *McGraw, K. L., Fuhler, G. M., & Johnson, J. O., et al. (2012). Erythropoietin receptor signaling is membrane raft dependent.* PloS One 7*(4), e34477. doi:10.1371/journal.pone.0034477. Available under a Creative Commons Attribution (CC BY 4.0) license.*

fractions (Fig. 5). This was true in both cultured cell lines and primary erythroid progenitors. This was confirmed by raft fractionation by ultracentrifugation and western blotting for Epo-R. Although there was basal colocalization of the receptor in the raft fractions, stimulation of the cells by rhEpo significantly increased receptor residence in raft fractions, similar to that observed in lymphocytes after TCR engagement. Importantly, upon Epo stimulation we observed not only recruitment of Epo-R to rafts but also recruitment of the downstream signaling intermediates Jak2, STAT5, MAPK, and Lyn kinase, shown in Fig. 6. Notably, the negative-regulating phosphatase, CD45, is repartitioned out of raft fractions upon Epo stimulation that further strengthens signal fidelity, analogous to that in T-cell signaling. As mentioned earlier, receptor turnover after ligand engagement is a rapid process and vital to assure extinction of the Epo-R signal. Epo-R is recruited to raft fractions within the first minute of stimulation and peaks at approximately 10 min. This time point coincides with the point at which the maximal activation of the Jak/STAT pathway occurs. After 30 min, however, the level of receptor residing in rafts is nearly

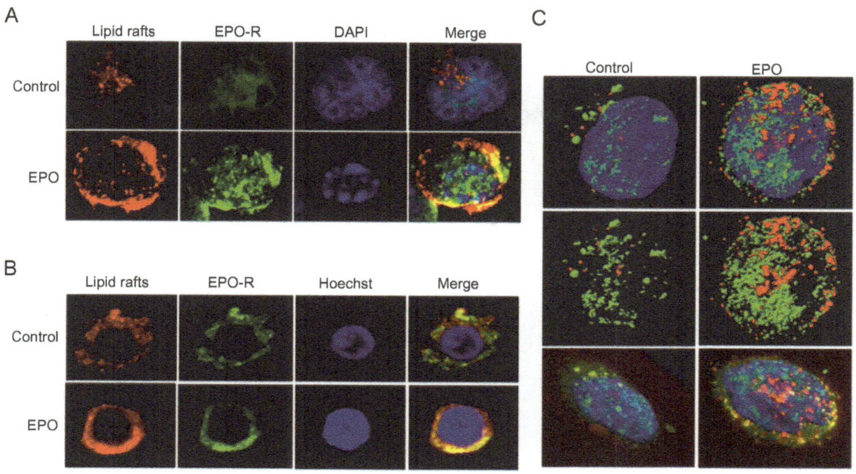

Fig. 5 Epo-R colocalizes with lipid rafts. (A) Confocal immunofluorescence of UT-7 cells untreated or after Epo stimulation, lipid rafts: *red*, Epo-R: *green*, DAPI/Hoechst: *blue*. *Right panel* is a merged image showing lipid raft and Epo-R colocalization (*yellow*). (B) Confocal immunofluorescence in human primary erythroid progenitors. (C) Three-dimensional rendering of UT7 cells either untreated (*left*) or after Epo treatment (*right*). *Top two rows* display isosurfacing of the rafts (*red*), Epo-R (*green*), and nucleus (Dapi, *blue*). Dapi was removed from the *middle row* to further visualize association of the receptor with rafts in the *second row* of panels. The *bottom row* displays volume rendering of the same cells to illustrate membrane colocalization (*yellow*). *McGraw, K. L., Fuhler, G. M., & Johnson, J. O., et al. (2012). Erythropoietin receptor signaling is membrane raft dependent.* PloS One *7(4), e34477. doi:10.1371/journal.pone.0034477. Available under a Creative Commons Attribution (CC BY 4.0) license.*

undetectable, demonstrating a return to the steady state level (Fig. 6). We next investigated whether lipid rafts were integral to Epo-R signaling. In order to study this, we utilized two methods of raft disruption prior to Epo stimulation in the UT-7 cell line, and then studied downstream signal activation. Methyl, beta-cyclodextran (MβCD) is a raft disrupter that removes cholesterol from the membrane, thereby destroying integrity of the raft fraction. Nystatin, on the other hand, intercalates membrane cholesterol and is an alternative, albeit less effective raft disrupter. Importantly, upon disruption of lipid rafts by either method, Epo signaling is severely impaired as demonstrated by decreased phosphorylation of Jak2, STAT5, MAPK, and Akt. We next validated these findings in primary bone marrow erythroid progenitors isolated from healthy donors. Indeed, raft disruption in primary erythroid progenitors significantly diminished signaling initiated by Epo stimulation (McGraw et al., 2012). These results demonstrate that

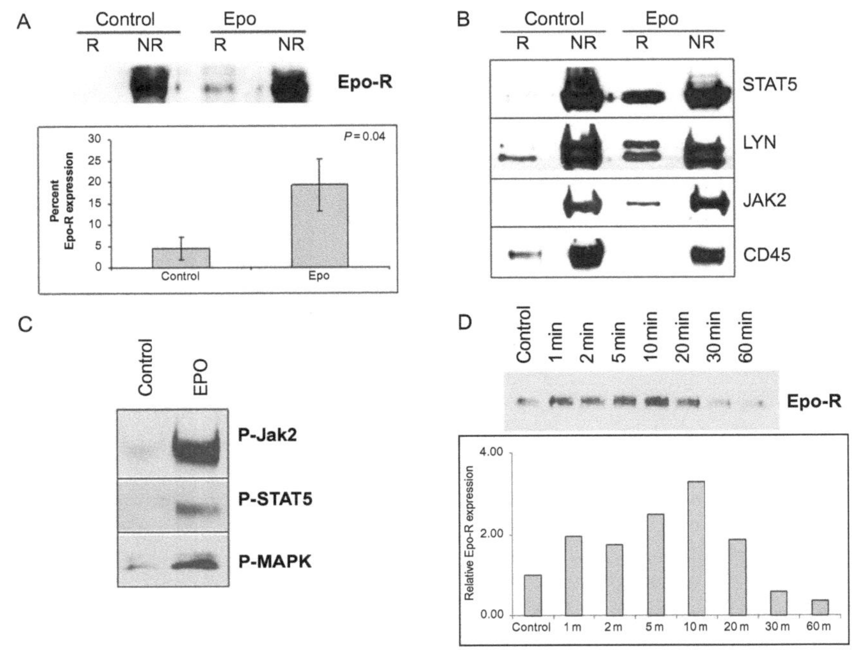

Fig. 6 Epo stimulation recruits signal effectors into raft fractions. (A) Raft fractions (R) were separated from nonraft fractions (NR) by gradient centrifugation and immunoblotted for Epo-R to investigate receptor translocation into rafts after Epo stimulation. Corresponding quantitation represents the mean ± SE of receptor recruitment. (B) UT-7 cells were starved overnight then treated with Epo for 10 min. After fractionation, the nonraft (NR) fractions and raft (R) fractions were pooled and immunoblotted for the indicated proteins showing recruitment of these effector to the raft fractions after stimulation. (C) Activated (phosphorylated) forms of Jak2, STAT5, and MAPK were also increased in the raft fractions after Epo stimulation. (D) Raft fractions were isolated after stimulation with Epo at the indicated time points and immunoblotted for Epo-R. Results show that Epo-R is recruited into rafts within 1 min of Epo stimulation reaching maximum loading at 10 min, followed by gradual redistribution thereafter. Accompanying graphic quantitation of the representative experiment. *McGraw, K. L., Fuhler, G. M., & Johnson, J. O., et al. (2012). Erythropoietin receptor signaling is membrane raft dependent.* PloS One 7*(4), e34477. doi:10.1371/journal.pone.0034477. Available under a Creative Commons Attribution (CC BY 4.0) license.*

lipid rafts are critical for maximum signaling initiated by Epo-R engagement. But how does this occur? We know that aggregation of rafts and recruitment of Epo-R and signaling intermediates is dependent on ROCK and the Rho GTPases; however, the precise mechanisms by which this happens and the method by which rafts are aggregated remain unclear. Several potential mechanisms have been postulated to explain this. One proposal is that as rafts aggregate, there is increased cholesterol affinity, potentiating coalescence and recruitment of the receptors, and intermediates contributing to signal coordination. Alternatively, as receptors dimerize in response to ligand engagement, this may induce the aggregation of lipid raft fractions to support these growing protein complexes. Others have postulated that rafts themselves activate enzymes that drive downstream effectors and signal initiation (Brown & London, 1998). Further investigation is required to determine the precise mechanisms by which these processes occur.

7. CLINICAL IMPLICATIONS

As discussed earlier, the MDSs include a spectrum of disorders characterized by ineffective hematopoiesis and diminished Epo-R signaling. Despite normal levels of circulating cytokine and comparable levels of cell surface receptor, Epo signaling is diminished as demonstrated by decreased STAT5 activation. Previous studies showed that neutrophils isolated from MDS patients had decreased density of lipid rafts, leading to diminished activation by granulocyte-macrophage colony-stimulating factor (GM-CSF) (Fuhler, Blom, Coffer, Drayer, & Vellenga, 2007). Based on this observation, and the requirements for intact lipid rafts for Epo signal fidelity, it reasons that lipid rafts may play a major role in the deficient signaling of MDS patients. And in fact, this is precisely the case. MDS erythroid progenitor cells have a marked reduction in lipid raft fractions compared to their normal counterparts (McGraw, Basiorka, Johnson, et al., 2014). In addition, the rafts are smaller in size, limiting recruitment of key signaling molecules and spatially impairing aggregation. As rafts are indispensable for Epo signal fidelity, lipid raft deficiency in MDS cells diminishes activation of STAT5 in response to Epo-R activation. Therapeutic strategies that promote accumulation of cholesterol-rich, lipid rafts, and recruitment of signaling intermediates may rescue anemia caused by deficient Epo signaling.

Lenalidomide is one of the most effective agents used to treat MDS patients. It is part of a family of compounds known as immunomodulatory (IMiDs) and a second generation analog of the known teratogen, thalidomide. Lenalidomide was approved for the treatment of transfusion-dependent patients with lower risk MDS and deletion in the long arm of chromosome 5 [del(5q)] in 2005, making it the first and only targeted therapy for these disorders. In del(5q) MDS, lenalidomide is cytotoxic to the clone, in part due to haploinsufficiency of the cereblon substrate casein kinase 1α, and another beta-catenin destruction complex member, protein phosphatase 2a, genes that are both located within or near the commonly deleted region on 5q (Kronke, Fink, Hollenbach, et al., 2015; Wei, Chen, Rocha, et al., 2009). Up to 75% of all del(5q) patients will respond to lenalidomide, with most experiencing sustained and prolonged transfusion independence and cytogenetic response. In non-del(5q) MDS patients, however, lenalidomide is able to restore erythropoiesis in approximately 25% of patients. The mechanism of how this occurs has been the focus of much research during the last decade. Our findings of the dependence of Epo-R signaling on lipid rafts led us to investigate whether lenalidomide had an effect on lipid rafts. Raft coalescence is dependent upon activation of GTPases, and it is well established that lenalidomide activates the Rho GTPases albeit through unknown mechanisms (Xu, Li, Ferguson, et al., 2009). In addition, lenalidomide augments Epo-induced signal activation evidenced by increased activation of signaling intermediates Jak2 and STAT5, and increased DNA binding of STAT5 to promote transcription of erythroid differentiation genes. Indeed, we found that treatment of erythroid cell lines and primary erythroid progenitors with lenalidomide-induced aggregation and accumulation of lipid raft fractions. In addition, lenalidomide fostered the recruitment of Jak2 and STAT5 into raft microdomains. And similar to Epo stimulation, lenalidomide treatment repartitioned CD45 outside of the raft fractions. Interestingly, the increase in raft accumulation in response to lenalidomide treatment was less in erythroid cells isolated from healthy donors compared to those from patients with lower risk non-del(5q) MDS. In the latter patient specimens, there was a striking and statistically significant upregulation of raft aggregation, correcting the deficiency to levels comparable to normal counterparts (McGraw et al., 2014). Furthermore, lenalidomide not only fosters lipid raft accumulation augmenting Epo signaling but also upregulates the protein expression of Epo-R. It does this via inhibition of the E3 ubiquitin ligase, RNF41 (Basiorka, McGraw, De Ceuninck, et al., 2016). RNF41 regulates expression of Jak-2 associated, Type-I receptors via ubiquitination and subsequent proteasomal degradation. Lenalidomide blocks RNF41 ubiquitinating function resulting in upregulation of Epo-R, another mechanism by which Epo signaling is augmenting by this IMiD agent. These data indicate novel mechanisms by which lenalidomide acts, as well as providing a rationale for the development and use of molecules targeting lipid rafts.

The role of lipid rafts in Epo signaling has led to the study of other disorders characterized by aberrant signaling. In contrast to diminished Epo-induced Jak2 and STAT5 activation in MDS, the MPNs have constitutive activation of these proteins. Although we know that Jak2 resides in the rafts, particularly after receptor stimulation, it has recently been shown that the Jak2 V617F mutant similarly is positioned within raft fractions. Importantly, raft disruption results in diminished Jak2 V617F constitutive signaling. Similar to the use of nystatin explained earlier in this chapter, other statins may also be utilized to disrupt rafts. Statins are highly prescribed to lower cholesterol in individuals with hyperlipidemia, and those with increased risk of coronary artery disease or stroke. The ability of statins to inhibit aberrant signaling caused by the constitutively active Jak2 V617F mutant, and growth inhibition of primary MPN cells, with no discernible effect on normal counterparts is an intriguing finding that may be exploitable in these disorders (Griner et al., 2013).

Recently, lipid rafts have been shown to play a major role in the pathogenesis paroxysmal nocturnal hemoglobinuria (PNH). PNH is a stem cell disorder in patients that harbor a mutation in *PIG-A*, which results in a decrease in expression of GPI-A anchored proteins on hematopoietic stem and progenitor cells (HSPCs), including CXCR4 and VLA4, adhesion receptors that facilitate stromal adherence of HSPCs in the bone marrow. It has recently been shown that CXCR4 and VLA-4 reside in lipid rafts that are diminished in PNH CD34+ bone marrow and peripheral blood cells. Raft deficiencies result in decreased stromal attachment via diminished CXCR4 ligand binding, increased mobility and clonal expansion. This results in differentiated cells including erythrocytes that are more susceptible to complement-mediated intravascular lysis resulting in the anemia observed in these patients (Ratajczak, Borkowska, Mierzejewska, et al., 2015). It reasons that compounds capable of manipulating rafts may offer benefit for this disorder.

8. CONCLUSION AND FUTURE DIRECTIONS

In conclusion, the findings discussed here demonstrate the importance of plasma membrane lipid rafts and the recruitment of receptors and signaling intermediates to these platforms for optimal Epo-R signal fidelity. These results explain, in part, differences in receptor signal potential in patients with MDS, where impaired Epo signaling is attributed in part to diminished raft fractions in the plasma membrane. Furthermore, the dependency of Epo signaling on rafts provides the possibility for new therapeutic strategies to treat these disorders, such as lenalidomide that, in addition to several other functions, induces lipid raft coalescence as well as the recruitment of signaling intermediates. Alternatively, in MPNs, disruption of rafts may be a key in abrogating the constitutive activation of Jak2 V617F that resides in raft fractions. Here, the use of statins has shown promising in vitro activity in decreasing growth of MPN cell models and primary cells. Together, as the roles of lipid rafts in disease pathogenesis continue to be discovered, we must look to future investigation to unravel the mechanisms by which lipid rafts assembly is controlled. Furthermore, we need to examine more closely the interaction between GTPases and lipid rafts to develop methods to exploit these interactions for future therapeutic benefit.

REFERENCES

Basiorka, A., McGraw, K., De Ceuninck, L., et al. (2016). Lenalidomide stabilizes the erythropoietin receptor by inhibiting the E3 ubiquitin ligase RNF41. *Cancer Research*, *76*, 3531–3540. http://dx.doi.org/10.1158/0008-5472. Published Online First April 6, 2016; CAN-15-1756.

Brown, D. A., & London, E. (1998). Functions of lipid rafts in biological membranes. *Annual Review of Cell and Developmental Biology*, *14*, 111–136.

D'Andrea, A. D., & Zon, L. I. (1990). Erythropoietin receptor. Subunit structure and activation. *The Journal of Clinical Investigation*, *86*(3), 681–687.

Ebert, B. L., & Bunn, H. F. (1999). Regulation of the erythropoietin gene. *Blood*, *94*(6), 1864–1877.

Fisher, J. W. (2003). Erythropoietin: Physiology and pharmacology update. *Experimental Biology and Medicine*, *228*(1), 1–14.

Fuhler, G. M., Blom, N. R., Coffer, P. J., Drayer, A. L., & Vellenga, E. (2007). The reduced GM-CSF priming of ROS production in granulocytes from patients with myelodysplasia is associated with an impaired lipid raft formation. *Journal of Leukocyte Biology*, *81*(2), 449–457.

Genovese, G., Kahler, A. K., Handsaker, R. E., et al. (2014). Clonal hematopoiesis and blood-cancer risk inferred from blood DNA sequence. *The New England Journal of Medicine*, *371*(26), 2477–2487.

Griner, L. N., McGraw, K. L., Johnson, J. O., List, A. F., & Reuther, G. W. (2013). JAK2-V617F-mediated signalling is dependent on lipid rafts and statins inhibit JAK2-V617F-dependent cell growth. *British Journal of Haematology*, *160*(2), 177–187.

Hoefsloot, L. H., van Amelsvoort, M. P., Broeders, L. C., et al. (1997). Erythropoietin-induced activation of STAT5 is impaired in the myelodysplastic syndrome. *Blood*, *89*(5), 1690–1700.

Huang, L. J., Constantinescu, S. N., & Lodish, H. F. (2001). The N-terminal domain of Janus kinase 2 is required for Golgi processing and cell surface expression of erythropoietin receptor. *Molecular Cell*, *8*(6), 1327–1338.

Jaiswal, S., Fontanillas, P., Flannick, J., et al. (2014). Age-related clonal hematopoiesis associated with adverse outcomes. *The New England Journal of Medicine*, *371*(26), 2488–2498.

Jegalian, A. G., & Wu, H. (2002). Differential roles of SOCS family members in EpoR signal transduction. *Journal of Interferon & Cytokine Research*, *22*(8), 853–860.

Jing, X., Infante, J., Nachtman, R. G., & Jurecic, R. (2008). E3 ligase FLRF (Rnf41) regulates differentiation of hematopoietic progenitors by governing steady-state levels of cytokine and retinoic acid receptors. *Experimental Hematology*, *36*(9), 1110–1120.

Kabouridis, P. S., & Jury, E. C. (2008). Lipid rafts and T-lymphocyte function: Implications for autoimmunity. *FEBS Letters*, *582*(27), 3711–3718.

Kalfa, T. A., & Zheng, Y. (2014). Rho GTPases in erythroid maturation. *Current Opinion in Hematology*, *21*(3), 165–171.

Krebs, D. L., & Hilton, D. J. (2001). SOCS proteins: Negative regulators of cytokine signaling. *Stem Cells*, *19*(5), 378–387.

Kronke, J., Fink, E. C., Hollenbach, P. W., et al. (2015). Lenalidomide induces ubiquitination and degradation of CK1alpha in del(5q) MDS. *Nature*, *523*(7559), 183–188.

Lacombe, C., & Mayeux, P. (1998). Biology of erythropoietin. *Haematologica*, *83*(8), 724–732.

Lewinski, U. H., Floru, S., Cohen, A. M., & Mittelmann, M. (1994). Recombinant human erythropoietin in the treatment of myelodysplastic syndromes—Response patterns. *Leukemia & Lymphoma*, *15*(1–2), 149–152.

Mantovani, L., Lentini, G., Hentschel, B., et al. (2000). Treatment of anaemia in myelodysplastic syndromes with prolonged administration of recombinant human granulocyte colony-stimulating factor and erythropoietin. *British Journal of Haematology*, *109*(2), 367–375.

McGraw, K. L., Basiorka, A. A., Johnson, J. O., et al. (2014). Lenalidomide induces lipid raft assembly to enhance erythropoietin receptor signaling in myelodysplastic syndrome progenitors. *PloS One*, *9*(12), e114249.

McGraw, K. L., Fuhler, G. M., Johnson, J. O., et al. (2012). Erythropoietin receptor signaling is membrane raft dependent. *PloS One*, 7(4), e34477.

McLornan, D., Percy, M., & McMullin, M. F. (2006). JAK2 V617F: A single mutation in the myeloproliferative group of disorders. *The Ulster Medical Journal*, *75*(2), 112–119.

McMullin, M. F., & Percy, M. J. (1999). Erythropoietin receptor and hematological disease. *American Journal of Hematology*, *60*(1), 55–60.

Meyer, L., Deau, B., Forejtnikova, H., et al. (2007). beta-Trcp mediates ubiquitination and degradation of the erythropoietin receptor and controls cell proliferation. *Blood*, *109*(12), 5215–5222.

Mikhalyov, I., & Samsonov, A. (2011). Lipid raft detecting in membranes of live erythrocytes. *Biochimica et Biophysica Acta*, *1808*(7), 1930–1939.

Miura, Y., Miura, O., Ihle, J. N., & Aoki, N. (1994). Activation of the mitogen-activated protein kinase pathway by the erythropoietin receptor. *The Journal of Biological Chemistry*, *269*(47), 29962–29969.

Miyake, T., Kung, C. K., & Goldwasser, E. (1977). Purification of human erythropoietin. *The Journal of Biological Chemistry*, *252*(15), 5558–5564.

Mulcahy, L. (2001). The erythropoietin receptor. *Seminars in Oncology*, *28*(2 Suppl. 8), 19–23.

Murphy, S. C., Fernandez-Pol, S., Chung, P. H., et al. (2007). Cytoplasmic remodeling of erythrocyte raft lipids during infection by the human malaria parasite Plasmodium falciparum. *Blood*, *110*(6), 2132–2139.

Murphy, S. C., Samuel, B. U., Harrison, T., et al. (2004). Erythrocyte detergent-resistant membrane proteins: Their characterization and selective uptake during malarial infection. *Blood*, *103*(5), 1920–1928.

Ratajczak, M. Z., Borkowska, S., Mierzejewska, K., et al. (2015). Further evidence that paroxysmal nocturnal haemoglobinuria is a disorder of defective cell membrane lipid rafts. *Journal of Cellular and Molecular Medicine*, *19*(9), 2193–2201.

Singbartl, G. (1994). Adverse events of erythropoietin in long-term and in acute/short-term treatment. *The Clinical Investigator*, *72*(6 Suppl.), S36–S43.

Slavova-Azmanova, N. S., Kucera, N., Louw, A., et al. (2014). Lyn kinase plays important roles in erythroid expansion, maturation and erythropoietin receptor signalling by regulating inhibitory signalling pathways that control survival. *The Biochemical Journal*, *459*(3), 455–466.

Sulahian, R., Cleaver, O., & Huang, L. J. (2009). Ligand-induced EpoR internalization is mediated by JAK2 and p85 and is impaired by mutations responsible for primary familial and congenital polycythemia. *Blood, 113*(21), 5287–5297.
Sundaravel, S., Duggan, R., Bhagat, T., et al. (2015). Reduced DOCK4 expression leads to erythroid dysplasia in myelodysplastic syndromes. *Proceedings of the National Academy of Sciences of the United States of America, 112*(46), E6359–E6368.
Watowich, S. S., Xie, X., Klingmuller, U., et al. (1999). Erythropoietin receptor mutations associated with familial erythrocytosis cause hypersensitivity to erythropoietin in the heterozygous state. *Blood, 94*(7), 2530–2532.
Wei, S., Chen, X., Rocha, K., et al. (2009). A critical role for phosphatase haplodeficiency in the selective suppression of deletion 5q MDS by lenalidomide. *Proceedings of the National Academy of Sciences of the United States of America, 106*(31), 12974–12979.
Xu, Y., Li, J., Ferguson, G. D., et al. (2009). Immunomodulatory drugs reorganize cytoskeleton by modulating Rho GTPases. *Blood, 114*(2), 338–345.
Yi, T., Zhang, J., Miura, O., & Ihle, J. N. (1995). Hematopoietic cell phosphatase associates with erythropoietin (Epo) receptor after Epo-induced receptor tyrosine phosphorylation: Identification of potential binding sites. *Blood, 85*(1), 87–95.
Zhao, W., Kitidis, C., Fleming, M. D., Lodish, H. F., & Ghaffari, S. (2006). Erythropoietin stimulates phosphorylation and activation of GATA-1 via the PI3-kinase/AKT signaling pathway. *Blood, 107*(3), 907–915.

CHAPTER SIX

Erythropoietin and Hypothalamic–Pituitary Axis

Soumyadeep Dey, Constance T. Noguchi[1]
National Institute of Diabetes and Digestive and Kidney Diseases, National Institutes of Health, Bethesda, MD, United States
[1]Corresponding author: e-mail address: connien@helix.nih.gov

Contents

Abstract

Erythropoietin (EPO), known primarily for its erythropoietic activity, is commonly used clinically to treat anemia of chronic kidney disease. However, the expression of EPO receptor (EpoR) beyond erythroid tissue provides for potential extrahematopoietic effects of EPO, including EPO regulation of metabolic homeostasis (Zhang et al., 2014). Small clinical studies have shown that EPO treatment in patients with end-stage renal disease improved glycemic control and insulin sensitivity. Studies in animal models have shown that EPO regulation of metabolism is mainly attributed to its response in fat, and the hypothalamus–pituitary axis (Dey et al., 2016; Dey, Scullen, & Noguchi, 2015; Teng, Gavrilova, et al., 2011; Wang et al., 2013) and is not dependent on its hematopoietic activity. EpoR expression in the hypothalamus is localized to the neurons expressing proopiomelanocortin (POMC) in the arcuate nucleus region, the most important site in the brain for the regulation of physiological energy expenditure. EPO treatment increases POMC production in anorexigenic POMC neurons in the hypothalamus. In the pituitary, EPO modulates the secretion of the POMC-derived peptide, adrenocorticotropic hormone (ACTH) that regulates physiological and metabolic stress response. With EPO produced by cells in the brain, such as astrocytes, and with EPO-stimulated POMC expression in the hypothalamus and EPO-inhibited ACTH

Vitamins and Hormones, Volume 105
ISSN 0083-6729
http://dx.doi.org/10.1016/bs.vh.2017.02.007
© 2017 Elsevier Inc.
All rights reserved.

secretion in the pituitary, EPO signaling contributes to the hypothalamic–pituitary axis as a major regulator of glucose metabolism and energy homeostasis.

1. ERYTHROPOIETIN

Erythropoietin (EPO) is produced in a hypoxia-dependent manner primarily in the fetal liver and adult kidney, and regulates red blood cell production. During hypoxia or ischemic stress, EPO is induced to stimulate erythropoiesis in response to the higher demand for oxygen (Bunn, 2013). EPO acts through its cell surface receptor (EpoR) on erythroid progenitor cells. EpoR is a homodimeric transmembrane receptor with an extracellular domain that binds EPO and a cytosolic domain that activates downstream signaling. EpoR has no intrinsic tyrosine kinase activity and associates with a nonreceptor tyrosine kinase, Janus kinase 2 (JAK2), that binds the EpoR Box1/Box2 region of the cytosolic domain (Funakoshi-Tago, Pelletier, Moritake, Parganas, & Ihle, 2008). EPO binding to the extracellular domain results in a conformational change that brings the JAK2 proteins in close proximity and results in transphosphorylation and activation of JAK2 and EpoR (Livnah et al., 1999). This initiates the activation of different possible signaling pathways including signal transducer and activator of transcription (STAT), phosphoinositide 3-kinase/AKT, and mitogen-activated protein kinase (MAPK) (Quelle et al., 1996; Zhao, Kitidis, Fleming, Lodish, & Ghaffari, 2006).

Knockout of the *EPO* ($EPO^{-/-}$) or *EpoR* ($EpoR^{-/-}$) gene in mice leads to embryonic lethality and is associated with severe anemia (Lin, Lim, D'Agati, & Costantini, 1996; Wu, Liu, Jaenisch, & Lodish, 1995), abnormalities in blood vessel formation (Kertesz, Wu, Chen, Sucov, & Wu, 2004), a thinner neuroepithelium, and a reduction in neural progenitor cells and increased apoptosis (Yu et al., 2002). The effect of EPO and EpoR signaling in endothelial cells is related to its ability to regulate blood flow and vascular tone by modulating endothelial nitric oxide (NO) synthase expression and NO production under reduced oxygen condition (Beleslin-Cokic et al., 2011; Beleslin-Cokic et al., 2004). This is responsible for the ventricular hyperplasia seen in $EpoR^{-/-}$ embryos (Wu, Lee, Gao, Liu, & Iruela-Arispe, 1999; Yu et al., 2002) and also for the cardioprotective effects associated with EPO in cardiac ischemia/reperfusion injury models (Burger et al., 2006; Teng, Calvert, et al., 2011).

During embryonic development, *EpoR* mRNA expression in mouse brain is high during mid-gestation coinciding with regions of neurogenesis (Liu, Chin, & Noguchi, 1994; Yu et al., 2002) and decreases to minimal level in the whole brain as development progresses to adulthood, becoming localized to distinct brain areas (Digicaylioglu et al., 1995; Liu, Shen, Liu, & Noguchi, 1997). However, EPO is also produced in the adult brain, mainly by astrocytes. EPO level in the brain does not follow circulatory levels, consistent with EPO production on both sides of the blood–brain barrier (Bernaudin et al., 2000; Marti et al., 1997; Masuda et al., 1994). EPO in the brain is neuroprotective in animal models, and hypoxia-induced EPO production and EPO treatment are associated with neural progenitor cell proliferation and differentiation (Shingo, Sorokan, Shimazaki, & Weiss, 2001; Studer et al., 2000). Conversely, loss of endogenous EPO signaling in brain is associated with increased neural progenitor cell apoptosis (Chen, Asavaritikrai, Prchal, & Noguchi, 2007; Tsai et al., 2006).

2. ERYTHROPOIETIN AND BRAIN DEVELOPMENT

In brain, EpoR expression is associated with regions of neurogenesis and the level of EpoR expression determines the EPO response to EPO in the brain (Chen et al., 2007; Sanchez et al., 2009; Tsai et al., 2006). In mouse embryonic development between E9.5 and E10.5, lack of *EpoR* gene expression results in marked decrease in proportion of neural progenitor cells as measured by *nestin* mRNA expression (Yu et al., 2002). Cultures of embryonic neural cell exhibit a proliferative response to EPO and fibroblast growth factor (FGF), but the combination of EPO with FGF is not more effective than FGF alone (Chen et al., 2007). Loss of *EpoR* in brain reduces neural progenitor cell proliferation in embryonic and adult brain, and culture of neurons with targeted deletion of *EpoR* exhibits reduced survival to hypoxic challenge (Chen et al., 2007; Tsai et al., 2006; Yu et al., 2002). The level of *EpoR* expression at E10.5 in the developing mouse brain is comparable with that in adult erythroid tissue, and is downregulated in the whole brain by 3 orders of magnitude at birth (Liu et al., 1997).

In humans, EpoR expression is widespread during the first two trimesters of embryo development (Juul, Yachnis, & Christensen, 1998), and EPO and EpoR expression could be detected in brain and spinal cord as early as 7–8 weeks of gestation (Juul, Anderson, Li, & Christensen, 1998; Li, Juul, Morris-Wiman, Calhoun, & Christensen, 1996). Primary cultures prepared from embryonic human brains at 12–15 weeks of gestation showed *EPO* mRNA expression in astrocytes, while EpoR expression was found in neurons, astrocytes, and microglial cells (Nagai et al., 2001). At 20 weeks, EpoR expression in brain was localized on neurons, astrocytes, and choroids plexus, and the expression of EPO and EpoR persists after birth (Juul, Yachnis, Rojiani, & Christensen, 1999).

3. EFFECTS OF ERYTHROPOIETIN RECEPTOR GENE KNOCKOUT IN METABOLISM

Gene knockout of either *EPO* or *EpoR* in mice is embryonic lethal at around embryonic day 13.5 (E13.5) (Lin et al., 1996; Wu et al., 1995), but the expression of an erythroid-specific *EpoR* transgene under the control of GATA-1 hematopoietic regulatory domain rescues the lethal phenotype of the $EpoR^{-/-}$ mice (Suzuki et al., 2002). These mice ($\Delta EpoR_E$) show no gross morphologic abnormalities, but gain excess weight from the first week after birth and become obese by 4 months in female mice and 6 months in male mice (Fig. 1), with significantly higher fat mass but no difference in lean mass (Teng, Gavrilova, et al., 2011). With age, the $\Delta EpoR_E$ mice become insulin resistant, and show decreased energy expenditure and physical activity. Moreover, EPO treatment of wild-type C57Bl/6 mice, but not $\Delta EpoR_E$ mice, is able to reduce body weight, while hematocrit levels are increased in both groups of mice (Fig. 2) (Teng, Gavrilova, et al., 2011). Hence the metabolic effect of EPO reflected in change of body/fat mass is independent of its erythropoietic effect and has been attributed to EpoR signaling primarily

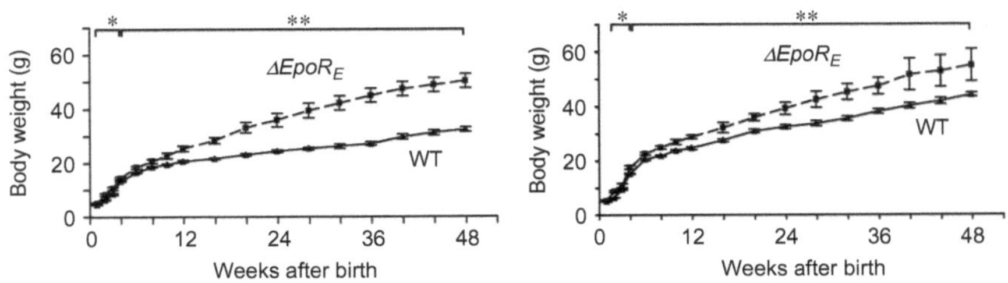

Fig. 1 EpoR expression in nonhematopoietic tissue regulates body weight gain. Female (A) and male (B) mice with *EpoR* gene expression restricted to hematopoietic tissue ($\Delta EpoR_E$) gain more weight (*upper dashed line*) compared to wild-type (WT) control (*lower solid line*), starting as early as 1 week of age, and the effect is more prominent with increasing age. $^*P < 0.05$ and $^{**}P < 0.01$. *From Teng, R., Gavrilova, O., Suzuki, N., Chanturiya, T., Schimel, D., Hugendubler, L., et al., (2011). Disrupted erythropoietin signalling promotes obesity and alters hypothalamus proopiomelanocortin production.* Nature Communications, *2, 520.*

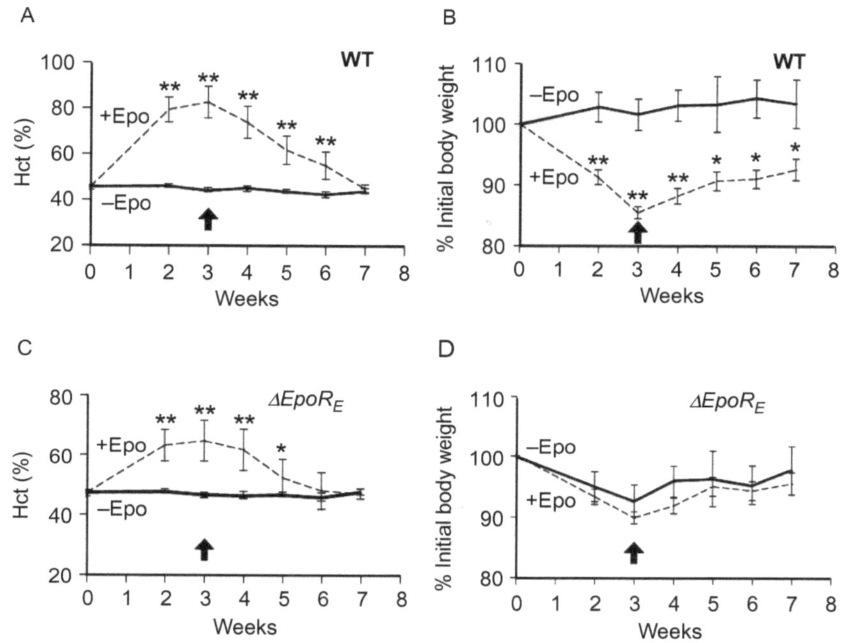

Fig. 2 EPO-mediated decrease in body weight is independent of change in hematocrit. Wild-type (WT) mice (A, B) and mice with *EpoR* gene expression restricted to hematopoietic tissue ($\Delta EpoR_E$) (C, D) on normal chow treated with EPO (3000 U/kg body weight, *dashed line*), or saline (*solid line*), show an increase in hematocrit (A, C). EPO-mediated reduction of body weight is observed in only WT mice, and not in $\Delta EpoR_E$ mice (B, D). *Bold arrow* shows the end of EPO treatment period after which hematocrit is normalized and body weight increases, but more gradually. $^*P < 0.05$ and $^{**}P < 0.01$. *From Teng, R., Gavrilova, O., Suzuki, N., Chanturiya, T., Schimel, D., Hugendubler, L., et al., (2011). Disrupted erythropoietin signalling promotes obesity and alters hypothalamus proopiomelanocortin production.* Nature Communications, 2, *520.*

in the white adipose tissue (WAT) and the hypothalamus (Dey et al., 2016; Teng, Gavrilova, et al., 2011; Wang et al., 2013).

In WAT from mice, *EpoR* mRNA expression is at 60% levels compared to hematopoietic tissues, and the $\Delta EpoR_E$ mice show abnormal preadipocyte differentiation with reduced phosphorylation of p38MAPK, ERK42/44, and PPARγ (Teng, Gavrilova, et al., 2011). Targeted deletion of *EpoR* gene in the WAT of mice by crossing *aP2-Cre* mice with *EpoR-floxed* mice on C57Bl/6 background, increased body weight and fat mass, and by 30 weeks, the mice had 65% more fat with reduced oxygen consumption and total respiratory exchange ratio (Wang et al., 2013). Of note, metabolic difference compared to control mice was not observed in *EpoR* gene knockout in fat cells of mice in a mixed background (Luk et al., 2013). This suggests that genetic background of the mouse model that affects susceptibility to obesity may also be a confounding variable in metabolic response to EPO.

In the hypothalamus, *EpoR* expression is localized specifically in the neurons of the arcuate nucleus (ARC) that produce proopiomelanocortin (POMC), an anorexigenic neuropeptide, and lack of *EpoR* results in decrease in POMC expression (Teng, Gavrilova, et al., 2011). EpoR expression is not detectable in other neurons of the ARC, and lack of *EpoR* did not affect neuropeptides such as neuropeptide Y/agouti-related peptide (NPY/AgRP), melanin-concentrating hormone (pro-MCH) or prepro-orexin, or expression of POMC-processing enzymes prohormone convertases-1 and -2 (PC1 or PC2) (Dey et al., 2016; Teng, Gavrilova, et al., 2011).

4. EFFECTS OF ERYTHROPOIETIN TREATMENT AND OVEREXPRESSION IN METABOLISM

EPO treatment in wild-type mice and mice expressing high transgenic levels of EPO have reduced body weight and fat mass, and exhibit improved glucose tolerance, increased physical activity, and lower food intake (Foskett, Alnaeeli, Wang, Teng, & Noguchi, 2011; Katz et al., 2010; Teng, Gavrilova, et al., 2011). EPO treatments of two classical obese mouse models have also been studied; mice lacking either the gene encoding leptin (*ob/ob* mouse model) or its receptor (*db/db* mouse model) have also been examined. Leptin shares structural similarities with EPO; both are members of the class I cytokine superfamily, and both signal through the JAK/STAT pathway. Leptin is secreted by adipocytes, acts on the hypothalamus, and induces POMC expression to suppress food intake. The *ob/ob* and *db/db* mice become grossly overweight with age, hyperphagic, and develop severe insulin resistance (Schwartz & Baskin, 2013). Leptin treatment in *ob/ob* mice, but not *db/db* mice, reduced body weight, body fat, food intake, and serum glucose and insulin (Campfield, Smith, Guisez, Devos, & Burn, 1995; Halaas et al., 1995; Pelleymounter et al., 1995). EPO treatment of *ob/ob* mice was also able to reduce glucose levels, improve glucose tolerance, and prevent weight gain (Katz et al., 2010), and reduced fat mass even when hematocrit was maintained at normal levels by phlebotomy, suggesting that increased erythropoiesis was not responsible for this effect (Teng, Gavrilova, et al., 2011).

Chronic EPO overexpression studies have been carried out in transgenic mice by using a human platelet-derived growth factor B-chain promoter to drive human *EPO* cDNA (*tg6*) that achieves 12-fold higher circulating EPO (Ruschitzka et al., 2000). These mice show high hematocrit (>80%), and have lower blood glucose and serum insulin levels, improved glucose tolerance and insulin sensitivity, and lower body weight and fat mass (Katz et al., 2010). Mice with brain-specific overexpression of human *EPO* (*tg21*) showed improved ventilator response and acclimatization to severe acute and chronic hypoxia (Soliz et al., 2005). These mice also exhibit improved exercise performance with increased maximal aerobic capacity and prolonged time to exhaustion that was also observed in mice receiving single bolus dose of EPO that did not increase erythroid response (Schuler et al., 2012). Moreover, intrathecal EPO administration at mid-cervical C4 in rats contributed to neural system respiratory motor control and elicited long-lasting phrenic motor facilitation mediated via ERK and AKT activation (Dale, Satriotomo, & Mitchell, 2012). These various studies in rodents with EPO treatment and overexpression models provide strong evidence for EPO regulation of metabolism through peripheral and neuroendocrine regulation of energy expenditure and respiration.

Clinical studies of small groups of patients have reported improvement in insulin sensitivity and glycemic control with EPO treatment. For example, young end-stage renal disease patients on peritoneal dialysis treated with EPO for 6 months and assessed by hyperglycemic clamp showed improvement in insulin and lipid abnormalities (Mak, 1996, 1998).

5. THE HYPOTHALAMIC–PITUITARY AXIS

The hypothalamic–pituitary (HP) axis is the most important neuroendocrine control over physiological stress response and energy metabolism (Lemche, Chaban, & Lemche, 2016). The hypothalamus and pituitary gland function toward maintaining a precise balance between energy intake and energy expenditure. The hypothalamus and pituitary gland exert influence over our endocrine system and are active sites of processing signals from peripheral energy stores. The HP axis receives input from key adiposity- and/or nutrient-related signals and integrates this information for subsequent regulation of energy homeostasis, food intake, and fuel metabolism through secretion of several peptide hormones that instruct other endocrine organs like the adrenal gland and the thyroid to subsequently perform their endocrine functions. In times of energy excess, new energy intake is inhibited, energy expenditure is increased, and fat stores are mobilized. On the other hand, during energy deficiency, food intake is increased, energy expenditure is reduced, and hepatic glucose production is raised until a metabolic homeostasis is reached (Obici, Feng, Arduini, Conti, & Rossetti, 2003; Seeley & Woods, 2003). The hypothalamus and pituitary gland play a key function in metabolic balance, and disruption of this regulation of energy homeostasis leads to weight gain and obesity.

Rodent models demonstrate the nonhematopoietic activity of endogenous EPO and of EPO administration including regulation of body weight and energy expenditure. To assess the relationship between plasma EPO levels and energy expenditure in humans, a study was carried out in full-heritage Southwestern Native Americans that have a high prevalence of obesity and type 2 diabetes mellitus (Astrup et al., 1994; Smith et al., 1996). In a clinical study for cause and consequences of obesity in full-heritage Southwestern Native Americans from the Gila River Indian Community, a secondary analysis using assessment of plasma EPO levels adjusted for covariates showed a positive association with 24 h energy expenditure, sleeping energy expenditure, and weight loss per year in males (Reinhardt et al., 2016). This observation lends support to a potential role for EPO regulation of metabolism and body weight.

6. HYPOTHALAMUS

The hypothalamus is a master regulatory site for energy metabolism. It communicates with and responds to peripheral energy stores and secretes several neuropeptides and chemical neurotransmitters. These are released to act upon the anterior pituitary gland or are transported into the circulation to target distant tissues. Historically, lesion models have demarcated the hypothalamus to be comprised of the "satiety center" localized in the ventromedial nucleus of the hypothalamus, and the "hunger center" that is localized in the lateral hypothalamus (Saper, Chou, & Elmquist, 2002). Advancements have provided mechanistic insights into the hypothalamic function (Fan, Boston, Kesterson, Hruby, & Cone, 1997; Morton, Cummings, Baskin, Barsh, & Schwartz, 2006; Tartaglia et al., 1995; Zhang et al., 1994). These studies have identified and shed light on functions of the adipose-derived hormone, leptin, and its receptor, gut hormones like ghrelin, as well as hypothalamic peptides such as POMC, cocaine and amphetamine-related peptide (CART), NPY, and AgRP.

Peripheral satiety signals, like leptin and insulin, signal to the ARC region of the hypothalamus to inhibit anabolic NPY/AgRP neurons while stimulating catabolic POMC/CART neurons to promote an anorexic response (Schwartz, Woods, Porte, Seeley, & Baskin, 2000). On the other hand, hunger signals like ghrelin are produced from an empty stomach and incite the opposite effects in these areas of the hypothalamus and stimulate appetite (Patterson, Bloom, & Gardiner, 2011). EPO is also a potent regulator of these centrally acting signals as shown in recent mouse studies; EpoR is expressed in the ARC region, EPO treatment decreased food intake, and *EpoR* knockout affects energy expenditure (Dey et al., 2016; Teng, Gavrilova, et al., 2011). These signals are further integrated with other sensory, circadian, and hedonic drives that can also affect appetite and is highly modulated by external factors such as stress as a result of diet, lifestyle, or other environmental factors. Stress can be defined as a deviation from homeostasis and a challenge to the physiological harmony arising from intrinsic or extrinsic factors. Interestingly, $\Delta EpoR_E$ mice with loss of EpoR expression in nonhematopoietic tissue including the hypothalamus and with an age-dependent onset of obesity exhibit reduced endogenous POMC production and STAT3 signaling in hypothalamus (Dey et al., 2016).

To counteract the challenges posed by stress factors, several physiological and behavioral responses are triggered that are collectively known as the "adaptive stress response." One of the earliest responses are the release of corticotropin-releasing hormone (CRH) and arginine vasopressin (AVP) from the paraventricular nucleus region of the hypothalamus (Fig. 3) and their subsequent release into the hypophyseal portal system, the blood vessels connecting the hypothalamus to the anterior pituitary gland. AVP and CRH can stimulate the hypothalamic secretion of each other, and both induce the release of adrenocorticotropic hormone (ACTH) from the anterior pituitary gland. However, while CRH can directly stimulate pituitary ACTH secretion, AVP exhibits only synergistic or additive effects (Korbonits et al., 1999). In absence of any stress factor, the release of CRH and AVP occurs in a circadian fashion, reaching a peak at early morning hours, declining through the day to a nadir around midnight and then begin to rise again (Horrocks et al., 1990). Changes in the physiological circadian rhythm, or any acute stressor that perturbs the amplitude and synchronization of CRH and AVP pulsations, disturb this variation of secretion with induction in CRH and AVP secretion in the PVH (Chrousos, 1998).

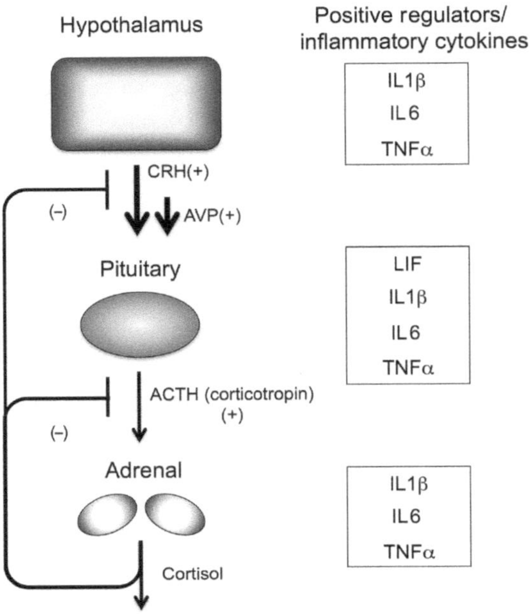

Fig. 3 Regulation of pituitary production of ACTH. Corticotropin-releasing hormone (CRH) from the hypothalamus, together with arginine vasopressin (AVP) that synergizes with CRH, acts on the pituitary gland to secrete adrenocorticotropic hormone (ACTH). ACTH targets the adrenal gland to produce cortisol which provides a negative feedback to CRH, AVP production from hypothalamus, and also ACTH production from the pituitary. Inflammatory cytokines like interleukin 1β (IL1β), interleukin 6, tumor necrosis factor α (TNFα), and leukemia inhibitory factor (LIF) also provide positive stimulus for these hormone secretions.

7. HYPOTHALAMUS AND PITUITARY GLAND CROSS TALK

The pituitary gland is considered the "master" endocrine gland; the hormones synthesized here control other endocrine glands, such as the adrenal gland, thyroid gland, ovaries, and testes. The communication between the hypothalamus and pituitary gland was first recognized in the early half of the last century (Harris, 1948) that laid the foundation for the field of neuroendocrinology. The adult pituitary is linked to the hypothalamus through the hypophyseal portal system harbored in the pituitary stalk. The pituitary gland consists of three lobes: posterior, intermediate, and anterior. The posterior lobe is an extension of the hypothalamus, while the intermediate lobe consists of melanotroph cells producing α-melanocyte-stimulating hormone. The intermediate lobe is present in many species including rodents, but in human it regresses at about week 15 of gestation and appears as a thin layer of cells between anterior and posterior lobes, nearly indistinguishable from the anterior lobe. The anterior lobe is a site of synthesis and release of several hormones including ACTH, luteinizing hormone, follicle-stimulating hormone, growth hormone, prolactin, and thyroid-stimulating hormone. These hormones constitute the hypothalamus–pituitary–target organ axes and the major brain–endocrine connection transducing neuronal messages into humoral signals. The target endocrine organs also provide feedback regulation by specific hormones that target the hypothalamus and the pituitary. Importantly, *EpoR* gene is expressed in the mouse hypothalamus and pituitary gland at 50% of the level that is in spleen, and makes these two tissues a target for EPO signaling (Dey et al., 2016; Dey, Scullen, & Noguchi, 2015).

8. POSTTRANSLATIONAL PROCESSING OF POMC: ACTH SYNTHESIS AND RELEASE

ACTH, a 39-amino acid polypeptide, is derived from the POMC precursor. POMC is a 267-amino acid polypeptide product that is recognized and cleaved by the prohormone cleavage enzymes (Fig. 4). POMC gives rise

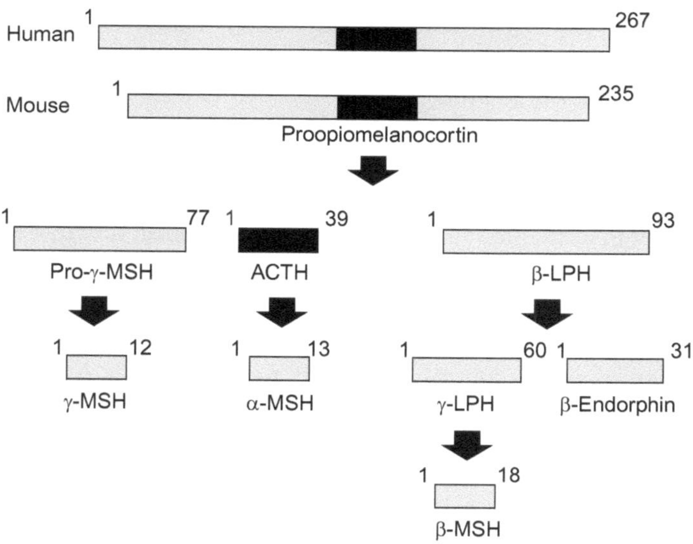

Fig. 4 POMC processing in humans and mice. The *black filled box* shows the region of POMC leading to ACTH synthesis. Further cleaving of ACTH gives rise to α-melanocyte-stimulating hormone (α-MSH), while γ-MSH is formed from pro-γ-MSH. β-Lipotropin (β-LPH) is cleaved into γ-LPH and β-endorphin, and γ-LPH is further cleaved into β-MSH.

to several smaller, biologically active peptide products including ACTH by the action of prohormone convertase endopeptidase enzymes, which cleave the precursor on the C-terminal side of regions of two or more basic amino acid residues (lys-lys or lys-arg) (Bicknell, 2008). The final end product of this processing depends on tissue specificity and differential distribution of the processing enzymes in the tissues that produce POMC. Proteolytic digestion of human POMC is mediated by either of the two related endopeptidases, PC1 and PC2. Distribution of *PC1* and *PC2* mRNA is distinct. While *PC1* mRNA is present in 20% of anterior pituitary cells comprised of mainly corticotrophic cells, and supraoptic nucleus of the hypothalamus, *PC2* is absent from the anterior pituitary, but highly expressed in cerebral cortex, hippocampus, and thalamus (Korner, Chun, Harter, & Axel, 1991; Seidah et al., 1990; Smeekens, Avruch, LaMendola, Chan, & Steiner, 1991). These suggest that PC1 is responsible for the POMC cleavage in anterior pituitary, whereas PC2 cleaves elsewhere in the brain.

Signal transduction pathways that control the release of ACTH from the pituitary involve protein kinase A (PKA), protein kinase C (PKC), JAK/STAT, or glucocorticoids. CRH and AVP can activate cAMP-dependent PKA or PKC, respectively, by increasing cAMP by inducing adenylate cyclase activity (Antoni, 1993). These pathways can act both on release of a ready pool of ACTH, followed by later release of newly synthesized protein (DeBold et al., 1983). During inflammation, leukemia inhibitory factor can increase gene expression of *POMC* by activating JAK/STAT3 signaling (Bousquet, Zatelli, & Melmed, 2000). Glucocorticoids, on the other hand, are the primary negative regulators of ACTH secretion, and act on transcriptional activators of POMC as well as reduce production of cAMP (Dallman et al., 1992). EPO-stimulated production of POMC in hypothalamic neurons and EpoR expression in the pituitary raise the possibility that EPO may also modulate pituitary expression of POMC.

Corticotropic cells are not easily available for in vivo studies or grown in culture; therefore, a pituitary tumor cell line AtT20 was used to study the mechanisms of Ca^{2+} control of cAMP formation and ACTH release. AtT20 cells secrete ACTH in response both to secretagogues that stimulate cAMP synthesis (CRH, AVP, forskolin) and to secretagogues that work independently of cAMP (potassium, calcium ionophores, phorbol esters), and, hence, provide a useful tool to study such mechanisms. CRH stimulation of AtT20 cells leads to the activation of transmembrane adenylyl cyclase, followed by voltage-operated Ca^{2+} channels (VOCC) activation, Ca^{2+} influx, and enhanced secretion of ACTH. At a basal level, there is a low oscillation of intracellular Ca^{2+}, dependent on extracellular levels

(Antoni, Hoyland, Woods, & Mason, 1992), while the addition of CRH leads to an increase in frequency and amplitude of oscillation, and an overall rise in intracellular Ca^{2+} (Luini, Lewis, Guild, Corda, & Axelrod, 1985). Interestingly, the *POMC* gene regulatory regions in mammals and amphibians lack a cAMP-response element or a Ca^{2+}-response element. However, transcription factors that drive *POMC* gene expression are induced by intracellular Ca^{2+} (Murakami, Takeuchi, Kudo, Sutou, & Takahashi, 2007), together with a mechanism by which exocytosis is triggered by depolarization caused by intracellular Ca^{2+} uptake (Tse, Lee, & Tse, 2012). Regulation of POMC expression in the hypothalamus was found to be distinct from that in the pituitary AtT20 cells.

9. ERYTHROPOIETIN EFFECTS IN HYPOTHALAMUS AND PITUITARY

The ARC of the hypothalamus is a critical site for energy regulation and the fact that the POMC neurons express EpoR makes it a direct target of EPO. Wild-type mice treated with EPO show an increase in POMC expression and also an increase in physical activity and a reduction in food intake, resulting in a reduction in body weight (Foskett et al., 2011; Teng, Gavrilova, et al., 2011). The $\Delta EpoR_E$ mice, on the other hand, showed a 50% reduction in POMC expression and reduced physical activity, and developed obesity, but without detectable increase in food intake (Teng, Gavrilova, et al., 2011). The uncoupling of food intake and energy expenditure has been observed in POMC neurons, and selective deletion of leptin receptor in POMC neurons fails to affect food intake (Balthasar et al., 2004). It has been argued that neural control over feeding behavior has several layers of redundancy (Schwartz et al., 2003). In this context, it should be noted that although the knockout of anorexigenic neuropeptide-expressing *POMC* gene causes increased food intake (Yaswen, Diehl, Brennan, & Hochgeschwender, 1999), the knockout of orexigenic *NPY* gene does not affect daily food intake (Erickson, Clegg, & Palmiter, 1996).

Leptin and EPO can both activate POMC neurons and act through the JAK2/STAT3 signaling pathway (Constantinescu et al., 2001; Mancour et al., 2012). In the hypothalamus, EPO-stimulated increase in POMC is dependent on STAT3 activation, and leptin's ability to regulate POMC is also dependent on active EpoR signaling (Dey et al., 2016). This explains the lower level of POMC expression and the blunted leptin response in the hypothalamus of $\Delta EpoR_E$ mice in terms of POMC induction

(Dey et al., 2016). However, the absence of EpoR signaling does not affect leptin-mediated inhibition of *NPY* and *AgRP* mRNA expression, and therefore EPO regulation is specific to POMC neurons only.

In addition to the feeding circuit, EPO can also regulate the stress-response pathways that are triggered under conditions of chronic metabolic stress or more acute brain ischemic–hypoxic stress. In rat hypothalamic explant model, CRH release by two secretagogues, potassium chloride and veratridine, was significantly reduced by EPO (Tringali, Pozzoli, Lisi, & Navarra, 2007). This effect was not due to decreasing *CRH* gene expression, but through modulating CRH release and led to an increase in intrahypothalamic CRH content. Although this suggests a possible indirect effect on ACTH release from the pituitary, EPO can also directly regulate pituitary function in terms of ACTH secretion. In the AtT20 pituitary cell line, EPO can suppress ACTH secretion into the supernatant and increase ACTH intracellular content without affecting the *POMC* gene expression (Dey et al., 2015). One contributory factor for this effect is the drop in intracellular Ca^{2+} content that occurs optimally at 10 U/mL of EPO that also caused maximum STAT3 activation (Dey et al., 2015). The *EpoR* mRNA expression level in the mouse pituitary is comparable to that in the hypothalamus, and wild-type mice treated with EPO showed a reduction in serum ACTH levels. Conversely, the $\Delta EpoR_E$ mice had a higher serum ACTH content compared to wild-type mice (Dey et al., 2015). This suggests that in rodents, endogenous and exogenous EPO regulate the serum ACTH levels probably through regulating the CRH release from the hypothalamus and also ACTH secretion from the pituitary (Dey et al., 2015; Tringali et al., 2007).

Clinical studies in renal failure patients undergoing hemodialysis have provided evidence of EPO-mediated regulation of HP axis. Renal failure patients with EPO deficiency show several complications including endocrine abnormalities, and kidney failure is associated with HP axis dysfunction (Holley, 2004). Patients with renal failure have elevated ACTH (Clodi et al., 1998; Kokot et al., 1995). A small clinical study compared ACTH and cortisol levels in five hemodialyzed patients treated with EPO for 3 months, to a control group of six hemodialyzed patients who received no EPO (Kokot, Wiecek, Grzeszczak, & Klin, 1990). The EPO-treated group had similar hematocrit and hemoglobin levels to the control group at the end of treatment, but showed a significant suppression of ACTH and a moderate suppression of cortisol (Kokot et al., 1990). In a larger study carried out with 34 hemodialyzed patients divided equally to receive EPO or not, the EPO-treated group showed a suppression of ACTH 3 months after EPO treatment (Kokot et al., 1995). The reduction in ACTH following EPO treatment was comparable to levels in healthy volunteers 3 months after EPO treatment, but this effect reverted back to elevated values at later time points of 6, 9, and 12 months after EPO treatment (Kokot et al., 1995). Cortisol levels, on the other hand, showed a prolonged suppression until 9 months after EPO treatment, and increased to no treatment control group levels only after 12 months.

Taken together, EPO's effect in the hypothalamus and the pituitary appears to be reversed in terms of POMC production and ACTH secretion. EPO-mediated POMC expression in the hypothalamus suppresses food intake and increases energy expenditure, while EPO inhibition of ACTH secretion from the pituitary is expected to reduce metabolic stress via cortisol inhibition. This suggests that the EPO effect on the HP axis serves to maintain a systemic metabolic homeostasis.

10. CONCLUSIONS

In summary, EPO's role in regulating metabolic homeostasis is a combinatorial effect in nonhematopoietic tissues. The neuroendocrine regulation of energy metabolism is carried out through the HP regulation and EPO cross talks with leptin to regulate POMC expression in the hypothalamus. Since the introduction of recombinant EPO in the clinic in 1989, clinical studies with EPO have provided important information regarding metabolic effects in human. However, patients with renal disease have other underlying disorders such as diabetes, stress, and comorbidities and concomitant treatment with other therapeutic agents. Therefore, the study of EPO regulation in mouse models, including genetically engineered mice, is useful to better understand endogenous and exogenous EPO function in neuroendocrine regulation, and to distinguish the direct and indirect effects of EPO as well as identify specific target tissues in direct EPO responses. Such studies have the potential to discover new regulatory pathways and identify therapeutic benefits of EPO beyond anemia.

REFERENCES

Antoni, F. A. (1993). Vasopressinergic control of pituitary adrenocorticotropin secretion comes of age. *Frontiers in Neuroendocrinology*, *14*, 76–122.

Antoni, F. A., Hoyland, J., Woods, M. D., & Mason, W. T. (1992). Glucocorticoid inhibition of stimulus-evoked adrenocorticotrophin release caused by suppression of intracellular calcium signals. *The Journal of Endocrinology*, *133*, R13–R16.

Astrup, A., Buemann, B., Western, P., Toubro, S., Raben, A., & Christensen, N. J. (1994). Obesity as an adaptation to a high-fat diet: Evidence from a cross-sectional study. *The American Journal of Clinical Nutrition*, *59*, 350–355.

Balthasar, N., Coppari, R., McMinn, J., Liu, S. M., Lee, C. E., Tang, V., et al. (2004). Leptin receptor signaling in POMC neurons is required for normal body weight homeostasis. *Neuron, 42*, 983–991.

Beleslin-Cokic, B. B., Cokic, V. P., Wang, L., Piknova, B., Teng, R., Schechter, A. N., et al. (2011). Erythropoietin and hypoxia increase erythropoietin receptor and nitric oxide levels in lung microvascular endothelial cells. *Cytokine, 54*, 129–135.

Beleslin-Cokic, B. B., Cokic, V. P., Yu, X., Weksler, B. B., Schechter, A. N., & Noguchi, C. T. (2004). Erythropoietin and hypoxia stimulate erythropoietin receptor and nitric oxide production by endothelial cells. *Blood, 104*, 2073–2080.

Bernaudin, M., Bellail, A., Marti, H. H., Yvon, A., Vivien, D., Duchatelle, I., et al. (2000). Neurons and astrocytes express EPO mRNA: Oxygen-sensing mechanisms that involve the redox-state of the brain. *Glia, 30*, 271–278.

Bicknell, A. B. (2008). The tissue-specific processing of pro-opiomelanocortin. *Journal of Neuroendocrinology, 20*, 692–699.

Bousquet, C., Zatelli, M. C., & Melmed, S. (2000). Direct regulation of pituitary pro-opiomelanocortin by STAT3 provides a novel mechanism for immuno-neuroendocrine interfacing. *The Journal of Clinical Investigation, 106*, 1417–1425.

Bunn, H. F. (2013). Erythropoietin. *Cold Spring Harbor Perspectives in Medicine, 3*, a011619.

Burger, D., Lei, M., Geoghegan-Morphet, N., Lu, X., Xenocostas, A., & Feng, Q. (2006). Erythropoietin protects cardiomyocytes from apoptosis via up-regulation of endothelial nitric oxide synthase. *Cardiovascular Research, 72*, 51–59.

Campfield, L. A., Smith, F. J., Guisez, Y., Devos, R., & Burn, P. (1995). Recombinant mouse OB protein: Evidence for a peripheral signal linking adiposity and central neural networks. *Science, 269*, 546–549.

Chen, Z. Y., Asavaritikrai, P., Prchal, J. T., & Noguchi, C. T. (2007). Endogenous erythropoietin signaling is required for normal neural progenitor cell proliferation. *The Journal of Biological Chemistry, 282*, 25875–25883.

Chrousos, G. P. (1998). Ultradian, circadian, and stress-related hypothalamic-pituitary-adrenal axis activity—A dynamic digital-to-analog modulation. *Endocrinology, 139*, 437–440.

Clodi, M., Riedl, M., Schmaldienst, S., Vychytil, A., Kotzmann, H., Kaider, A., et al. (1998). Adrenal function in patients with chronic renal failure. *American Journal of Kidney Diseases, 32*, 52–55.

Constantinescu, S. N., Keren, T., Socolovsky, M., Nam, H., Henis, Y. I., & Lodish, H. F. (2001). Ligand-independent oligomerization of cell-surface erythropoietin receptor is mediated by the transmembrane domain. *Proceedings of the National Academy of Sciences of the United States of America, 98*, 4379–4384.

Dale, E. A., Satriotomo, I., & Mitchell, G. S. (2012). Cervical spinal erythropoietin induces phrenic motor facilitation via extracellular signal-regulated protein kinase and Akt signaling. *The Journal of Neuroscience, 32*, 5973–5983.

Dallman, M. F., Akana, S. F., Scribner, K. A., Bradbury, M. J., Walker, C. D., Strack, A. M., et al. (1992). Stress, feedback and facilitation in the hypothalamo-pituitary-adrenal axis. *Journal of Neuroendocrinology, 4*, 517–526.

DeBold, C. R., DeCherney, G. S., Jackson, R. V., Sheldon, W. R., Alexander, A. N., Island, D. P., et al. (1983). Effect of synthetic ovine corticotropin-releasing factor: Prolonged duration of action and biphasic response of plasma adrenocorticotropin and cortisol. *The Journal of Clinical Endocrinology and Metabolism, 57*, 294–298.

Dey, S., Li, X., Teng, R., Alnaeeli, M., Chen, Z., Rogers, H., et al. (2016). Erythropoietin regulates POMC expression via STAT3 and potentiates leptin response. *Journal of Molecular Endocrinology, 56*, 55–67.

Dey, S., Scullen, T., & Noguchi, C. T. (2015). Erythropoietin negatively regulates pituitary ACTH secretion. *Brain Research, 1608*, 14–20.

Digicaylioglu, M., Bichet, S., Marti, H. H., Wenger, R. H., Rivas, L. A., Bauer, C., et al. (1995). Localization of specific erythropoietin binding sites in defined areas of the mouse brain. *Proceedings of the National Academy of Sciences of the United States of America, 92*, 3717–3720.

Erickson, J. C., Clegg, K. E., & Palmiter, R. D. (1996). Sensitivity to leptin and susceptibility to seizures of mice lacking neuropeptide Y. *Nature, 381*, 415–421.

Fan, W., Boston, B. A., Kesterson, R. A., Hruby, V. J., & Cone, R. D. (1997). Role of melanocortinergic neurons in feeding and the agouti obesity syndrome. *Nature, 385*, 165–168.

Foskett, A., Alnaeeli, M., Wang, L., Teng, R., & Noguchi, C. T. (2011). The effects of erythropoietin dose titration during high-fat diet-induced obesity. *Journal of Biomedicine & Biotechnology, 2011*, 373781.

Funakoshi-Tago, M., Pelletier, S., Moritake, H., Parganas, E., & Ihle, J. N. (2008). Jak2 FERM domain interaction with the erythropoietin receptor regulates Jak2 kinase activity. *Molecular and Cellular Biology, 28*, 1792–1801.

Halaas, J. L., Gajiwala, K. S., Maffei, M., Cohen, S. L., Chait, B. T., Rabinowitz, D., et al. (1995). Weight-reducing effects of the plasma protein encoded by the obese gene. *Science, 269*, 543–546.

Harris, G. W. (1948). Neural control of the pituitary gland. *Physiological Reviews, 28*, 139–179.

Holley, J. L. (2004). The hypothalamic-pituitary axis in men and women with chronic kidney disease. *Advances in Chronic Kidney Disease, 11*, 337–341.

Horrocks, P. M., Jones, A. F., Ratcliffe, W. A., Holder, G., White, A., Holder, R., et al. (1990). Patterns of ACTH and cortisol pulsatility over twenty-four hours in normal males and females. *Clinical Endocrinology (Oxford), 32*, 127–134.

Juul, S. E., Anderson, D. K., Li, Y., & Christensen, R. D. (1998). Erythropoietin and erythropoietin receptor in the developing human central nervous system. *Pediatric Research, 43*, 40–49.

Juul, S. E., Yachnis, A. T., & Christensen, R. D. (1998). Tissue distribution of erythropoietin and erythropoietin receptor in the developing human fetus. *Early Human Development, 52*, 235–249.

Juul, S. E., Yachnis, A. T., Rojiani, A. M., & Christensen, R. D. (1999). Immunohistochemical localization of erythropoietin and its receptor in the developing human brain. *Pediatric and Developmental Pathology, 2*, 148–158.

Katz, O., Stuible, M., Golishevski, N., Lifshitz, L., Tremblay, M. L., Gassmann, M., et al. (2010). Erythropoietin treatment leads to reduced blood glucose levels and body mass: Insights from murine models. *The Journal of Endocrinology, 205*, 87–95.

Kertesz, N., Wu, J., Chen, T. H., Sucov, H. M., & Wu, H. (2004). The role of erythropoietin in regulating angiogenesis. *Developmental Biology, 276*, 101–110.

Kokot, F., Wiecek, A., Grzeszczak, W., & Klin, M. (1990). Influence of erythropoietin treatment on function of the pituitary-adrenal axis and somatotropin secretion in hemodialyzed patients. *Clinical Nephrology, 33*, 241–246.

Kokot, F., Wiecek, A., Schmidt-Gayk, H., Marcinkowski, W., Gilge, U., Heidland, A., et al. (1995). Function of endocrine organs in hemodialyzed patients of long-term erythropoietin therapy. *Artificial Organs, 19*, 428–435.

Korbonits, M., Kaltsas, G., Perry, L. A., Putignano, P., Grossman, A. B., Besser, G. M., et al. (1999). The growth hormone secretagogue hexarelin stimulates the hypothalamo-pituitary-adrenal axis via arginine vasopressin. *The Journal of Clinical Endocrinology and Metabolism, 84*, 2489–2495.

Korner, J., Chun, J., Harter, D., & Axel, R. (1991). Isolation and functional expression of a mammalian prohormone processing enzyme, murine prohormone convertase 1. *Proceedings of the National Academy of Sciences of the United States of America, 88*, 6834–6838.

Lemche, E., Chaban, O. S., & Lemche, A. V. (2016). Neuroendorine and epigentic mechanisms subserving autonomic imbalance and HPA dysfunction in the metabolic syndrome. *Frontiers in Neuroscience, 10*, 142.

Li, Y., Juul, S. E., Morris-Wiman, J. A., Calhoun, D. A., & Christensen, R. D. (1996). Erythropoietin receptors are expressed in the central nervous system of mid-trimester human fetuses. *Pediatric Research, 40*, 376–380.

Lin, C. S., Lim, S. K., D'Agati, V., & Costantini, F. (1996). Differential effects of an erythropoietin receptor gene disruption on primitive and definitive erythropoiesis. *Genes & Development, 10*, 154–164.

Liu, Z. Y., Chin, K., & Noguchi, C. T. (1994). Tissue specific expression of human erythropoietin receptor in transgenic mice. *Developmental Biology, 166*, 159–169.

Liu, C., Shen, K., Liu, Z., & Noguchi, C. T. (1997). Regulated human erythropoietin receptor expression in mouse brain. *The Journal of Biological Chemistry, 272*, 32395–32400.

Livnah, O., Stura, E. A., Middleton, S. A., Johnson, D. L., Jolliffe, L. K., & Wilson, I. A. (1999). Crystallographic evidence for preformed dimers of erythropoietin receptor before ligand activation. *Science, 283*, 987–990.

Luini, A., Lewis, D., Guild, S., Corda, D., & Axelrod, J. (1985). Hormone secretagogues increase cytosolic calcium by increasing cAMP in corticotropin-secreting cells. *Proceedings of the National Academy of Sciences of the United States of America, 82*, 8034–8038.

Luk, C. T., Shi, S. Y., Choi, D., Cai, E. P., Schroer, S. A., & Woo, M. (2013). In vivo knockdown of adipocyte erythropoietin receptor does not alter glucose or energy homeostasis. *Endocrinology, 154*, 3652–3659.

Mak, R. H. (1996). Effect of recombinant human erythropoietin on insulin, amino acid, and lipid metabolism in uremia. *The Journal of Pediatrics, 129*, 97–104.

Mak, R. H. (1998). Metabolic effects of erythropoietin in patients on peritoneal dialysis. *Pediatric Nephrology, 12*, 660–665.

Mancour, L. V., Daghestani, H. N., Dutta, S., Westfield, G. H., Schilling, J., Oleskie, A. N., et al. (2012). Ligand-induced architecture of the leptin receptor signaling complex. *Molecular Cell, 48*, 655–661.

Marti, H. H., Gassmann, M., Wenger, R. H., Kvietikova, I., Morganti-Kossmann, M. C., Kossmann, T., et al. (1997). Detection of erythropoietin in human liquor: Intrinsic erythropoietin production in the brain. *Kidney International, 51*, 416–418.

Masuda, S., Okano, M., Yamagishi, K., Nagao, M., Ueda, M., & Sasaki, R. (1994). A novel site of erythropoietin production. Oxygen-dependent production in cultured rat astrocytes. *The Journal of Biological Chemistry, 269*, 19488–19493.

Morton, G. J., Cummings, D. E., Baskin, D. G., Barsh, G. S., & Schwartz, M. W. (2006). Central nervous system control of food intake and body weight. *Nature, 443*, 289–295.

Murakami, I., Takeuchi, S., Kudo, T., Sutou, S., & Takahashi, S. (2007). Corticotropin-releasing hormone or dexamethasone regulates rat proopiomelanocortin transcription through Tpit/Pitx-responsive element in its promoter. *The Journal of Endocrinology, 193*, 279–290.

Nagai, A., Nakagawa, E., Choi, H. B., Hatori, K., Kobayashi, S., & Kim, S. U. (2001). Erythropoietin and erythropoietin receptors in human CNS neurons, astrocytes, microglia, and oligodendrocytes grown in culture. *Journal of Neuropathology and Experimental Neurology, 60*, 386–392.

Obici, S., Feng, Z., Arduini, A., Conti, R., & Rossetti, L. (2003). Inhibition of hypothalamic carnitine palmitoyltransferase-1 decreases food intake and glucose production. *Nature Medicine, 9*, 756–761.

Patterson, M., Bloom, S. R., & Gardiner, J. V. (2011). Ghrelin and appetite control in humans–potential application in the treatment of obesity. *Peptides, 32*, 2290–2294.

Pelleymounter, M. A., Cullen, M. J., Baker, M. B., Hecht, R., Winters, D., Boone, T., et al. (1995). Effects of the obese gene product on body weight regulation in ob/ob mice. *Science, 269*, 540–543.
Quelle, F. W., Wang, D., Nosaka, T., Thierfelder, W. E., Stravopodis, D., Weinstein, Y., et al. (1996). Erythropoietin induces activation of Stat5 through association with specific tyrosines on the receptor that are not required for a mitogenic response. *Molecular and Cellular Biology, 16*, 1622–1631.
Reinhardt, M., Dey, S., Noguchi, C. T., Zhang, Y., Krakoff, J., & Thearle, M. S. (2016). Non-hematopoietic effects of endogenous erythropoietin on lean mass and body weight regulation. *Obesity, 24*, 1530–1536.
Ruschitzka, F. T., Wenger, R. H., Stallmach, T., Quaschning, T., de Wit, C., Wagner, K., et al. (2000). Nitric oxide prevents cardiovascular disease and determines survival in polyglobulic mice overexpressing erythropoietin. *Proceedings of the National Academy of Sciences of the United States of America, 97*, 11609–11613.
Sanchez, P. E., Fares, R. P., Risso, J. J., Bonnet, C., Bouvard, S., Le-Cavorsin, M., et al. (2009). Optimal neuroprotection by erythropoietin requires elevated expression of its receptor in neurons. *Proceedings of the National Academy of Sciences of the United States of America, 106*, 9848–9853.
Saper, C. B., Chou, T. C., & Elmquist, J. K. (2002). The need to feed: Homeostatic and hedonic control of eating. *Neuron, 36*, 199–211.
Schuler, B., Vogel, J., Grenacher, B., Jacobs, R. A., Arras, M., & Gassmann, M. (2012). Acute and chronic elevation of erythropoietin in the brain improves exercise performance in mice without inducing erythropoiesis. *The FASEB Journal, 26*, 3884–3890.
Schwartz, M. W., & Baskin, D. G. (2013). Leptin and the brain: Then and now. *The Journal of Clinical Investigation, 123*, 2344–2345.
Schwartz, M. W., Woods, S. C., Porte, D., Jr., Seeley, R. J., & Baskin, D. G. (2000). Central nervous system control of food intake. *Nature, 404*, 661–671.
Schwartz, M. W., Woods, S. C., Seeley, R. J., Barsh, G. S., Baskin, D. G., & Leibel, R. L. (2003). Is the energy homeostasis system inherently biased toward weight gain? *Diabetes, 52*, 232–238.
Seeley, R. J., & Woods, S. C. (2003). Monitoring of stored and available fuel by the CNS: Implications for obesity. *Nature Reviews. Neuroscience, 4*, 901–909.
Seidah, N. G., Gaspar, L., Mion, P., Marcinkiewicz, M., Mbikay, M., & Chretien, M. (1990). cDNA sequence of two distinct pituitary proteins homologous to Kex2 and furin gene products: Tissue-specific mRNAs encoding candidates for pro-hormone processing proteinases. *DNA and Cell Biology, 9*, 789.
Shingo, T., Sorokan, S. T., Shimazaki, T., & Weiss, S. (2001). Erythropoietin regulates the in vitro and in vivo production of neuronal progenitors by mammalian forebrain neural stem cells. *The Journal of Neuroscience, 21*, 9733–9743.
Smeekens, S. P., Avruch, A. S., LaMendola, J., Chan, S. J., & Steiner, D. F. (1991). Identification of a cDNA encoding a second putative prohormone convertase related to PC2 in AtT20 cells and islets of Langerhans. *Proceedings of the National Academy of Sciences of the United States of America, 88*, 340–344.
Smith, C. J., Nelson, R. G., Hardy, S. A., Manahan, E. M., Bennett, P. H., & Knowler, W. C. (1996). Survey of the diet of Pima Indians using quantitative food frequency assessment and 24-hour recall. Diabetic Renal Disease Study. *Journal of the American Dietetic Association, 96*, 778–784.
Soliz, J., Joseph, V., Soulage, C., Becskei, C., Vogel, J., Pequignot, J. M., et al. (2005). Erythropoietin regulates hypoxic ventilation in mice by interacting with brainstem and carotid bodies. *The Journal of Physiology, 568*, 559–571.

Studer, L., Csete, M., Lee, S. H., Kabbani, N., Walikonis, J., Wold, B., et al. (2000). Enhanced proliferation, survival, and dopaminergic differentiation of CNS precursors in lowered oxygen. *The Journal of Neuroscience*, *20*, 7377–7383.

Suzuki, N., Ohneda, O., Takahashi, S., Higuchi, M., Mukai, H. Y., Nakahata, T., et al. (2002). Erythroid-specific expression of the erythropoietin receptor rescued its null mutant mice from lethality. *Blood*, *100*, 2279–2288.

Tartaglia, L. A., Dembski, M., Weng, X., Deng, N., Culpepper, J., Devos, R., et al. (1995). Identification and expression cloning of a leptin receptor, OB-R. *Cell*, *83*, 1263–1271.

Teng, R., Calvert, J. W., Sibmooh, N., Piknova, B., Suzuki, N., Sun, J., et al. (2011). Acute erythropoietin cardioprotection is mediated by endothelial response. *Basic Research in Cardiology*, *106*, 343–354.

Teng, R., Gavrilova, O., Suzuki, N., Chanturiya, T., Schimel, D., Hugendubler, L., et al. (2011). Disrupted erythropoietin signalling promotes obesity and alters hypothalamus proopiomelanocortin production. *Nature Communications*, *2*, 520.

Tringali, G., Pozzoli, G., Lisi, L., & Navarra, P. (2007). Erythropoietin inhibits basal and stimulated corticotropin-releasing hormone release from the rat hypothalamus via a non-transcriptional mechanism. *Endocrinology*, *148*, 4711–4715.

Tsai, P. T., Ohab, J. J., Kertesz, N., Groszer, M., Matter, C., Gao, J., et al. (2006). A critical role of erythropoietin receptor in neurogenesis and post-stroke recovery. *The Journal of Neuroscience*, *26*, 1269–1274.

Tse, A., Lee, A. K., & Tse, F. W. (2012). Ca2+ signaling and exocytosis in pituitary corticotropes. *Cell Calcium*, *51*, 253–259.

Wang, L., Teng, R., Di, L., Rogers, H., Wu, H., Kopp, J. B., et al. (2013). PPARalpha and Sirt1 mediate erythropoietin action in increasing metabolic activity and browning of white adipocytes to protect against obesity and metabolic disorders. *Diabetes*, *62*, 4122–4131.

Wu, H., Lee, S. H., Gao, J., Liu, X., & Iruela-Arispe, M. L. (1999). Inactivation of erythropoietin leads to defects in cardiac morphogenesis. *Development*, *126*, 3597–3605.

Wu, H., Liu, X., Jaenisch, R., & Lodish, H. F. (1995). Generation of committed erythroid BFU-E and CFU-E progenitors does not require erythropoietin or the erythropoietin receptor. *Cell*, *83*, 59–67.

Yaswen, L., Diehl, N., Brennan, M. B., & Hochgeschwender, U. (1999). Obesity in the mouse model of pro-opiomelanocortin deficiency responds to peripheral melanocortin. *Nature Medicine*, *5*, 1066–1070.

Yu, X., Shacka, J. J., Eells, J. B., Suarez-Quian, C., Przygodzki, R. M., Beleslin-Cokic, B., et al. (2002). Erythropoietin receptor signalling is required for normal brain development. *Development*, *129*, 505–516.

Zhang, Y., Proenca, R., Maffei, M., Barone, M., Leopold, L., & Friedman, J. M. (1994). Positional cloning of the mouse obese gene and its human homologue. *Nature*, *372*, 425–432.

Zhang, Y., Wang, L., Dey, S., Alnaeeli, M., Suresh, S., Rogers, H., et al. (2014). Erythropoietin action in stress response, tissue maintenance and metabolism. *International Journal of Molecular Sciences*, *15*, 10296–10333.

Zhao, W., Kitidis, C., Fleming, M. D., Lodish, H. F., & Ghaffari, S. (2006). Erythropoietin stimulates phosphorylation and activation of GATA-1 via the PI3-kinase/AKT signaling pathway. *Blood*, *107*, 907–915.

CHAPTER SEVEN

Erythropoietin-Mediated Regulation of Central Respiratory Command

Tommy Seaborn*, Céline Caravagna†,1
*Faculty of Medicine, Université Laval, Quebec City, QC, Canada
†Institut de Neurosciences de la Timone, Marseille, France
1Corresponding author: e-mail address: celine.caravagna@univ-amu.fr

Contents

Abstract

Erythropoietin (Epo) is a cytokine expressed throughout the body, including in the central nervous system where it can act as a breathing modulator in the central respiratory network. In vitro, Epo allows maintaining the activity of respiratory neurons during acute

Vitamins and Hormones, Volume 105
ISSN 0083-6729
http://dx.doi.org/10.1016/bs.vh.2017.02.006
© 2017 Elsevier Inc.
All rights reserved.

hypoxia, resulting in inhibition of the hypoxia-induced rhythm depression. In vivo, Epo action on the central respiratory command results in enhancement of the acute hypoxic ventilatory response, allowing a better oxygenation of the body by improvement of gases exchanges in the lungs. Importantly, this effect of Epo is age-dependent, being observed at adulthood and at both early and late postnatal ages, but not at middle postnatal ages, when an important setup of the central respiratory command occurs. Epo regulation of the central respiratory command involves at least two intracellular signaling pathways, PI3K–Akt and MEK–ERK pathways. However, the exact mechanism underlying the action of Epo on the central respiratory control remains to be deciphered, as well as the exact cell types and nuclei involved in this control. Epo-mediated effect on the central respiratory command is regulated by several factors, including hypoxia, sex hormones, and an endogen antagonist. Although more knowledge is needed before reaching the clinical trial step, Epo seems to be a promising therapeutic treatment, notably against newborn breathing disorders.

1. ERYTHROPOIETIN

Erythropoietin (Epo) is a cytokine expressed throughout the body and having numerous roles. Among these roles, Epo acts as breathing modulator. This regulatory action takes place in different parts of the organism, including brain, blood, and carotid bodies. In this review, the current knowledge on the Epo roles in the modulation of the respiratory process is reviewed, and further directions to explore are proposed to improve the understanding of Epo-modulated regulation of breathing.

1.1 History

The Epo molecule was discovered in 1906 by Paul Carnot, who observed that injection of blood from hemorrhagic rabbit induced erythrocytes production in nonhemorrhagic rabbits (Jelkmann, 1986). Since then, Epo has been characterized as a glycoprotein hormone of 34kDa (Lai, Everett, Wang, Arakawa, & Goldwasser, 1986) and as the main molecule controlling erythrocytes production (Reissmann, 1950). Epo has been synthetized for the first time in 1985 (Egrie, Browne, Lai, & Lin, 1985), permitting the multiplication of studies on this molecule. In human, Epo is mainly produced by the liver in the fetus (Zanjani, Poster, Burlington, Mann, & Wasserman, 1977) and the kidney at adulthood (Jacobson, Goldwasser, Fried, & Plzak, 1957), and is released in the blood to stimulate erythropoiesis in bone marrow. Epo is also synthesized by other organs, including lung, heart, and brain, and has different roles depending on its site of expression, e.g., erythropoiesis in bone marrow and neuroprotection in central nervous system (CNS).

1.2 Ligands and Receptors

Epo binds the Epo receptor (EpoR), which is expressed by a wide variety of cell types throughout the body. EpoR belongs to the cytokine receptor superfamily, and Epo binding induces EpoR homodimerization then tyrosine-kinase activity leading to the activation of various intracellular pathways (cf. Section 1.3), depending on the cell type concerned (Witthuhn et al., 1993). Epo affinity to EpoR depends on the targeted cells, some expressing membrane-truncated forms of EpoR with various levels of Epo affinity (Masuda et al., 1993) (Table 1). Moreover, Epo-induced enhancement of EpoR expression has been characterized in rat cardiomyocytes (Salisch, Klar, Thurisch, Bungert, & Dame, 2011), vascular system (Beleslin-Cokic et al., 2004), and, more relevantly to the context of this review, in the CNS (Hellewell, Yan, Alwis, Bye, & Morganti-Kossmann, 2013).

There is also a soluble truncated form of EpoR, lacking both the transmembrane and intracellular domains. This EpoR isoform is called soluble

Table 1 Epo Affinity for Various EpoR Forms

Epo Receptor	Dissociation Constant (K_d)	References
EpoR	High-affinity binding site: $0.092\,nM < K_d < 1.5\,nM$	Ahaded, Winchenne, Cartron, Lambin, and Lopez (1999), Hatayama and Ide (2015), and Avedissian, Poola, and Spivak (1995)
sEpoR	High-affinity binding sites: $0.25\,nM < K_d < 3.4\,nM$	Ahaded et al. (1999)
	Low-affinity binding sites: $0.85\,\mu M < K_d < 1.35\,\mu M$	Avedissian et al. (1995) and Philo, Aoki, Arakawa, Narhi, and Wen (1996)
Various EpoR isoforms[a]	High-affinity binding sites: $0.095\,nM < K_d < 0.58\,nM$	Jarsch, Brandt, Lanzendörfer, and Haselbeck (2008)
	Low-affinity binding sites: $1.9\,nM < K_d < 16\,nM$	Masuda et al. (1993)

[a]Several studies characterized different affinities of Epo to EpoR, but did not precisely sequence the concerned EpoR isoform. The affinity could also depend on the cell type expressing EpoR and the Epo molecule used during the experimentation (e.g., labeled or not, recombinant human).

EpoR (sEpoR) and acts as endogen Epo antagonist by binding Epo in the extracellular space without inducing intracellular signaling (Nagao, Masuda, Abe, Ueda, & Sasaki, 1992), creating a competition with active membrane EpoR (Table 1). sEpoR has been detected in various organs, including the brain (Soliz, Gassmann, & Joseph, 2007).

Epo can also bind a heterodimeric complex composed of one EpoR subunit and one β-common receptor (βCR) subunit (Brines et al., 2004). The expression level of both βCR and EpoR is enhanced by Epo treatment after spinal cord injury in rat (King et al., 2007). However, βCR and EpoR usually do not colocalize in studied structures (i.e., hippocampus, neocortex, and hypothalamus) (Sanchez et al., 2009), and neuroprotective role of Epo is also observed in cells where βCR is not been expressed (Nadam et al., 2007; Um, Gross, & Lodish, 2007). Thus, the implication of βCR in Epo signaling needs to be focused on, especially in the brain, where potential neuroprotective agents are required.

1.3 Signaling

Epo binding on EpoR can stimulate various intracellular signaling pathways, depending on which cell type the activated receptor is expressed. In the CNS, Epo-induced neuroprotection involves antiapoptotic kinase (Akt), janus kinase (JAK), and extracellular signal-regulated protein kinase (ERK) pathways (Jiang, Haustein, Sofroniew, & Khakh, 2014; Kilic et al., 2005). The implication of these pathways in the Epo activity has been shown in different neural contexts, from Parkinson disease (Jiang et al., 2014) to cerebral focal ischemia (Kilic et al., 2005). The implication of these signaling pathways in the Epo effect on respiratory modulation is more precisely discussed in Section 4.5.

2. CONTROL OF THE RESPIRATORY SYSTEM

2.1 Carotid Bodies

Breathing is a vital function that requires precise modulation to fit the body energy requirements. Blood gases (dioxygen—O_2 and carbon dioxide—CO_2) levels are continuously monitored by the carotid bodies to ensure maintenance of appropriate body oxygenation (Madurga, Mizíková, Ruiz-Camp, & Morty, 2013). When blood gases reach inappropriate levels, carotid bodies send a specific signal to the central respiratory network via the carotid sinus and the glossopharyngeal (IX) nerves. Breathing is thus modulated by the central respiratory command until blood gases levels come back to physiological values.

2.2 Central Respiratory Command

The central respiratory network is composed of numerous neuronal groups located in the brainstem (Richter, 1982). More precisely, the respiratory rhythm is generated in the medulla oblongata, while adaptation and control of the rhythm are handled by respiratory groups located in both medulla oblongata and pons. In this regulatory network, different types of neurons are involved, e.g., pacemakers and nonpacemakers rhythmic neurons, tonically activated neurons, and neurons integrating the peripheral afferences (Feldman & Del Negro, 2006; Feldman, Del Negro, & Gray, 2013). In addition, many neurotransmitters are involved, including gamma-aminobutyric acid (Paton & Richter, 1995), glycine (Janczewski, Tashima, Hsu, Cui, & Feldman, 2013), and catecholamines (Guyenet et al., 2013). The complex functioning of this system is not yet fully understood, but some of its fundamental functions are well defined. For example, the pre-Bötzinger complex is necessary for proper generation of the inspiratory rhythm (Arata, Onimaru, & Homma, 1990; Smith, Ellenberger, Ballanyi, Richter, & Feldman, 1991), while the nucleus tractus solitarius is the site of integration for the peripheral afferences (Jean, 1991). Also, catecholamines nuclei A1C1, A2C2, A5, and A7 are known to be activated during hypoxia (Elam, Yao, Thorén, & Svensson, 1981; Erickson & Millhorn, 1994; Soulier et al., 1997), and their lesions alter the hypoxic ventilatory response (Coles & Dick, 1996; Housley & Sinclair, 1988), demonstrating their necessary implication in proper response to hypoxia (Alheid & McCrimmon, 2008; Hilaire, Viemari, Coulon, Simonneau, & Bévengut, 2004; Roux, Dura, & Villard, 2008).

2.3 Responses to Hypoxia

Whole-body plethysmography is a technique allowing the recording of breathing parameters (frequency and tidal volume) on awake and nonrestricted animals. The major advantage of this technique is the very limited stress resulting from animal manipulation. There is thus only a weak influence of the experimental procedure on the recorded breathing parameters

(Seaborn, Gassmann, & Soliz, 2013) that can be considered as resting parameters. Another advantage of this technique is the possibility to expose the animal to various controlled levels of environmental gases. The ventilatory response of the animal can thus be precisely observed, monitored, and analyzed.

Whole-body plethysmography allowed demonstrating, among others, that when the body is submitted to hypoxic challenge, breathing is adjusted to enhance O_2 capture. This hypoxic ventilatory response is characterized by two phases: A first phase in which breathing frequency is rapidly enhanced (Kazemi & Hoop, 1991), and a second phase during which breathing frequency decreases before stabilizing as a plateau (Teppema & Dahan, 2010). Noteworthy, frequency value reached by this plateau in newborn is lower than its basal normoxic ventilation level (Nsegbe et al., 2004), making them more sensitive to hypoxia than adults.

The central respiratory network is located in the brainstem, and it is possible to isolate brainstem from the body and maintain it in an artificially oxygenated cerebrospinal fluid (Rousseau & Caravagna, 2015; Suzue, 1984). The rhythmic activity generated by the central respiratory command is then recordable from the fourth ventral root and associated to inspiratory activity (Rousseau & Caravagna, 2015; Suzue, 1984). This rhythm displays a lower frequency than the breathing frequency occurring in resting living animal, mainly due to the absence of stimuli from peripheral afferences to the brainstem (Kubin, Alheid, Zuperku, & McCrimmon, 2006). The hypoxic response of the isolated central respiratory network is also different from the hypoxic ventilatory response observed in living animal. Indeed, when isolated central respiratory network is submitted to hypoxia, the rhythm frequency decreases immediately and rises during posthypoxic recovery (Viemari, Burnet, Bévengut, & Hilaire, 2003). The rhythm could even disappear if hypoxia persists. This in vivo vs in vitro difference is likely due to the lack of peripheral afferences and rostral pons nuclei, and this explanation is now unanimously accepted by the scientists in the field (Siljehav, Shvarev, & Herlenius, 2014; Viemari et al., 2003). This in vitro approach has been widely used to investigate the functioning of the central respiratory command per se, including to study the effect of Epo on central respiratory command.

In Sections 3 and 4, the current knowledge of Epo action on breathing is further developed by overviewing the different levels at which Epo can modulate respiration, from the periphery of the body to the central respiratory command, in both in vitro and in vivo models.

3. PERIPHERAL EPO REGULATION OF RESPIRATORY PARAMETERS

3.1 Erythropoietic Function

There is an indirect Epo-mediated regulation of breathing that relies on the involvement of Epo on erythropoiesis (Reissmann, 1950). Indeed, erythrocytes are required to O_2 transport, and when O_2 is efficiently carried to the organs, there is no need for increased gases exchanges between blood and air in the lung alveoli, thus for increased breathing. In brief, accurate O_2 blood transport throughout the body thanks to Epo-induced erythropoiesis contributes to maintain an accurate breathing.

3.2 Action on the Carotid Bodies

Peripheral Epo allows regular breathing not only by maintaining appropriate erythrocytes levels but also by acting directly on the carotid bodies. In mice, intravenous injection of 2000 U/kg of synthetic Epo right before hypoxic challenge increases breathing frequency and decreases tidal volume during acute severe hypoxia (6% O_2), suggesting that Epo modifies carotid bodies gases detection or response to blood gases variations (Soliz et al., 2005; Soliz, Thomsen, Soulage, Lundby, & Gassmann, 2009). Indeed, this effect is likely too fast to involve gene transduction or to allow penetration of Epo into the CNS at appropriate doses to modulate the central respiratory command. Accordingly, this effect has been associated to the well blood-irrigated cells of the carotid bodies (Soliz et al., 2005). Moreover, chemosensitive cells of the carotid bodies express EpoR (Soliz et al., 2005), and the expression of both Epo and EpoR is enhanced in the carotid bodies during both chronic and intermittent hypoxia (Lam, Tipoe, & Fung, 2009), strongly supporting the hypothesis of Epo modulation on the blood gases detection by carotid bodies.

4. EPO REGULATION ON THE CENTRAL RESPIRATORY COMMAND

Since Epo can cross the blood–brain barrier in a dose-dependent manner (Banks, Jumbe, Farrell, Niehoff, & Heatherington, 2004), it represents a potential valuable agent to treat neural pathologies, including respiratory-related neural diseases.

4.1 Epo Effects on Breathing

Epo-mediated regulation of the central respiratory command is first observable at the physiological level, breathing modulation being directly dependent on the central respiratory command.

Intracisternal injection (i.c.i., injection in the cisterna magna), allowing selective injection in the CNS, has been used to induce Epo blockade in the brainstem of living animals (Ballot, Joseph, & Soliz, 2015; Ballot, Laouafa, et al., 2015). In adult mice, sEpoR i.c.i. decreases normoxic ventilation and hypoxic (12% O_2) ventilatory response. In newborn mice (postnatal day 10), sEpoR i.c.i. decreases normoxic ventilation, as a result of a decreased tidal volume. During moderate hypoxia (10% O_2), sEpoR i.c.i. decreases hypoxic ventilatory response, while during extreme hypoxia (6% O_2), sEpoR i.c.i. induces respiratory depression, signs of asphyxia (gasping), and a high mortality rate (Ballot, Joseph, et al., 2015). Taken together, these data show that endogenous Epo is required in the central respiratory network to maintain proper breathing in both adult and newborn mice.

The effect of chronic Epo overexpression on the central respiratory command has also been investigated, thanks to transgenic Tg21 mice constitutively overexpressing Epo in the CNS only (four times higher than in wild-type mice) (Wiessner et al., 2001). Such a chronic Epo overexpression in the brain increases hypoxic ventilatory response to acute severe hypoxia (15 min, 6% O_2) (Soliz et al., 2005). This beneficial effect is age dependent, being present at early and late postnatal ages (days 3 and 21, respectively) (Caravagna, Gasser, Ballot, Joseph, & Soliz, 2015) as well as in adult mice (Soliz et al., 2005), but not at middle postnatal ages (postnatal days 7 and 15) (Caravagna et al., 2015). Moreover, at postnatal day 15, Tg21 mice display hypoxic (6% O_2) ventilatory response that is lower than control mice under the same conditions. This difference is observed neither at postnatal day 7 nor at postnatal day 21. These results suggest that Epo effect on the central respiratory command is modified at postnatal day 15. This can be attributed to the modification occurring in the central respiratory network around postnatal days 12–13 (Wong-Riley, Liu, & Gao, 2013). The expression of many molecules is affected during this critical period (e.g., brain-derived neurotrophic factor (BDNF) (Gao, Liu, Nair, & Wong-Riley, 2014), serotonin (Wong-Riley et al., 2013), and glycine (Liu & Wong-Riley, 2013)). Interestingly, Epo has been shown to interact with these molecules. Indeed, BDNF is involved in Epo neuroprotective role (Solev et al., 2013), serotonin is implicated in Epo effects in various contexts (Azzadin et al., 1992), and glycine level is regulated by Epo in the blood of dialyzed patients (Riedel et al., 1989). It is thus conceivable that the Epo effect during this critical period can vary by itself or by its interaction with other molecules involved in the setting up of the central respiratory network. It is also relevant to underline that no Epo effect is observed on mouse ventilation at postnatal day 21 and in adulthood during normoxia or 10% O_2 hypoxia. In vivo, Epo effect appears at postnatal day 21 and in adulthood during 6% hypoxia. In vitro, Epo effect appears at both postnatal days 0 and 3, when isolated brainstems are directly exposed to hypoxia (described in Section 4.2). Thus, Epo effect seems to be observed only when the hypoxia directly reaches the brainstem. This stimulatory effect is actually a "maintaining" effect, in conditions where control mice display breathing depression, which could occur at least partially by modulating the integration of carotid bodies signaling. Tg21 mice also display improved ventilatory acclimatization to chronic hypoxic exposure (Soliz et al., 2005). Thus, chronic Epo overexpression in the brain improves hypoxic ventilatory response in mice in an age-dependent manner that would then need to be considered in the view of eventual therapeutic application of Epo. Furthermore, following bilateral transection of the carotid sinus nerves (chemodenervation) that uncouples the brain from the carotid bodies, Tg21 mice adjusted their ventilation to acute severe hypoxia (6% O_2), while chemodenervated wild-type mice developed a life-threatening apnea (Soliz et al., 2005). Taken together, these results imply that cerebral Epo is required for breathing modulation during both acute and chronic hypoxia, and may have a therapeutic use in case of chemosensitive system failure.

Finally, in vivo Epo-mediated modulation of the central respiratory command has been unexpectedly demonstrated in premature human neonates. Subcutaneous treatment with Epo (300 U/kg/dose, 3 times/week) to very premature neonates (gestational age <30 weeks) has been reported to improve both erythropoietic (higher hemoglobin, hematocrit, and reticulocytes) and ventilatory functions (reduced need for assisted ventilation and O_2 supplementation) (Tempera et al., 2011). Although a better hematological status clearly helps in stabilizing respiration (as mentioned in Section 3.1), the remarkable observed improvement of ventilatory function cannot be merely explained by a higher number of erythrocytes since the ventilatory improvement was not observed in preterm neonates who received only blood transfusion without Epo treatment (Kasat, Hendricks-Muñoz, & Mally, 2011). Moreover, Epo can easily cross the blood–brain barrier in neonates (Juul et al., 2004), allowing hypothesizing that Epo has concomitant direct effects on both central respiratory command and carotid bodies. Interestingly, repeated high-dose Epo treatment (7 daily injections of 30,000 U/kg) in newborn rats has no discernible adverse effects (Kellert, McPherson, & Juul, 2007), describing Epo as a promising therapeutic molecule against newborn breathing distress.

4.2 Epo Effects on the Brainstem

In order to study the effect of Epo on the central respiratory network itself, brainstem preparations have been isolated and the output signal from the respiratory network recorded under normoxic and hypoxic conditions. One-hour incubation of the brainstems with Epo attenuates hypoxia-induced central respiratory depression, resulting in maintained rhythm frequency (Khemiri, Seaborn, Gestreau, & Soliz, 2012). Moreover, one-hour incubation of the brainstems with antagonist sEpoR dramatically increased hypoxia-induced depression, confirming the importance of endogenous Epo under acute hypoxic conditions (Khemiri et al., 2012). This effect is dose- and time-dependent, highlighting the importance of appropriate administration regimen in eventual therapeutic applications.

Transgenic Tg21 mice overexpressing Epo in the brain have also been pivotal in the study of the effect of chronic Epo stimulation on the central respiratory command. On the opposite of brainstem isolated from wild-type mice, isolated brainstems of mice chronically overexpressing Epo display no rhythm attenuation during hypoxia (Caravagna, Kinkead, & Soliz, 2014). Thus, Epo allows the central respiratory command to continue to generate an inspiratory rhythm even when the central respiratory network is directly submitted to acute hypoxia. This effect is different from the modification of hypoxic ventilatory response previously described in living animal because isolated brainstems are directly submitted to hypoxia, while in whole-body plethysmography hypoxemia is moderated by breathing modulation and usually does not reach the CNS directly. Thus, these two experimental paradigms highlight two distinct mechanisms.

4.3 Epo Effects on Respiratory Groups

The respiratory groups targeted by Epo are not well described yet. However, it has been documented that EpoR is expressed in the nucleus tractus solitarius, the pre-Bötzinger complex, as well as in A1C1, A2C2, A5, and A7 nuclei in adult wild-type mice (Soliz et al., 2005). These descriptive observations strongly suggest that Epo plays a role in these respiratory groups, although the effect of Epo on a precise group has not been formally demonstrated to date. Nevertheless, significant differences were found in A2C2 and A5 cell groups between wild-type and Tg21 mice. Indeed, the activity of the tyrosine hydroxylase (TH), the enzyme allowing catecholamine neurotransmitters synthesis, and norepinephrine (NE) content were decreased in A2C2 of Tg21 mice, while NE content was increased in their A5 cell group (Soliz et al., 2005). Moreover, Tg6 female mice, another transgenic mice overexpressing Epo in the brain (26-fold than wild-type mice) (Wiessner et al., 2001), showed higher TH activity in A6 and A5 groups but lower NE content in A2C2 and A1C1 groups (Soliz et al., 2009). These data strongly suggest that cerebral Epo could regulate breathing by modulating catecholamine expression.

4.4 Epo Effects on Respiratory Cells

Epo expression has been first detected in cultured astrocytes (Masuda et al., 1994). On the other hand, cultured neurons express both EpoR (Ruscher et al., 2002) and Epo (Sun et al., 2013). In vivo, both cortical neurons and astrocytes have been shown to express Epo mRNA in response to hypoxic stress (Bernaudin et al., 2000). However, the expression level of Epo mRNA in normoxic condition has not been documented in this in vivo study, and the expression of Epo protein has not been assessed so far. In the central respiratory network, both Epo and EpoR are expressed (see Section 3.2), but the exact cell types expressing Epo in these neural groups are not yet identified. In fact, Epo and EpoR expression by the respiratory neurons and astrocytes has never been studied at the cellular level. This information would be pivotal in the understanding of the mechanisms underlying Epo-mediated regulation of the central respiratory command.

4.5 Epo Effects on Cell Signaling

Various pathways have been shown to be activated by Epo in various cell types, but this review focuses on those Epo-modulated specifically in the respiratory neural nuclei.

It is known that cerebral Epo protects against focal ischemia by activating both ERK and Akt (Kilic et al., 2005). In addition, Epo induces phrenic motor facilitation by activation of $ERK_{1/2}$ and Akt pathways (Dale, Satriotomo, & Mitchell, 2012). Using isolated brainstem–spinal cord preparations, the role of these pathways in the central respiratory network of Tg21 mice has been studied (Caravagna & Soliz, 2015). Exposition of brainstems isolated from Tg21 mice to 1 μM solution of $MEK_{1/2}$ or PI3K inhibitors (U0126 and LY294002, respectively) enhances bursts frequency during both normoxia and hypoxia, compared with preincubation frequency. This effect is dose-dependent and is not present at lower and higher tested doses. Moreover, the exposition of brainstems isolated from wild-type animal to $MEK_{1/2}$ or PI3K inhibitors solutions abolishes hypoxia-induced depression. This result reveals a complex interaction between Epo and $MEK_{1/2}$ and/or PI3K pathways. Indeed, inhibition of $MEK_{1/2}$ or PI3K does not restore control phenotype on Tg21 brainstems, but mimic chronic Epo overexpression effect on control brainstems. Thus, $MEK_{1/2}$ and PI3K are necessary to the hypoxia-induced frequency depression to appear. This data can be interpreted in different ways. For instance, Epo may not activate $MEK_{1/2}$ or PI3K in the central respiratory command and act via others pathways to stimulate the brainstem, perhaps the JAK–STAT pathway, which has been shown to be implicated in the Epo neuroprotective effect (Miljus et al., 2014). Thus, $MEK_{1/2}$ and PI3K pathways would be recruited by a molecule other than Epo to allow hypoxic frequency depression. Alternatively, Epo may activate $MEK_{1/2}$ and/or PI3K pathways in respiratory groups involved in the hypoxia-induced depression, such as the retrotrapezoid nucleus associated to the parafacial respiratory group (RTN/pFRG) (Gourine et al., 2010), or in different respiratory groups, the resulting effect on the respiratory output being an increased frequency. Finally, a potential compensatory phenomenon between $MEK_{1/2}$ and PI3K pathways is also conceivable. In any case, the observed results allow the conclusion that $MEK_{1/2}$ or PI3K alone is not sufficient to support Epo stimulation under hypoxic conditions.

5. FACTORS INFLUENCING EPO-MEDIATED REGULATION OF THE RESPIRATORY COMMAND

5.1 Sex Hormones

Epo interaction with sex hormones has already been demonstrated in other systems than the central respiratory network. For example, long-term Epo therapy in hemodialyzed patients decreases levels of follicle-stimulating hormone and luteinizing hormone and increases plasma levels of estradiol and testosterone (Kokot et al., 1992). Moreover, plasma estradiol levels are higher in transgenic female mice presenting systemic overexpression of Epo (Gassmann, Pfistner, Doan, Vogel, & Soliz, 2010). Reciprocally, sex hormones modulate the erythropoietic functions of Epo. In rats, after acclimatization to chronic hypoxia, hemoglobin and hematocrit levels are lower in females than in males (Lumbroso & Joseph, 2009). This effect is likely due to the inhibiting effect of estradiol on the expression of Epo in females during hypoxia (Mukundan, Resta, & Kanagy, 2002), and the stimulating effect of testosterone on the expression of Epo reported in men (Coviello et al., 2008), male rats (Rencricca et al., 1969), and male mice (Fried & Gurney, 1965).

Regarding the central respiratory command in human, Epo enhances hypoxic (10% O_2) ventilation in women, while decreases hypoxic ventilation in men, as a result of increased tidal volume in women and decreased frequency in men (Soliz et al., 2009). However, this effect is observed in peripherally injected subjects and Epo action on the central respiratory command may be mediated by carotid bodies. Epo sex dimorphism directly in the central respiratory command has been described by sEpoR i.c.i. and plethysmography in adult male and female mice, where sEpoR decreases normoxic ventilation due to a decrease of tidal volume in males and breathing frequency in females (Ballot, Joseph, et al., 2015). Moreover, sEpoR injection decreases the hypoxic ventilatory response (12% O_2) in male but not in female mice (Ballot, Joseph, et al., 2015). In newborn mice exposed to normoxia, sEpoR decreases ventilation, tidal volume, and frequency in females only (Ballot, Joseph, et al., 2015). These results clearly show that Epo regulation of the central respiratory command is sex dependent. However, this sex dimorphism may also be due to the action of sex hormones directly on the central respiratory command, making the response to Epo different between males and females.

5.2 Hypoxia

In many cells, hypoxia induces the expression of the hypoxia-inducible factor (HIF), a transcription factor controlling the expression of many genes in response to hypoxia. Among these genes, HIF induces the transcription of Epo (Semenza, 1994). The expression of the HIF-α subunit is regulated by oxygen levels, whereas the expression of the HIF-β subunit is constitutive (Pugh, O'Rourke, Nagao, Gleadle, & Ratcliffe, 1997). There are three known isoforms of the α-subunit: HIF-1α, HIF-2α, and HIF-3α. A study suggested that the induction of HIF-2α rather than HIF-1α may be responsible for hypoxia-induced increase in Epo expression in astrocytes (Chavez, Baranova, Lin, & Pichiule, 2006).

Moreover, Epo concentrations in the brainstem of high-altitude newborn rats (harvested from high-altitude rats having developed hypoxia adaptation, as revealed by significantly higher hematocrit levels and basal ventilation than in sea-level controls) are twice higher than the Epo concentrations measured in the brainstem of sea-level newborn rats (Seaborn et al., 2011).

5.3 sEpoR

sEpoR i.c.i. in adult and newborn mice interferes with breathing frequency and tidal volume. Indeed, i.c.i. of sEpoR reduces normoxic and hypoxic ventilation in newborn and adult mice, resulting in high mortality rate in newborn mice exposed to severe (6% O_2) hypoxia (Ballot, Joseph, et al., 2015), demonstrating that sEpoR can regulate Epo-mediated effect on the central respiratory command. However, sEpoR expression itself in the central respiratory network neuronal groups has never been studied. Nevertheless, there is no evidence showing that sEpoR is not expressed in the central respiratory groups and then acts as Epo modulator. This would be an important part of Epo regulatory mechanism, deserving to be deciphered.

6. CONCLUSION

Despite the promising therapeutic potential, Epo studies are still limited, mainly by the lack of appropriate specific antibodies against Epo and EpoR. Indeed, not only the specificity of the available antibodies has been questioned (Brown et al., 2007; Elliott et al., 2006), but their sensitivity is limited (Matsubara et al., 1989), requiring important quantities of tissue that is very restricted in small structures as those forming the mice central respiratory network. Moreover, the implication of βCR in Epo signaling, especially in the central respiratory network, needs to be focused on. Indeed, the respiratory effect of Epo could involve βCR as well as EpoR, and this hypothesis should be studied to better understand how Epo modulates the central respiratory command.

On the other hand, the functioning of the central respiratory command itself is still to be fully deciphered. The understanding of the respiratory rhythm generation and modulating mechanisms is also required to perfectly evaluate the actions of Epo on the central respiratory command groups and cells.

We described here the current knowledge about Epo-mediated regulation on the central respiratory command (Fig. 1). Epo is required for proper breathing, as demonstrated by acute Epo blockade in the brain. Moreover, Epo treatment seems to be beneficial to improve the hypoxic ventilatory response in early newborns and adults in both mice and humans. However, the underlying mechanisms and precise Epo targets are still to be identified.

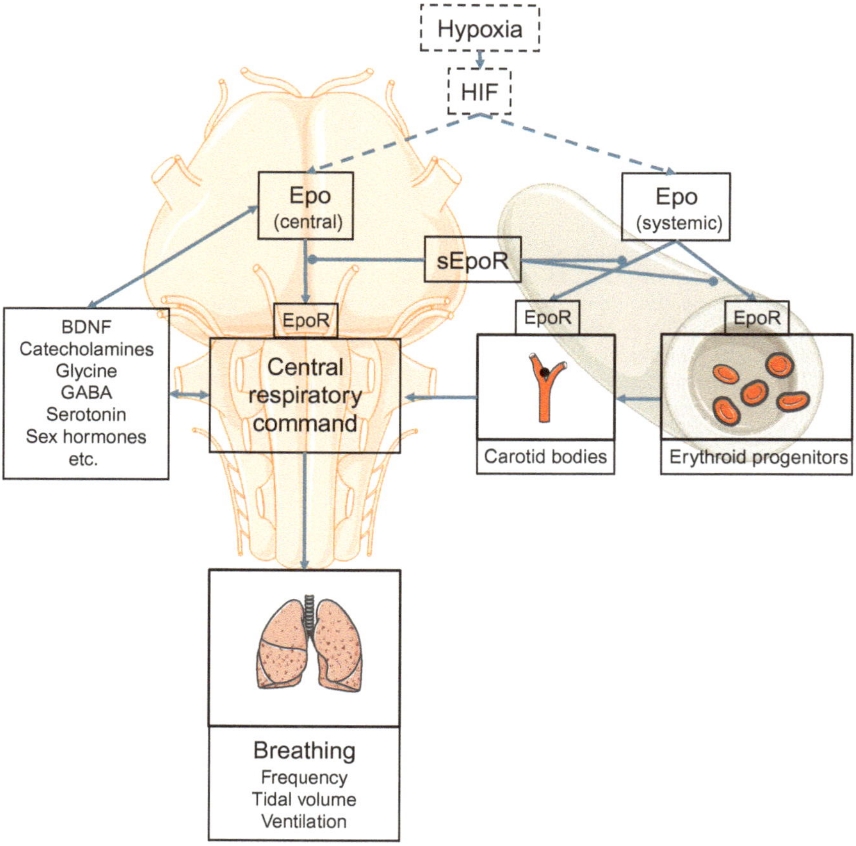

Fig. 1 Summary of Epo-mediated control of breathing. Epo modulates respiration in normoxic conditions and its expression is enhanced in response to hypoxia. This Epo modulation occurs by acting at both systemic level, i.e., on erythropoiesis and gases level detection in the carotid body, and central level, i.e., on the central respiratory network, where it could involve various molecules, including neurotransmitters and sex hormones. *BDNF*, brain-derived neurotrophic factor; *Epo*, erythropoietin; *EpoR*, erythropoietin receptor; *GABA*, gamma-aminobutyric acid; *HIF*, hypoxia-inducible factor; sEpoR, soluble erythropoietin receptor.

It would be very valuable to better understand the action of Epo on the central respiratory command, in order to use it as a breathing stimulant, for example, to treat premature newborns suffering from breathing disorders.

ACKNOWLEDGMENT

The authors thank Miss Orlane Ballot for helpful and constructive ideas and discussions.

REFERENCES

Ahaded, A., Winchenne, J. J., Cartron, J. P., Lambin, P., & Lopez, C. (1999). The extracellular domain of the human erythropoietin receptor: Expression as a fusion protein in *Escherichia coli*, purification, and biological properties. *Preparative Biochemistry & Biotechnology, 29*(2), 163–176. http://doi.org/10.1080/10826069908544888.

Alheid, G. F., & McCrimmon, D. R. (2008). The chemical neuroanatomy of breathing. *Respiratory Physiology & Neurobiology, 164*(1–2), 3–11. http://doi.org/10.1016/j.resp.2008.07.014.

Arata, A., Onimaru, H., & Homma, I. (1990). Respiration-related neurons in the ventral medulla of newborn rats in vitro. *Brain Research Bulletin, 24*(4), 599–604. Retrieved from http://www.ncbi.nlm.nih.gov/pubmed/2357591.

Avedissian, L. S., Poola, I., & Spivak, J. L. (1995). Ligand binding kinetics of a soluble full-length murine erythropoietin receptor. *Biochemical and Biophysical Research Communications, 216*(1), 62–68. http://doi.org/10.1006/bbrc.1995.2592.

Azzadin, A., Bialkowska, A., Pawlak, D., Chabielska, E., Mazerska, M., Mysliwiec, M., et al. (1992). Serotonin is involved in some effects of erythropoietin. *Thrombosis Research, 65*(4–5), 657–661. Retrieved from http://www.ncbi.nlm.nih.gov/pubmed/1615501.

Ballot, O., Joseph, V., & Soliz, J. (2015). Endogenous brain erythropoietin is a potent sex-specific respiratory stimulant in adult and newborn mice. *Journal of Applied Physiology (Bethesda, Md.: 1985), 118*(11), 1386–1395. http://doi.org/10.1152/japplphysiol.00143.2015.

Ballot, O., Laouafa, S., Elliot-Portal, E., Tam, R., Voituron, N., Joseph, V., et al. (2015). The central chemosensitivity is not altered by cerebral erythropoietin. *Neuroscience Letters, 609*, 63–68. http://doi.org/10.1016/j.neulet.2015.10.026.

Banks, W. A., Jumbe, N. L., Farrell, C. L., Niehoff, M. L., & Heatherington, A. C. (2004). Passage of erythropoietic agents across the blood-brain barrier: A comparison of human and murine erythropoietin and the analog darbepoetin alfa. *European Journal of Pharmacology, 505*(1–3), 93–101. http://doi.org/10.1016/j.ejphar.2004.10.035.

Beleslin-Cokic, B. B., Cokic, V. P., Yu, X., Weksler, B. B., Schechter, A. N., & Noguchi, C. T. (2004). Erythropoietin and hypoxia stimulate erythropoietin receptor and nitric oxide production by endothelial cells. *Blood, 104*(7), 2073–2080. http://doi.org/10.1182/blood-2004-02-0744.

Bernaudin, M., Bellail, A., Marti, H. H., Yvon, A., Vivien, D., Duchatelle, I., et al. (2000). Neurons and astrocytes express EPO mRNA: Oxygen-sensing mechanisms that involve the redox-state of the brain. *Glia, 30*(3), 271–278. Retrieved from http://www.ncbi.nlm.nih.gov/pubmed/10756076.

Brines, M., Grasso, G., Fiordaliso, F., Sfacteria, A., Ghezzi, P., Fratelli, M., et al. (2004). Erythropoietin mediates tissue protection through an erythropoietin and common beta-subunit heteroreceptor. *Proceedings of the National Academy of Sciences of the United States of America, 101*(41), 14907–14912. http://doi.org/10.1073/pnas.0406491101.

Brown, W. M., Maxwell, P., Graham, A. N. J., Yakkundi, A., Dunlop, E. A., Shi, Z., et al. (2007). Erythropoietin receptor expression in non-small cell lung carcinoma: A question of antibody specificity. *Stem Cells (Dayton, Ohio)*, *25*(3), 718–722. http://doi.org/10.1634/stemcells.2006-0687.

Caravagna, C., Gasser, E. M. S., Ballot, O., Joseph, V., & Soliz, J. (2015). Chronic overexpression of cerebral Epo improves the ventilatory response to acute hypoxia during the postnatal development. *International Journal of Developmental Neuroscience*, *44*, 84–91. http://doi.org/10.1016/j.ijdevneu.2015.06.001.

Caravagna, C., Kinkead, R., & Soliz, J. (2014). Post-natal hypoxic activity of the central respiratory command is improved in transgenic mice overexpressing Epo in the brain. *Respiratory Physiology & Neurobiology*, *200*, 64–71. http://doi.org/10.1016/j.resp.2014.06.003.

Caravagna, C., & Soliz, J. (2015). PI3K and MEK1/2 molecular pathways are involved in the erythropoietin-mediated regulation of the central respiratory command. *Respiratory Physiology & Neurobiology*, *206*, 36–40. http://doi.org/10.1016/j.resp.2014.11.012.

Chavez, J. C., Baranova, O., Lin, J., & Pichiule, P. (2006). The transcriptional activator hypoxia inducible factor 2 (HIF-2/EPAS-1) regulates the oxygen-dependent expression of erythropoietin in cortical astrocytes. *The Journal of Neuroscience: The Official Journal of the Society for Neuroscience*, *26*(37), 9471–9481. http://doi.org/10.1523/jneurosci.2838-06.2006.

Coles, S. K., & Dick, T. E. (1996). Neurones in the ventrolateral pons are required for post-hypoxic frequency decline in rats. *The Journal of Physiology*, *497*(Pt. 1)), 79–94. Retrieved from http://www.pubmedcentral.nih.gov/articlerender.fcgi?artid=1160914&tool=pmcentrez&rendertype=abstract.

Coviello, A. D., Kaplan, B., Lakshman, K. M., Chen, T., Singh, A. B., & Bhasin, S. (2008). Effects of graded doses of testosterone on erythropoiesis in healthy young and older men. *The Journal of Clinical Endocrinology and Metabolism*, *93*(3), 914–919. http://doi.org/10.1210/jc.2007-1692.

Dale, E. A., Satriotomo, I., & Mitchell, G. S. (2012). Cervical spinal erythropoietin induces phrenic motor facilitation via extracellular signal-regulated protein kinase and Akt signaling. *The Journal of Neuroscience: The Official Journal of the Society for Neuroscience*, *32*(17), 5973–5983. http://doi.org/10.1523/jneurosci.3873-11.2012.

Egrie, J. C., Browne, J., Lai, P., & Lin, F. K. (1985). Characterization of recombinant monkey and human erythropoietin. *Progress in Clinical and Biological Research*, *191*, 339–350. Retrieved from http://www.ncbi.nlm.nih.gov/pubmed/2996025.

Elam, M., Yao, T., Thorén, P., & Svensson, T. H. (1981). Hypercapnia and hypoxia: Chemoreceptor-mediated control of locus coeruleus neurons and splanchnic, sympathetic nerves. *Brain Research*, *222*(2), 373–381. Retrieved from http://www.ncbi.nlm.nih.gov/pubmed/6793212.

Elliott, S., Busse, L., Bass, M. B., Lu, H., Sarosi, I., Sinclair, A. M., et al. (2006). Anti-Epo receptor antibodies do not predict Epo receptor expression. *Blood*, *107*(5), 1892–1895. http://doi.org/10.1182/blood-2005-10-4066.

Erickson, J. T., & Millhorn, D. E. (1994). Hypoxia and electrical stimulation of the carotid sinus nerve induce Fos-like immunoreactivity within catecholaminergic and serotoninergic neurons of the rat brainstem. *The Journal of Comparative Neurology*, *348*(2), 161–182. http://doi.org/10.1002/cne.903480202.

Feldman, J. L., & Del Negro, C. A. (2006). Looking for inspiration: New perspectives on respiratory rhythm. *Nature Reviews Neuroscience*, 7(3), 232–242. http://doi.org/10.1038/nrn1871.

Feldman, J. L., Del Negro, C. A., & Gray, P. A. (2013). Understanding the rhythm of breathing: So near, yet so far. *Annual Review of Physiology*, *75*, 423–452. http://doi.org/10.1146/annurev-physiol-040510-130049.

Fried, W., & Gurney, C. W. (1965). Erythropoietic effect of plasma from mice receiving testosterone. *Nature*, *206*(989), 1160–1161. Retrieved from http://www.ncbi.nlm.nih.gov/pubmed/5866160.

Gao, X.-P., Liu, Q., Nair, B., & Wong-Riley, M. T. T. (2014). Reduced levels of brain-derived neurotrophic factor contribute to synaptic imbalance during the critical period of respiratory development in rats. *The European Journal of Neuroscience*, *40*(1), 2183–2195. http://doi.org/10.1111/ejn.12568.

Gassmann, M., Pfistner, C., Doan, V. D., Vogel, J., & Soliz, J. (2010). Impaired ventilatory acclimatization to hypoxia in female mice overexpressing erythropoietin: Unexpected deleterious effect of estradiol in carotid bodies. *American Journal of Physiology. Regulatory, Integrative and Comparative Physiology*, *299*(6), R1511–R1520. http://doi.org/10.1152/ajpregu.00205.2010.

Gourine, A. V., Kasymov, V., Marina, N., Tang, F., Figueiredo, M. F., Lane, S., et al. (2010). Astrocytes control breathing through pH-dependent release of ATP. *Science (New York, N.Y.)*, *329*(5991), 571–575. http://doi.org/10.1126/science.1190721.

Guyenet, P. G., Stornetta, R. L., Bochorishvili, G., Depuy, S. D., Burke, P. G. R., & Abbott, S. B. G. (2013). C1 neurons: The body's EMTs. *American Journal of Physiology Regulatory, Integrative and Comparative Physiology*, *305*(3), R187–R204. http://doi.org/10.1152/ajpregu.00054.2013.

Hatayama, K., & Ide, T. (2015). Engineering of erythropoietin receptor for use as an affinity ligand. *Protein Expression and Purification*, *111*, 1–8. http://doi.org/10.1016/j.pep.2015.03.005.

Hellewell, S. C., Yan, E. B., Alwis, D. S., Bye, N., & Morganti-Kossmann, M. C. (2013). Erythropoietin improves motor and cognitive deficit, axonal pathology, and neuroinflammation in a combined model of diffuse traumatic brain injury and hypoxia, in association with upregulation of the erythropoietin receptor. *Journal of Neuroinflammation*, *10*, 156. http://doi.org/10.1186/1742-2094-10-156.

Hilaire, G., Viemari, J.-C., Coulon, P., Simonneau, M., & Bévengut, M. (2004). Modulation of the respiratory rhythm generator by the pontine noradrenergic A5 and A6 groups in rodents. *Respiratory Physiology & Neurobiology*, *143*(2–3), 187–197. http://doi.org/10.1016/j.resp.2004.04.016.

Housley, G. D., & Sinclair, J. D. (1988). Localization by kainic acid lesions of neurones transmitting the carotid chemoreceptor stimulus for respiration in rat. *The Journal of Physiology*, *406*, 99–114. Retrieved from http://www.pubmedcentral.nih.gov/articlerender.fcgi?artid=1191090&tool=pmcentrez&rendertype=abstract.

Jacobson, L. O., Goldwasser, E., Fried, W., & Plzak, L. F. (1957). Studies on erythropoiesis. VII. The role of the kidney in the production of erythropoietin. *Transactions of the Association of American Physicians*, *70*, 305–317. Retrieved from http://www.ncbi.nlm.nih.gov/pubmed/13496139.

Janczewski, W. A., Tashima, A., Hsu, P., Cui, Y., & Feldman, J. L. (2013). Role of inhibition in respiratory pattern generation. *The Journal of Neuroscience: The Official Journal of the Society for Neuroscience*, *33*(13), 5454–5465. http://doi.org/10.1523/jneurosci.1595-12.2013.

Jarsch, M., Brandt, M., Lanzendörfer, M., & Haselbeck, A. (2008). Comparative erythropoietin receptor binding kinetics of C.E.R.A. and epoetin-beta determined by surface plasmon resonance and competition binding assay. *Pharmacology*, *81*(1), 63–69. http://doi.org/10.1159/000109166.

Jean, A. (1991). The nucleus tractus solitarius: Neuroanatomic, neurochemical and functional aspects. *Archives Internationales de Physiologie, de Biochimie et de Biophysique*, *99*(5), A3–52. Retrieved from http://www.ncbi.nlm.nih.gov/pubmed/1720691.

Jelkmann, W. (1986). Erythropoietin research, 80 years after the initial studies by Carnot and Deflandre. *Respiration Physiology*, *63*(3), 257–266. Retrieved from http://www.ncbi.nlm.nih.gov/pubmed/3515472.

Jiang, R., Haustein, M. D., Sofroniew, M. V., & Khakh, B. S. (2014). Imaging intracellular Ca^{2+} signals in striatal astrocytes from adult mice using genetically-encoded calcium indicators. *Journal of Visualized Experiments*, *93*, e51972. http://doi.org/10.3791/51972.

Juul, S. E., McPherson, R. J., Farrell, F. X., Jolliffe, L., Ness, D. J., & Gleason, C. A. (2004). Erytropoietin concentrations in cerebrospinal fluid of nonhuman primates and fetal sheep following high-dose recombinant erythropoietin. *Biology of the Neonate*, *85*(2), 138–144. http://doi.org/10.1159/000074970.

Kasat, K., Hendricks-Muñoz, K. D., & Mally, P. V. (2011). Neonatal red blood cell transfusions: Searching for better guidelines. *Blood Transfusion = Trasfusione del sangue*, *9*(1), 86–94. http://doi.org/10.2450/2010.0031-10.

Kazemi, H., & Hoop, B. (1991). Glutamic acid and gamma-aminobutyric acid neurotransmitters in central control of breathing. *Journal of Applied Physiology (Bethesda, Md.: 1985)*, *70*(1), 1–7. Retrieved from http://www.ncbi.nlm.nih.gov/pubmed/1672687.

Kellert, B. A., McPherson, R. J., & Juul, S. E. (2007). A comparison of high-dose recombinant erythropoietin treatment regimens in brain-injured neonatal rats. *Pediatric Research*, *61*(4), 451–455. http://doi.org/10.1203/pdr.0b013e3180332cec.

Khemiri, H., Seaborn, T., Gestreau, C., & Soliz, J. (2012). Erythropoietin and its antagonist regulate hypoxic fictive breathing in newborn mice. *Respiratory Physiology & Neurobiology*, *183*(2), 115–121. http://doi.org/10.1016/j.resp.2012.05.027.

Kilic, E., Kilic, U., Soliz, J., Bassetti, C. L., Gassmann, M., & Hermann, D. M. (2005). Brain-derived erythropoietin protects from focal cerebral ischemia by dual activation of ERK-1/-2 and Akt pathways. *FASEB Journal: Official Publication of the Federation of American Societies for Experimental Biology*, *19*(14), 2026–2028. http://doi.org/10.1096/fj.05-3941fje.

King, V. R., Averill, S. A., Hewazy, D., Priestley, J. V., Torup, L., & Michael-Titus, A. T. (2007). Erythropoietin and carbamylated erythropoietin are neuroprotective following spinal cord hemisection in the rat. *The European Journal of Neuroscience*, *26*(1), 90–100. http://doi.org/10.1111/j.1460-9568.2007.05635.x.

Kokot, F., Wiecek, A., Schmidt-Gayk, H., Marcinkowski, W., Nieszporek, T., Rudka, R., et al. (1992). Influence of long-term erythropoietin therapy on endocrine abnormalities in haemodialyzed patients. *Przegląd Lekarski*, *49*(1–2), 43–49. Retrieved from http://www.ncbi.nlm.nih.gov/pubmed/1455006.

Kubin, L., Alheid, G. F., Zuperku, E. J., & McCrimmon, D. R. (2006). Central pathways of pulmonary and lower airway vagal afferents. *Journal of Applied Physiology (Bethesda, Md.: 1985)*, *101*(2), 618–627. http://doi.org/10.1152/japplphysiol.00252.2006.

Lai, P. H., Everett, R., Wang, F. F., Arakawa, T., & Goldwasser, E. (1986). Structural characterization of human erythropoietin. *The Journal of Biological Chemistry*, *261*(7), 3116–3121. Retrieved from http://www.ncbi.nlm.nih.gov/pubmed/3949763.

Lam, S. Y., Tipoe, G. L., & Fung, M. L. (2009). Upregulation of erythropoietin and its receptor expression in the rat carotid body during chronic and intermittent hypoxia. *Advances in Experimental Medicine and Biology*, *648*, 207–214. http://doi.org/10.1007/978-90-481-2259-2_24.

Liu, Q., & Wong-Riley, M. T. T. (2013). Postnatal development of glycine receptor subunits α1, α2, α3, and β immunoreactivity in multiple brain stem respiratory-related nuclear groups of the rat. *Brain Research*, *1538*, 1–16. http://doi.org/10.1016/j.brainres.2013.09.028.

Lumbroso, D., & Joseph, V. (2009). Impaired acclimatization to chronic hypoxia in adult male and female rats following neonatal hypoxia. *American Journal of Physiology. Regulatory, Integrative and Comparative Physiology*, *297*(2), R421–R427. http://doi.org/10.1152/ajpregu.00068.2009.

Madurga, A., Mizíková, I., Ruiz-Camp, J., & Morty, R. E. (2013). Recent advances in late lung development and the pathogenesis of bronchopulmonary dysplasia. *American Journal of Physiology. Lung Cellular and Molecular Physiology*, *305*(12), L893–L905. http://doi.org/10.1152/ajplung.00267.2013.

Masuda, S., Nagao, M., Takahata, K., Konishi, Y., Gallyas, F., Tabira, T., et al. (1993). Functional erythropoietin receptor of the cells with neural characteristics. Comparison with receptor properties of erythroid cells. *The Journal of Biological Chemistry*, *268*(15), 11208–11216. Retrieved from http://www.ncbi.nlm.nih.gov/pubmed/7684373.

Masuda, S., Okano, M., Yamagishi, K., Nagao, M., Ueda, M., & Sasaki, R. (1994). A novel site of erythropoietin production. Oxygen-dependent production in cultured rat astrocytes. *The Journal of Biological Chemistry*, *269*(30), 19488–19493. Retrieved from http://www.ncbi.nlm.nih.gov/pubmed/8034718.

Matsubara, K., Yoshimura, T., Kamachi, S., Fukushima, M., Hino, M., & Morii, H. (1989). Radioimmunoassay for erythropoietin using anti-recombinant erythropoietin antibody with high affinity. *Clinica Chimica Acta; International Journal of Clinical Chemistry*, *185*(2), 177–184. Retrieved from http://www.ncbi.nlm.nih.gov/pubmed/2624997.

Miljus, N., Heibeck, S., Jarrar, M., Micke, M., Ostrowski, D., Ehrenreich, H., et al. (2014). Erythropoietin-mediated protection of insect brain neurons involves JAK and STAT but not PI3K transduction pathways. *Neuroscience*, *258*, 218–227. http://doi.org/10.1016/j.neuroscience.2013.11.020.

Mukundan, H., Resta, T. C., & Kanagy, N. L. (2002). 17Beta-estradiol decreases hypoxic induction of erythropoietin gene expression. *American Journal of Physiology. Regulatory, Integrative and Comparative Physiology*, *283*(2), R496–R504. http://doi.org/10.1152/ajpregu.00573.2001.

Nadam, J., Navarro, F., Sanchez, P., Moulin, C., Georges, B., Laglaine, A., et al. (2007). Neuroprotective effects of erythropoietin in the rat hippocampus after pilocarpine-induced status epilepticus. *Neurobiology of Disease*, *25*(2), 412–426. http://doi.org/10.1016/j.nbd.2006.10.009.

Nagao, M., Masuda, S., Abe, S., Ueda, M., & Sasaki, R. (1992). Production and ligand-binding characteristics of the soluble form of murine erythropoietin receptor. *Biochemical and Biophysical Research Communications*, *188*(2), 888–897. Retrieved from http://www.ncbi.nlm.nih.gov/pubmed/1445329.

Nsegbe, E., Wallén-Mackenzie, A., Dauger, S., Roux, J.-C., Shvarev, Y., Lagercrantz, H., et al. (2004). Congenital hypoventilation and impaired hypoxic response in Nurr1 mutant mice. *The Journal of Physiology*, *556*(Pt. 1), 43–59. http://doi.org/10.1113/jphysiol.2003.058560.

Paton, J. F., & Richter, D. W. (1995). Role of fast inhibitory synaptic mechanisms in respiratory rhythm generation in the maturing mouse. *The Journal of Physiology*, *484*(Pt. 2), 505–521. Retrieved from http://www.pubmedcentral.nih.gov/articlerender.fcgi?artid=1157911&tool=pmcentrez&rendertype=abstract.

Philo, J. S., Aoki, K. H., Arakawa, T., Narhi, L. O., & Wen, J. (1996). Dimerization of the extracellular domain of the erythropoietin (EPO) receptor by EPO: One high-affinity and one low-affinity interaction. *Biochemistry*, *35*(5), 1681–1691. http://doi.org/10.1021/bi9524272.

Pugh, C. W., O'Rourke, J. F., Nagao, M., Gleadle, J. M., & Ratcliffe, P. J. (1997). Activation of hypoxia-inducible factor-1; definition of regulatory domains within the alpha subunit. *The Journal of Biological Chemistry*, *272*(17), 11205–11214. Retrieved from http://www.ncbi.nlm.nih.gov/pubmed/9111021.

Reissmann, K. R. (1950). Studies on the mechanism of erythropoietic stimulation in parabiotic rats during hypoxia. *Blood*, *5*(4), 372–380. Retrieved from http://www.ncbi.nlm.nih.gov/pubmed/15411424.

Rencricca, N. J., Solomon, J., Fimian, W. J., Howard, D., Rizzoli, V., & Stohlman, F. (1969). The effect of testosterone on erythropoiesis. *Scandinavian Journal of Haematology*, *6*(6), 431–436. Retrieved from http://www.ncbi.nlm.nih.gov/pubmed/5372961.

Richter, D. W. (1982). Generation and maintenance of the respiratory rhythm. *The Journal of Experimental Biology*, *100*, 93–107. Retrieved from http://www.ncbi.nlm.nih.gov/pubmed/6757372.

Riedel, E., Nündel, M., Algermissen, B., Hampl, H., Scigalla, P., & Stabell, U. (1989). Changes in the concentrations of hydroxyproline, glycine and serine in the plasma of haemodialysis patients undergoing erythropoietin therapy. *Journal of Clinical Chemistry and Clinical Biochemistry. Zeitschrift Für Klinische Chemie Und Klinische Biochemie*, *27*(11), 851–856. Retrieved from http://www.ncbi.nlm.nih.gov/pubmed/2514250.

Rousseau, J., & Caravagna, C. (2015). Electrophysiology on isolated brainstem-spinal cord preparations from newborn rodents allows neural respiratory network output recording. *Journal of Visualized Experiments*, *105*, 1–7. http://doi.org/10.3791/53071.

Roux, J.-C., Dura, E., & Villard, L. (2008). Tyrosine hydroxylase deficit in the chemoafferent and the sympathoadrenergic pathways of the Mecp2 deficient mouse. *Neuroscience Letters*, *447*(1), 82–86. http://doi.org/10.1016/j.neulet.2008.09.045.

Ruscher, K., Freyer, D., Karsch, M., Isaev, N., Megow, D., Sawitzki, B., et al. (2002). Erythropoietin is a paracrine mediator of ischemic tolerance in the brain: Evidence from an in vitro model. *The Journal of Neuroscience: The Official Journal of the Society for Neuroscience*, *22*(23), 10291–10301. Retrieved from http://www.ncbi.nlm.nih.gov/pubmed/12451129.

Salisch, S. v., Klar, M., Thurisch, B., Bungert, J., & Dame, C. (2011). Gata4 and Sp1 regulate expression of the erythropoietin receptor in cardiomyocytes. *Journal of Cellular and Molecular Medicine*, *15*(9), 1963–1972. http://doi.org/10.1111/j.1582-4934.2010.01193.x.

Sanchez, P. E., Navarro, F. P., Fares, R. P., Nadam, J., Georges, B., Moulin, C., et al. (2009). Erythropoietin receptor expression is concordant with erythropoietin but not with common β chain expression in the rat brain throughout the life span. *The Journal of Comparative Neurology*, *514*(4), 403–414. http://doi.org/10.1002/cne.22020.

Seaborn, T., Gassmann, M., & Soliz, J. (2013). Using plethysmography to determine erythropoietin's impact on neural control of ventilation. *Methods in Molecular Biology (Clifton, N.J.)*, *982*, 303–314. http://doi.org/10.1007/978-1-62703-308-4_18.

Seaborn, T., Gonzales, M., Villalpando, G., Grenacher, B., Soria, R., & Soliz, J. (2011). Enhanced erythropoietin expression in the brainstem of newborn rats at high altitude. *Neuroscience Letters*, *502*(1), 33–36. http://doi.org/10.1016/j.neulet.2011.07.019.

Semenza, G. L. (1994). Regulation of erythropoietin production. New insights into molecular mechanisms of oxygen homeostasis. *Hematology/Oncology Clinics of North America*, *8*(5), 863–884. Retrieved from http://www.ncbi.nlm.nih.gov/pubmed/7852211.

Siljehav, V., Shvarev, Y., & Herlenius, E. (2014). Il-1β and prostaglandin E2 attenuate the hypercapnic as well as the hypoxic respiratory response via prostaglandin E receptor type 3 in neonatal mice. *Journal of Applied Physiology (Bethesda, Md.: 1985)*, *117*(9), 1027–1036. http://doi.org/10.1152/japplphysiol.00542.2014.

Smith, J. C., Ellenberger, H. H., Ballanyi, K., Richter, D. W., & Feldman, J. L. (1991). Pre-Bötzinger complex: A brainstem region that may generate respiratory rhythm in mammals. *Science (New York, N.Y.)*, *254*(5032), 726–729. Retrieved from http://www.pubmedcentral.nih.gov/articlerender.fcgi?artid=3209964&tool=pmcentrez&rendertype=abstract.

Solev, I. N., Balabanyan, V. Y., Volchek, I. A., Elizarova, O. S., Litvinova, S. A., Garibova, T. L., et al. (2013). Involvement of BDNF and NGF in the mechanism of neuroprotective effect of human recombinant erythropoietin nanoforms. *Bulletin of Experimental Biology and Medicine*, *155*(2), 242–244. Retrieved from http://www.ncbi.nlm.nih.gov/pubmed/24131000.

Soliz, J., Gassmann, M., & Joseph, V. (2007). Soluble erythropoietin receptor is present in the mouse brain and is required for the ventilatory acclimatization to hypoxia. *The Journal of Physiology*, *583*(Pt. 1), 329–336. http://doi.org/10.1113/jphysiol.2007.133454.

Soliz, J., Joseph, V., Soulage, C., Becskei, C., Vogel, J., Pequignot, J. M., et al. (2005). Erythropoietin regulates hypoxic ventilation in mice by interacting with brainstem and carotid bodies. *The Journal of Physiology*, *568*(Pt. 2), 559–571. http://doi.org/10.1113/jphysiol.2005.093328.

Soliz, J., Thomsen, J. J., Soulage, C., Lundby, C., & Gassmann, M. (2009). Sex-dependent regulation of hypoxic ventilation in mice and humans is mediated by erythropoietin. *American Journal of Physiology. Regulatory, Integrative and Comparative Physiology, 296*(6), R1837–R1846. http://doi.org/10.1152/ajpregu.90967.2008.

Soulier, V., Gestreau, C., Borghini, N., Dalmaz, Y., Cottet-Emard, J. M., & Pequignot, J. M. (1997). Peripheral chemosensitivity and central integration: Neuroplasticity of catecholaminergic cells under hypoxia. *Comparative Biochemistry and Physiology. Part A, Physiology, 118*(1), 1–7. Retrieved from http://www.ncbi.nlm.nih.gov/pubmed/9243809.

Sun, Y.-Y., Lin, S.-H., Lin, H.-C., Hung, C.-C., Wang, C.-Y., Lin, Y.-C., et al. (2013). Cell type-specific dependency on the PI3K/Akt signaling pathway for the endogenous Epo and VEGF induction by baicalein in neurons versus astrocytes. *PLoS One, 8*(7), e69019. http://doi.org/10.1371/journal.pone.0069019.

Suzue, T. (1984). Respiratory rhythm generation in the in vitro brain stem-spinal cord preparation of the neonatal rat. *The Journal of Physiology, 354*, 173–183. Retrieved from http://www.pubmedcentral.nih.gov/articlerender.fcgi?artid=1193406&tool=pmcentrez&rendertype=abstract.

Tempera, A., Stival, E., Piastra, M., De Luca, D., Ottaviano, C., Tramontozzi, P., et al. (2011). Early erythropoietin influences both transfusion and ventilation need in very low birth weight infants. *The Journal of Maternal-Fetal & Neonatal Medicine: The Official Journal of the European Association of Perinatal Medicine, the Federation of Asia and Oceania Perinatal Societies, the International Society of Perinatal Obstetricians, 24*(8), 1060–1064. http://doi.org/10.3109/14767058.2010.545917.

Teppema, L. J., & Dahan, A. (2010). The ventilatory response to hypoxia in mammals: Mechanisms, measurement, and analysis. *Physiological Reviews, 90*(2), 675–754. http://doi.org/10.1152/physrev.00012.2009.

Um, M., Gross, A. W., & Lodish, H. F. (2007). A "classical" homodimeric erythropoietin receptor is essential for the antiapoptotic effects of erythropoietin on differentiated neuroblastoma SH-SY5Y and pheochromocytoma PC-12 cells. *Cellular Signalling, 19*(3), 634–645. http://doi.org/10.1016/j.cellsig.2006.08.014.

Viemari, J.-C., Burnet, H., Bévengut, M., & Hilaire, G. (2003). Perinatal maturation of the mouse respiratory rhythm-generator: In vivo and in vitro studies. *The European Journal of Neuroscience, 17*(6), 1233–1244. Retrieved from http://www.ncbi.nlm.nih.gov/pubmed/12670311.

Wiessner, C., Allegrini, P. R., Ekatodramis, D., Jewell, U. R., Stallmach, T., & Gassmann, M. (2001). Increased cerebral infarct volumes in polyglobulic mice overexpressing erythropoietin. *Journal of Cerebral Blood Flow and Metabolism: Official Journal of the International Society of Cerebral Blood Flow and Metabolism, 21*(7), 857–864. http://doi.org/10.1097/00004647-200107000-00011.

Witthuhn, B. A., Quelle, F. W., Silvennoinen, O., Yi, T., Tang, B., Miura, O., et al. (1993). JAK2 associates with the erythropoietin receptor and is tyrosine phosphorylated and activated following stimulation with erythropoietin. *Cell, 74*(2), 227–236. Retrieved from http://www.ncbi.nlm.nih.gov/pubmed/8343951.

Wong-Riley, M. T. T., Liu, Q., & Gao, X. (2013). Peripheral-central chemoreceptor interaction and the significance of a critical period in the development of respiratory control. *Respiratory Physiology & Neurobiology, 185*(1), 156–169. http://doi.org/10.1016/j.resp.2012.05.026.

Zanjani, E. D., Poster, J., Burlington, H., Mann, L. I., & Wasserman, L. R. (1977). Liver as the primary site of erythropoietin formation in the fetus. *The Journal of Laboratory and Clinical Medicine, 89*(3), 640–644. Retrieved from http://www.ncbi.nlm.nih.gov/pubmed/839121.

CHAPTER EIGHT

Mutual Cross Talk Between Iron Homeostasis and Erythropoiesis

Ilona Rybinska, Gaetano Cairo[1]
University of Milan, Milano, Italy
[1]Corresponding author: e-mail address: gaetano.cairo@unimi.it

Contents

Abstract

Iron is necessary for physiological processes essential for the activity of all cells, but the erythropoietic compartment is a privileged iron consumer. In fact, a considerable amount of iron is daily required for hemoglobin synthesis and erythroid cell proliferation. Therefore, a tight link exists between iron metabolism and erythropoiesis. The iron needed for hemoglobin synthesis is mainly ensured by inhibiting hepcidin expression, thereby increasing both ferroportin-mediated iron export from the duodenal absorptive cells and iron release from the reticuloendothelial cells that process old and/or damaged red blood cells. This mechanism makes certain that sufficient iron availability to the erythropoietic compartment occurs. Recent studies established that hypoxia and/or hypoxia-induced erythropoietin are not direct regulators of hepcidin, which is indirectly inhibited by erythropoietic drive, in particular under pathological conditions characterized by expanded but ineffective erythropoiesis, such as β-thalassemia. Among the number of factors proposed as mediators linking erythropoiesis with liver hepcidin suppression, erythroferrone, a hormone produced and secreted by erythroid precursors, appears the best candidate.

Vitamins and Hormones, Volume 105
ISSN 0083-6729
http://dx.doi.org/10.1016/bs.vh.2017.01.001
© 2017 Elsevier Inc.
All rights reserved.

1. INTRODUCTION

Iron is a micronutrient that plays a fundamental role in metabolic processes and thus is indispensable for the growth, development, and well-being of living organisms. As iron exchanges single electrons easily with a wide range of substrates and is incorporated in heme and iron–sulfur clusters (ISC), it is widely used for enzymatic reactions required for a number of essential functions, such as cell respiration, DNA synthesis, and gene regulation (Cairo, Bernuzzi, & Recalcati, 2006), but its major role, at least from a quantitative viewpoint, is in oxygen transport. In fact, iron is incorporated in the structure of hemoglobin (oxygen transport protein) and myoglobin (oxygen storage protein) and erythrocytes account for about 50% of body iron content. Being a structural part of hemoglobin, iron is a limiting factor in red blood cell production and, starting from erythroblast maturation, the process of erythropoiesis is increasingly dependent on iron acquisition. The substantial amount of iron (20–25 mg) which must be supplied to bone marrow daily to produce erythrocytes (see Fig. 1)

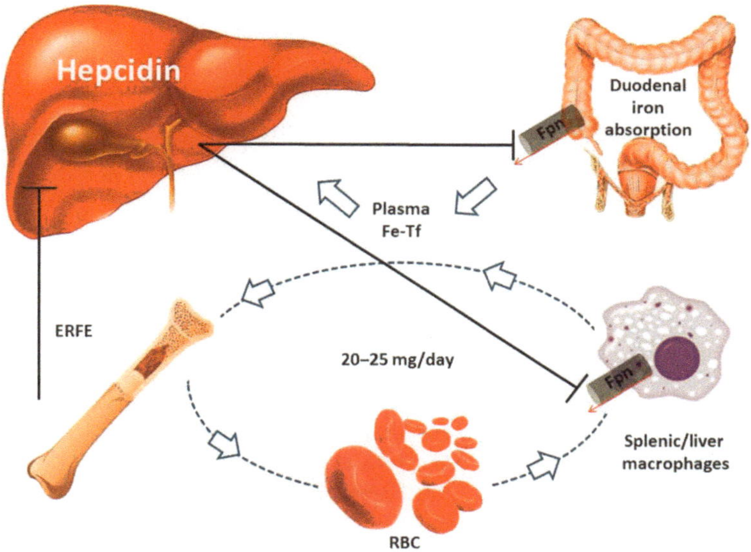

Fig. 1 Regulation of systemic iron homeostasis and effect of erythropoiesis. Plasma iron transferrin (Fe-Tf) delivers the iron supplied by ferroportin (Fpn)-mediated duodenal absorption and iron-recycling macrophages to the bone marrow, which incorporates the metal into red blood cells (RBC) hemoglobin. Liver-derived hepcidin causes Fpn degradation and thereby inhibits iron flow into circulation. Under conditions of enhanced erythropoiesis, inhibitory signals, such as erythroferrone (ERFE), suppress hepcidin production, thus permitting Fpn-dependent iron release into plasma and increasing iron availability.

is mainly provided by macrophage-mediated recycling of hemoglobin-derived iron from the breakdown of senescent erythrocytes, while about 2 mg iron daily derives from intestinal absorption of iron present in food. In the bone marrow, although the majority of iron in erythroid cells is incorporated in hemoglobin, it is also used for proteins that sustain the high proliferation rate of red cell precursors and for mitochondrial ISC synthesis. Since the function of erythropoietin (EPO) as stimulator of erythropoiesis would be meaningless in conditions of iron scarcity, regulation of iron traffic is comprised among the number of activities controlled by EPO, closely linking iron homeostasis and erythropoiesis.

2. SYSTEMIC IRON HOMEOSTASIS

As reported earlier, iron is important for processes essential for life, but its ability to easily gain or lose electrons also facilitates the generation of highly reactive oxygen species, which trigger oxidative stress, lipid peroxidation, and DNA damage and eventually lead to cell death (Gozzelino & Arosio, 2015; Paul & Lill, 2015; Zhang, 2014). Indeed, either too little or too much iron can have detrimental effects on health, and humans suffer from both iron deficiency (leading, for example, to iron deficiency anemia, which affects more than 10% of the world's population and is a huge global public health problem) (Sankaran & Weiss, 2015) and iron overload (e.g., hemochromatosis) (Gozzelino & Soares, 2014; Pietrangelo, 2010; Powell, Seckington, & Deugnier, 2016). Given the dramatic consequences of both iron deficiency and excess, organisms developed precise regulatory mechanisms to tightly control iron traffic in every single cell as well as systemic homeostasis.

Body iron balance is maintained by means of a network of proteins that regulates iron absorption (duodenal enterocytes), continuous recycling (splenic macrophages), utilization (erythroid and proliferating cells), storage (parenchymal liver cells and reticuloendothelial macrophages), and trafficking (Fig. 1). In contrast to iron uptake and recycling, there is no regulated pathway for iron efflux from the body, which mainly results from mucosal lining and blood loss from the gastrointestinal tract (Andrews, 2008; Hentze, Muckenthaler, Galy, & Camaschella, 2010). In the absence of a dedicated mechanism for iron dismissal, intestinal iron absorption is a critical determinant of body iron homeostasis and thus must be highly controlled (Andrews, 2008; Hentze et al., 2010). In fact, excessive iron uptake results in iron overload, like in hereditary hemochromatosis which leads to excess iron accumulation in tissues and consequent damage, manifesting as fatigue, liver disease, and cardiomyopathy (Pietrangelo, 2010; Powell et al., 2016).

2.1 Duodenal Iron Absorption

The molecular pathway of nonheme iron uptake is based on the reduction of the highly insoluble and poorly available nonheme ferric iron (Fe^{3+}) present in the diet by membrane-associated ferric reductase (Dcytb), and the subsequent transport of Fe^{2+} by divalent metal transporter 1 (DMT-1) across the luminal side of the duodenum (Fig. 2; Hentze et al., 2010).

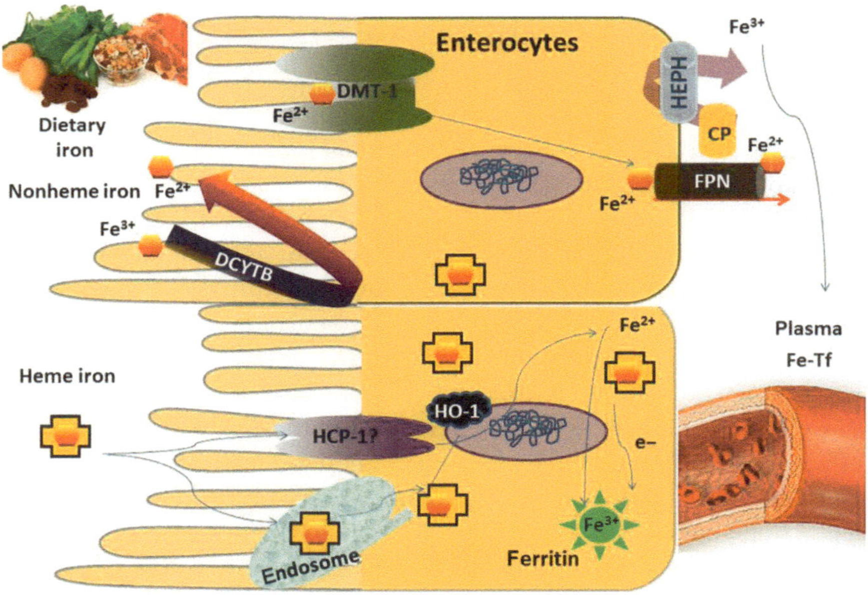

Fig. 2 Duodenal iron absorption. The molecular pathway of nonheme iron uptake is based on the reduction of the highly insoluble and poorly available dietary nonheme ferric iron (Fe^{3+}) by membrane-associated ferric reductase (Dcytb) and transport of Fe^{2+} by divalent metal transporter 1 (DMT-1) across the apical side of the duodenum. Heme iron, possibly absorbed through HCP-1, is released by the action of heme oxygenase (HO-1). Iron derived from both sources that is not stored in ferritin is then exported from the enterocyte into circulation by the basolateral iron exporter ferroportin (Fpn), assisted by the activity of oxidases (circulating ceruloplasmin, CP, and membrane-bound hephaestin, HEPH) that convert ferrous iron to ferric iron and thus permit the binding of iron to transferrin (Tf), which supplies iron to all cells.

Heme provides an important source of dietary iron for mammals because its absorption is efficient and independent on other dietary components. However, although various transporters have been proposed as candidates of intestinal heme uptake, such as HCP-1, the precise mechanism of heme iron absorption needs to be elucidated (Gulec, Anderson, & Collins, 2014). Both the iron liberated from heme through the action of heme oxygenase 1 (HO-1) and the nonheme iron transported by DMT-1 are then released from the enterocyte into circulation by the basolateral iron exporter ferroportin (Fpn) (Donovan et al., 2005), assisted by the activity of oxidases (circulating ceruloplasmin and membrane-bound hephaestin) that convert ferrous iron to ferric iron and thus permit the binding of iron to transferrin, which supplies iron to all cells through the interaction with the transferrin receptor 1 (TfR1) (Andrews, 2008; Hentze et al., 2010; McKie et al., 2000; Pantopoulos, Porwal, Tartakoff, & Devireddy, 2012). Erythroid cells are avid acceptors of transferrin-bound iron thanks to the very high number of TfR1 molecules (about 800,000/cell) expressed following the induction of their differentiation. Interestingly, it has been shown that the EPO receptor–Jak2–Stat5 pathway, which regulates proliferation, differentiation, and survival of erythroid cells, also controls the transcription of TfR1 and IRP2, the posttranscriptional regulator of TfR1 mRNA stability (Kerenyi et al., 2008). Mice with ablation of Stat5, which suffer from microcytic anemia, show a strong decrease of TfR1-mediated iron uptake and globin synthesis. This study established a direct EPO-dependent mechanistic link between iron metabolism and erythropoiesis.

Under physiological conditions, intestinal iron uptake is influenced principally by body iron content and erythropoiesis (Finch, 1994). Iron absorption can be accelerated in case of increased erythropoiesis or suppressed in response to iron overload. Cells across the intestine mucosal layer have their own iron- and hypoxia-sensing machinery based on iron regulatory proteins (IRPs), which posttranscriptionally regulate the expression of the major proteins of iron metabolism (Cairo & Recalcati, 2007), as well as that of hypoxia-inducible factor HIF2α, which induces the transcription of a large number of genes involved in the response to hypoxia, including EPO. In fact, the IRP1/HIF2α axis coordinates duodenal iron absorption according to systemic oxygen and iron availability (Recalcati, Gammella, & Cairo, 2015). However, the major player in the communication of the body iron stores to the duodenal iron absorption sites is hepcidin (Fleming & Sly, 2001).

Hepcidin was discovered at the beginning of the 21st century by independent groups and described as cysteine-rich peptide from hepatic origin with a characteristic typical for an antimicrobial activity, thus it was named hepcidin (hepatic bactericidal protein). Hepcidin is a short (25 amino acids) peptide hormone generated from an 84-amino acid propeptide by furin cleavage, produced and secreted mainly by hepatocytes. Much lower amounts are produced by macrophages and adipocytes (Ganz & Nemeth, 2012).

The peptide exerts its regulatory function by binding to Fpn, the only cellular iron exporter known to date, and leading to its internalization and degradation (Nemeth et al., 2003), consequently blocking the iron egress from cells, particularly reticuloendothelial cells and absorptive enterocytes, to plasma and extracellular fluids (Ganz, 2005). This molecular circuit maintains plasma iron in a nearly constant range, and both iron overload and deficiency conditions can be attributed to defects or imbalances in the hepcidin–Fpn axis. The prominence of hepcidin in systemic iron homeostasis was recognized by the progressive tissue iron overload caused by the lack of the hepcidin gene (Nicolas et al., 2001), as well as by the finding that transgenic mice overexpressing hepcidin suffered from severe and often lethal anemia (Nicolas, Bennoun, et al., 2002). Moreover, dysregulated hepcidin expression underlies iron overload conditions in humans (Pietrangelo, 2015).

Hepcidin synthesis is controlled mostly at the transcriptional level by various and sometimes opposite stimuli, the major players being the erythroid regulator (i.e., iron demand for erythropoiesis) and the store regulator (i.e., body iron deposits) (Fig. 3). In fact, plasma transferrin-iron, tissue iron stores, inflammation/infection, and ER stress induce hepcidin expression, whereas erythropoietic activity and hypoxia have an opposite effect (Andrews, 2008). Here, we will briefly summarize the mechanisms at the basis of hepcidin stimulation and focus more in detail on the pathways underlying erythropoiesis-mediated hepcidin inhibition.

3. REGULATION OF HEPCIDIN EXPRESSION

3.1 Hepcidin Regulation by Iron Availability

Increased iron levels activate the bone morphogenic proteins 6 (BMP6)–BMPR–SMAD1/5/8 transduction pathway which triggers hepcidin transcription. At the molecular level, in response to increased liver iron stores,

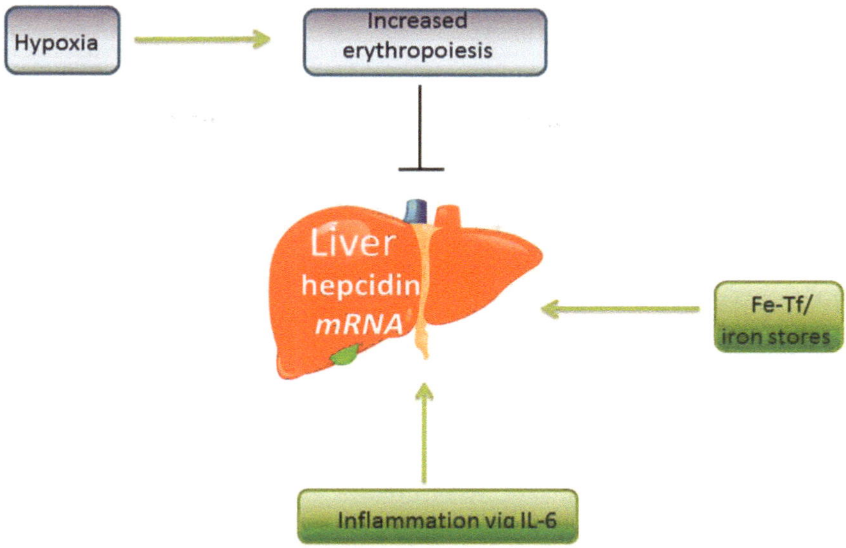

Fig. 3 Major regulatory pathways of hepcidin expression. Liver hepcidin expression is raised by increased iron stores and transferrin saturation, as well as by inflammation. Conversely, hepcidin is inhibited by erythropoietin-dependent increased erythroid activity (and iron deficiency).

BMP6, predominantly produced by liver nonparenchymal cells, binds to the coreceptor hemojuvelin and BMP type I and type II receptors (BMPR) to induce phosphorylation of SMAD1/5/8 proteins, which form heterodimers with SMAD4 and reach the nucleus to bind the hepcidin promoter (reviewed by Core, Canali, & Babitt, 2014; Ganz & Nemeth, 2012; Gozzelino & Arosio, 2016; Zhao, Zhang, & Enns, 2013). High levels of serum iron activate SMAD1/5/8 phosphorylation downstream or independent of BMP6. BMP6-dependent activation of hepcidin transcription is modulated by a complex comprising HJV, a BMP coreceptor, human hemochromatosis protein (HFE), and transferrin receptor 2 (TfR2), three proteins which are mutated in patients with distinct types of hemochromatosis (Zhang, West, Wyman, Bjorkman, & Enns, 2005). Conversely, in response to iron deficiency, hepcidin is controlled in a negative way by matriptase-2 (TMPRSS6), a hepatocyte serine protease which cleaves HJV from the plasma membrane. In vitro studies have shown that TMPRSS6 determines the levels of membrane-bound HJV (m-HJV) by cleaving the protein into an inactive soluble fragment (s-HJV) which is released and acts as a decoy molecule for BMPs, thus inhibiting hepcidin

(Du et al., 2008; Folgueras et al., 2008; Silvestri, Pagani, Nai, et al., 2008). The role of TMPRSS6 was shown by studies demonstrating that its inactivation in mice leads to hepcidin increase, block of iron absorption, and anemia (Core et al., 2014; Ganz & Nemeth, 2012; Gozzelino & Arosio, 2016; Zhao et al., 2013). Moreover, a rare form of anemia (refractory iron deficiency anemia, IRIDA) can be caused by mutations in the gene encoding TMPRSS6/matriptase-2 (De Falco et al., 2013).

Furin (Silvestri, Pagani, & Camaschella, 2008) and neogenin, an HJV-interacting protein which assists in the proper assembly of BMPs/BMPR complexes (Lee et al., 2010), are other factors coexpressed in hepatocytes which influence hepcidin regulation (Zhao et al., 2013). HJV complexed with neogenin is much more susceptible for cleavage by the proprotein convertase furin, and the cleaved HJV is rapidly released from cells. This action of furin is enhanced in iron deficiency and hypoxia and results in generation of s-HJV, which antagonizes BMP-dependent activation of hepcidin (Silvestri, Pagani, Nai, et al., 2008). Since furin participates also in processing of prohepcidin into the mature protein (Valore & Ganz, 2008), the effect of furin on hepcidin expression may be complex, resulting in activation or inhibition depending on the conditions (Poli et al., 2010). The precise molecular mechanisms underlying the interaction of these pathways and proteins remain incompletely understood. Since the detailed discussion of the control of hepcidin transcription is outside the scope of this chapter, readers are referred to recent reviews (Core et al., 2014; Ganz & Nemeth, 2012; Gozzelino & Arosio, 2016; Zhao et al., 2013).

3.2 Hepcidin Synthesis Regulation by Inflammatory Stimuli

As iron restriction can limit microbial growth, the increase in hepcidin levels in response to inflammation and infection is thought to be related to defense mechanisms protecting the host from infections (Casanovas, Mleczko-Sanecka, Altamura, Hentze, & Muckenthaler, 2009; Wang et al., 2005). The first insight into the mechanism at the basis of the well-known notion that patients with chronic inflammation have lower serum iron levels, which can contribute to the development of anemia of chronic disease, was provided by the discovery that hepatocytes treated with lipopolysaccharide activate hepcidin transcription (Pigeon et al., 2001). Further studies on human primary hepatocytes showed that hepcidin is induced by inflammatory mediators, in particular interleukin 6 (IL-6) (Nemeth et al., 2003), which triggers hepcidin transcription mainly through the JAK–STAT3 signaling pathway, although the BMP signaling pathway also takes part in the inflammatory response via SMAD4 (Wang et al., 2005). A cross talk between iron-sensing and inflammatory pathways regulating hepcidin expression seems to exist, as the type I BMPR is an essential element for the induction of hepcidin transcription via IL-6 (Mayeur et al., 2014). Moreover, inflammatory stimuli also induce another TGF-β/BMP superfamily ligand, activin B, which can be the missing part of the cross talk between the BMP–SMAD pathway and inflammation-mediated regulation of hepcidin (Besson-Fournier et al., 2012).

3.3 Hepcidin Regulation by Erythropoiesis

Stimulation of erythropoiesis via induction of EPO synthesis is a paradigmatic systemic adaptation to hypoxia, which eventually leads to the reestablishment of optimal oxygen availability (Nicolas, Chauvet, et al., 2002). HIF is the main mediator of cellular adaptation to low oxygen. Conditional knockout studies have provided clear evidence that liver and renal hypoxic induction of EPO expression in adult is mainly controlled by HIF2α (Gruber et al., 2007; Scortegagna, Morris, Oktay, Bennett, & Garcia, 2003). When an organism is exposed to systemic hypoxia or in case of hemorrhage, hemolysis, or other conditions that raise EPO levels and trigger stress erythropoiesis, a mutual mechanism assuring an adequate quantity of iron is needed to sustain the enhanced red blood cell production. Conversely, low iron availability, on the one hand, suppresses EPO production via the IRP1–HIF2α axis to prevent excessive iron usage by erythropoiesis during systemic iron restriction, on the other hand, also impairs erythroid maturation by inhibiting aconitase (Zhang, Ghosh, & Rouault, 2014).

Hepcidin coordinates erythropoietic activity and iron homeostasis; in fact, when erythropoiesis is enhanced, both intestinal iron absorption and release of iron from macrophages and liver stores are increased as a result of hepcidin downregulation. The inhibition of hepcidin synthesis, when erythropoiesis is stimulated, was observed both in animal models (Nicolas, Chauvet, et al., 2002) and in thalassemic patients (Gardenghi et al., 2007). In patients with β-thalassemia major, hepcidin levels are relatively suppressed and are associated at each time point with EPO concentrations and biomarkers of erythroid activity. Interestingly, erythropoietic drive can overcome both inflammatory and iron-sensing pathways, as indicated by studies in mice showing the suppression of STAT3 and SMAD4 signaling in vivo (Huang, Constante, Layoun, & Santos, 2009), as well as by the observation that the dysregulated erythropoiesis observed in patients with iron-loading anemias is able to suppress hepcidin even under conditions of iron overload (Kautz, Jung, et al., 2014), thus contributing to the development of iron overload in these subjects. Studies in mice lacking the second transferrin receptor (TfR2) in erythropoietic cells indicated that, by sensing the level of circulating iron, TfR2 modulates the sensitivity of erythroblasts to EPO (Nai et al., 2016), but the role of this pathway in the cross talk between iron homeostasis and red blood cells production requires further clarification.

In consideration of the interrelated physiological functions of hepcidin and EPO, the hypothesis that EPO could directly regulate hepcidin was plausible and at first supported by experimental evidence. In fact, the enhanced iron absorption necessary to face the elevated need of erythropoietic cells under conditions like hypoxia or anemia is accompanied by a strong inhibition of hepcidin expression (Kautz & Nemeth, 2014). Furthermore, a strong downregulation of hepcidin synthesis occurred in mice injected with EPO (reviewed by Kautz & Nemeth, 2014; Recalcati et al., 2015) and a direct inhibition of hepcidin transcription was found in EPO-treated cell lines (Fein, Merle, Ehehalt, Herrmann, & Kulaksiz, 2007; Pinto et al., 2008). Notably, EPO administration to human healthy subjects was followed by a prompt and strong decrease in hepcidin levels (Ashby et al., 2010; Robach et al., 2013, 2009) in the absence of significant changes of other hematological parameters. However, these results were challenged by experiments showing that EPO cannot inhibit hepcidin in mice with inactive or suppressed bone marrow activity, thus suggesting that although EPO can regulate hepcidin, it does not act directly (Pak, Lopez, Gabayan, Ganz, & Rivera, 2006; Sasaki, Noguchi-Sasaki, Yasuno, Yorozu, & Shimonaka, 2012). Furthermore, it has been demonstrated that by increasing liver iron content it is possible to stimulate hepcidin expression even under conditions of permanently high EPO serum levels and sustained erythropoiesis (Díaz et al., 2013), thus suggesting that hepcidin is not directly controlled by EPO in these settings. More recent findings showing that the inhibitory effect of EPO on hepcidin synthesis was observed also in mice lacking EPO receptor expression in hepatic cells (Gammella et al., 2015), conclusively demonstrated that the direct binding of EPO to liver receptors is not necessary to suppress hepcidin. These results obtained in experimental models are supported by human studies showing that anemia alone, even when EPO levels are elevated, is not able to suppress hepcidin. In fact, hepcidin levels are not decreased in patients with Diamond–Blackfan anemia, which is characterized by erythroid hypoplasia (Pospíšilová et al., 2012), thus strongly indicating that hepcidin downregulation requires an active erythroid marrow.

Therefore, the search for direct effectors connecting erythropoiesis with liver hepcidin downregulation has been pursued. Two proteins produced by erythroid cells: growth differentiation factor 15 (GDF15), a member of the transforming growth factor-β superfamily (Tanno et al., 2007), and twisted gastrulation factor 1 (TWSG1) (Tanno et al., 2009) were identified as potential "erythroid factors" suppressing hepcidin in ineffective erythropoiesis. Despite promising preliminary data, later studies did not find significant changes of GDF15 levels in humans injected with EPO (Robach et al., 2013). In addition, GDF15 knockout mice showed that in condition of increased erythropoietic activity due to blood loss, GDF15 does not diminish hepatic hepcidin mRNA expression, and it is not an essential factor balancing iron homeostasis (Casanovas et al., 2013). Moreover, evidence for the contribution of TWSG1 to regulation of hepcidin in physiology and pathology remains inconclusive (Kautz & Nemeth, 2014).

Recently, the hormone erythroferrone (ERFE) has been identified and characterized as suppressor of hepcidin expression (Fig. 1). ERFE is expressed by erythroid precursors in response to EPO treatment via the JAK2–STAT5 pathway and directly mediates hepcidin suppression during stress erythropoiesis (Kautz, Jung, et al., 2014). ERFE is also known as myonectin, a member of the C1Q/tumor necrosis factor (TNF)-related protein family (CTRP15) expressed in skeletal muscle and adipose tissue (Seldin, Peterson, Byerly, Wei, & Wong, 2012). Although the precise molecular mechanism underlying the action of ERFE on hepcidin repression needs to be elucidated, it seems to be not related to the iron-sensing or BMP–SMAD pathway (Kautz, Jung, et al., 2014). Recent studies proved that ERFE plays a role in mouse models of β-thalassemia and anemia of inflammation. In fact, deletion of ERFE improved iron overload and ineffective erythropoiesis in thalassemic mice, although it did not improve the anemia (Kautz et al., 2015). Moreover, when injected with heat-killed *Brucella abortus*, mice lacking ERFE also had higher hepcidin levels and iron sequestration, with more severe anemia than wild-type animals, suggesting that ERFE favors the recovery from anemia of inflammation (Kautz, Jung, Nemeth, & Ganz, 2014). ERFE knockout mice failed to suppress hepcidin and exhibited delayed recovery from acute blood loss anemia, thus indicating that this factor is important for stress erythropoiesis, but did not show abnormalities in iron and hematological parameters at baseline. Therefore, the lack of an evident phenotype suggests that there are additional unidentified regulators of hepcidin that maintain the connection between the erythroid compartment and the liver under physiological conditions. In this regard, it has been recently shown that the TfR1 expressed on erythroid cells is an ERFE-independent mediator of the inhibition of hepcidin expression in response to expanded erythropoiesis (Keel et al., 2015). Altogether, the available data indicate that the role of EPO is to stimulate the synthesis of ERFE in erythroblasts, which eventually downregulates hepcidin production in the liver.

3.4 Hepcidin Regulation by Hypoxia

Since oxygen tension regulates erythropoietic activity, it is conceivable that hypoxia can regulate hepcidin by a direct mechanism, but available evidence is against a biologically relevant direct effect of hypoxia/HIF1 on hepcidin transcription. Although a study proposed that the increased levels of HIF1α present in livers of iron-deficient mice might repress hepcidin transcription (Peyssonnaux et al., 2007), subsequent studies using liver-restricted deletion of various components of the HIF complex (Liu, Davidoff, Niss, & Haase, 2012; Mastrogiannaki et al., 2012) did not confirm a direct transcriptional suppression of hepcidin and showed that the real role of HIF in this context is to induce EPO. The consequent stimulation of erythropoiesis then leads to hepcidin inhibition, probably through ERFE. In line with the hypothesis that suppression of hepcidin in response to hypoxia is mediated via EPO or/and erythropoiesis-dependent pathways, serum hepcidin levels rapidly decreased in healthy lowlanders who were abruptly (Talbot et al., 2012) or more gradually (Piperno et al., 2011) subjected to high-altitude hypoxia, inversely mirroring EPO production. On the other hand, indigenous populations chronically adapted to high altitude by a pattern not involving increased erythropoiesis do not have diminished hepcidin levels in comparison to their lowland counterparts (Lundgrin et al., 2013).

While these findings ruled out the possibility of a direct inhibitory effect of hypoxia on hepcidin in vivo, studies reporting that activation of HIF by hypoxia resulted in hepcidin downregulation in cultured cells suggest that hypoxia may play a complementary role in this context. In fact, it has been demonstrated that the expression of proteins that negatively control hepcidin expression, such as matriptase-2 (Lakhal et al., 2011) and furin (McMahon, Grondin, McDonald, Richard, & Dubois, 2005; Silvestri, Pagani, & Camaschella, 2008) is increased in hypoxia via HIF1/2α.

Moreover, the recent demonstration that platelet-derived growth factor BB, which is stored in several cell types (e.g., platelets) and promptly released in response to EPO, is increased by hypoxia and inhibits hepcidin (Sonnweber et al., 2014) may explain how EPO administration in humans leads to early and considerable decrease of hepcidin levels that preceded any change in potentials mediators (Ashby et al., 2010; Robach et al., 2013, 2009). The role of these pathways in hepcidin downregulation under physiological and pathological circumstances and whether hierarchy or cooperation between them exists remains to be completely understood.

4. CONCLUSION AND FUTURE DIRECTIONS

The elevated heme production associated with erythropoiesis requires high entry of iron into erythroid precursor cells. Therefore, control of iron trafficking by the erythroid compartment, which is the main consumer, is a key homeostatic mechanism for iron regulation. Over the last years, the characterization of the mechanism based on the hepcidin–Fpn axis by which systemic iron homeostasis is maintained has proceeded fast, but knowledge of the pathways by which erythropoiesis suppresses hepcidin has lagged behind. The demonstration that the roles of HIF and EPO are indirect and mediated by erythropoiesis prompted the search for the so-called erythroid regulator(s) which link bone marrow iron demand with liver hepcidin expression. ERFE is currently the best-characterized candidate to mediate hepcidin inhibition, at least in the context of massively increased erythropoiesis. However, important questions about its mechanism of action and its role in human physiology and disease remain unresolved. Advances in the understanding of the molecular basis of the interaction between iron metabolism and erythropoiesis should hopefully translate into improved therapies aimed to correct iron imbalance in patients with disordered erythropoiesis.

ACKNOWLEDGMENT

This work was supported by grants from MIUR.

REFERENCES

Andrews, N. C. (2008). Forging a field: The golden age of iron biology. *Blood*, *112*(2), 219–230. http://dx.doi.org/10.1182/blood-2007-12-077388.

Ashby, D. R., Gale, D. P., Busbridge, M., Murphy, K. G., Duncan, N. D., Cairns, T. D., … Choi, P. (2010). Erythropoietin administration in humans causes a marked and prolonged reduction in circulating hepcidin. *Haematologica, 95*(3), 505–508. http://dx.doi.org/10.3324/haematol.2009.013136.

Besson-Fournier, C., Latour, C., Kautz, L., Bertrand, J., Ganz, T., Roth, M. P., & Coppin, H. (2012). Induction of activin B by inflammatory stimuli up-regulates expression of the iron-regulatory peptide hepcidin through Smad1/5/8 signaling. *Blood, 120*(2), 431–439. http://dx.doi.org/10.1182/blood-2012-02-411470.

Cairo, G., Bernuzzi, F., & Recalcati, S. (2006). A precious metal: Iron, an essential nutrient for all cells. *Genes & Nutrition, 1*(1), 25–39. http://dx.doi.org/10.1007/bf02829934.

Cairo, G., & Recalcati, S. (2007). Iron-regulatory proteins: Molecular biology and pathophysiological implications. *Expert Reviews in Molecular Medicine, 9*(33), 1–13. http://dx.doi.org/10.1017/S1462399407000531.

Casanovas, G., Mleczko-Sanecka, K., Altamura, S., Hentze, M. W., & Muckenthaler, M. U. (2009). Bone morphogenetic protein (BMP)-responsive elements located in the proximal and distal hepcidin promoter are critical for its response to HJV/BMP/SMAD. *Journal of Molecular Medicine (Berlin, Germany), 87*(5), 471–480. http://dx.doi.org/10.1007/s00109-009-0447-2.

Casanovas, G., Vujic Spasic, M., Casu, C., Rivella, S., Strelau, J., Unsicker, K., & Muckenthaler, M. U. (2013). The murine growth differentiation factor 15 is not essential for systemic iron homeostasis in phlebotomized mice. *Haematologica, 98*(3), 444–447. http://dx.doi.org/10.3324/haematol.2012.069807.

Core, A. B., Canali, S., & Babitt, J. L. (2014). Hemojuvelin and bone morphogenetic protein (BMP) signaling in iron homeostasis. *Frontiers in Pharmacology, 5*, 104. http://dx.doi.org/10.3389/fphar.2014.00104.

De Falco, L., Sanchez, M., Silvestri, L., Kannengiesser, C., Muckenthaler, M. U., Iolascon, A., … Beaumont, C. (2013). Iron refractory iron deficiency anemia. *Haematologica, 98*(6), 845–853. http://dx.doi.org/10.3324/haematol.2012.075515.

Díaz, V., Gammella, E., Recalcati, S., Santambrogio, P., Naldi, A. M., Vogel, J., … Cairo, G. (2013). Liver iron modulates hepcidin expression during chronically elevated erythropoiesis in mice. *Hepatology, 58*(6), 2122–2132. http://dx.doi.org/10.1002/hep.26550.

Donovan, A., Lima, C. A., Pinkus, J. L., Pinkus, G. S., Zon, L. I., Robine, S., & Andrews, N. C. (2005). The iron exporter ferroportin/Slc40a1 is essential for iron homeostasis. *Cell Metabolism, 1*(3), 191–200. http://dx.doi.org/10.1016/j.cmet.2005.01.003.

Du, X., She, E., Gelbart, T., Truksa, J., Lee, P., Xia, Y., … Beutler, B. (2008). The serine protease TMPRSS6 is required to sense iron deficiency. *Science, 320*(5879), 1088–1092. http://dx.doi.org/10.1126/science.1157121.

Fein, E., Merle, U., Ehehalt, R., Herrmann, T., & Kulaksiz, H. (2007). Regulation of hepcidin in HepG2 and RINm5F cells. *Peptides, 28*(5), 951–957. http://dx.doi.org/10.1016/j.peptides.2007.01.016.

Finch, C. (1994). Regulators of iron balance in humans. *Blood, 84*(6), 1697–1702.

Fleming, R. E., & Sly, W. S. (2001). Hepcidin: A putative iron-regulatory hormone relevant to hereditary hemochromatosis and the anemia of chronic disease. *Proceedings of the National Academy of Sciences of the United States of America, 98*(15), 8160–8162. http://dx.doi.org/10.1073/pnas.161296298.

Folgueras, A. R., de Lara, F. M., Pendas, A. M., Garabaya, C., Rodriguez, F., Astudillo, A., … Velasco, G. (2008). Membrane-bound serine protease matriptase-2 (Tmprss6) is an essential regulator of iron homeostasis. *Blood, 112*(6), 2539–2545. http://dx.doi.org/10.1182/blood-2008-04-149773.

Gammella, E., Diaz, V., Recalcati, S., Buratti, P., Samaja, M., Dey, S., … Cairo, G. (2015). Erythropoietin's inhibiting impact on hepcidin expression occurs indirectly. *American Journal of Physiology. Regulatory, Integrative and Comparative Physiology, 308*, R330–R335. http://dx.doi.org/10.1152/ajpregu.00410.2014. ajpregu.00410.02014.

Ganz, T. (2005). Hepcidin—A regulator of intestinal iron absorption and iron recycling by macrophages. *Best Practice & Research. Clinical Haematology*, *18*(2), 171–182. http://dx.doi.org/10.1016/j.beha.2004.08.020.

Ganz, T., & Nemeth, E. (2012). Hepcidin and iron homeostasis. *Biochimica et Biophysica Acta*, *1823*(9), 1434–1443. http://dx.doi.org/10.1016/j.bbamcr.2012.01.014.

Gardenghi, S., Marongiu, M. F., Ramos, P., Guy, E., Breda, L., Chadburn, A., … Rivella, S. (2007). Ineffective erythropoiesis in beta-thalassemia is characterized by increased iron absorption mediated by down-regulation of hepcidin and up-regulation of ferroportin. *Blood*, *109*(11), 5027–5035. http://dx.doi.org/10.1182/blood-2006-09-048868.

Gozzelino, R., & Arosio, P. (2015). The importance of iron in pathophysiologic conditions. *Frontiers in Pharmacology*, *6*, 26. http://dx.doi.org/10.3389/fphar.2015.00026.

Gozzelino, R., & Arosio, P. (2016). Iron homeostasis in health and disease. *International Journal of Molecular Sciences*, *17*(1), 130. http://dx.doi.org/10.3390/ijms17010130.

Gozzelino, R., & Soares, M. P. (2014). Coupling heme and iron metabolism via ferritin H chain. *Antioxidants & Redox Signaling*, *20*(11), 1754–1769. http://dx.doi.org/10.1089/ars.2013.5666.

Gruber, M., Hu, C. J., Johnson, R. S., Brown, E. J., Keith, B., & Simon, M. C. (2007). Acute postnatal ablation of Hif-2alpha results in anemia. *Proceedings of the National Academy of Sciences of the United States of America*, *104*(7), 2301–2306. http://dx.doi.org/10.1073/pnas.0608382104.

Gulec, S., Anderson, G. J., & Collins, J. F. (2014). Mechanistic and regulatory aspects of intestinal iron absorption. *American Journal of Physiology. Gastrointestinal and Liver Physiology*, *307*(4), G397–G409. http://dx.doi.org/10.1152/ajpgi.00348.2013.

Hentze, M. W., Muckenthaler, M. U., Galy, B., & Camaschella, C. (2010). Two to tango: Regulation of mammalian iron metabolism. *Cell*, *142*(1), 24–38. http://dx.doi.org/10.1016/j.cell.2010.06.028.

Huang, H., Constante, M., Layoun, A., & Santos, M. M. (2009). Contribution of STAT3 and SMAD4 pathways to the regulation of hepcidin by opposing stimuli. *Blood*, *113*(15), 3593–3599. http://dx.doi.org/10.1182/blood-2008-08-173641.

Kautz, L., Jung, G., Du, X., Gabayan, V., Chapman, J., Nasoff, M., … Ganz, T. (2015). Erythroferrone contributes to hepcidin suppression and iron overload in a mouse model of beta-thalassemia. *Blood*, *126*(17), 2031–2037. http://dx.doi.org/10.1182/blood-2015-07-658419.

Kautz, L., Jung, G., Nemeth, E., & Ganz, T. (2014a). Erythroferrone contributes to recovery from anemia of inflammation. *Blood*, *124*(16), 2569–2574. http://dx.doi.org/10.1182/blood-2014-06-584607.

Kautz, L., Jung, G., Valore, E. V., Rivella, S., Nemeth, E., & Ganz, T. (2014b). Identification of erythroferrone as an erythroid regulator of iron metabolism. *Nature Genetics*, *46*(7), 678–684. http://dx.doi.org/10.1038/ng.2996.

Kautz, L., & Nemeth, E. (2014). Molecular liaisons between erythropoiesis and iron metabolism. *Blood*, *124*(4), 479–482. http://dx.doi.org/10.1182/blood-2014-05-516252.

Keel, S. B., Doty, R., Liu, L., Nemeth, E., Cherian, S., Ganz, T., & Abkowitz, J. L. (2015). Evidence that the expression of transferrin receptor 1 on erythroid marrow cells mediates hepcidin suppression in the liver. *Experimental Hematology*, *43*(6). http://dx.doi.org/10.1016/j.exphem.2015.03.001. 469–478.e466.

Kerenyi, M. A., Grebien, F., Gehart, H., Schifrer, M., Artaker, M., Kovacic, B., … Müllner, E. W. (2008). Stat5 regulates cellular iron uptake of erythroid cells via IRP-2 and TfR-1. *Blood*, *112*(9), 3878–3888. http://dx.doi.org/10.1182/blood-2008-02-138339.

Lakhal, S., Schodel, J., Townsend, A. R., Pugh, C. W., Ratcliffe, P. J., & Mole, D. R. (2011). Regulation of type II transmembrane serine proteinase TMPRSS6 by hypoxia-inducible factors: New link between hypoxia signaling and iron homeostasis.

The Journal of Biological Chemistry, *286*(6), 4090–4097. http://dx.doi.org/10.1074/jbc.M110.173096.

Lee, D. H., Zhou, L. J., Zhou, Z., Xie, J. X., Jung, J. U., Liu, Y., … Xiong, W. C. (2010). Neogenin inhibits HJV secretion and regulates BMP-induced hepcidin expression and iron homeostasis. *Blood*, *115*(15), 3136–3145. http://dx.doi.org/10.1182/blood-2009-11-251199.

Liu, Q., Davidoff, O., Niss, K., & Haase, V. H. (2012). Hypoxia-inducible factor regulates hepcidin via erythropoietin-induced erythropoiesis. *The Journal of Clinical Investigation*, *122*(12), 4635–4644. http://dx.doi.org/10.1172/JCI63924.

Lundgrin, E. L., Janocha, A. J., Koch, C. D., Gebremedhin, A., Di Rienzo, A., Alkorta-Aranburu, G., … Beall, C. M. (2013). Plasma hepcidin of Ethiopian highlanders with steady-state hypoxia. *Blood*, *122*, 1989–1991. United States.

Mastrogiannaki, M., Matak, P., Mathieu, J. R., Delga, S., Mayeux, P., Vaulont, S., & Peyssonnaux, C. (2012). Hepatic hypoxia-inducible factor-2 down-regulates hepcidin expression in mice through an erythropoietin-mediated increase in erythropoiesis. *Haematologica*, *97*(6), 827–834. http://dx.doi.org/10.3324/haematol.2011.056119.

Mayeur, C., Lohmeyer, L. K., Leyton, P., Kao, S. M., Pappas, A. E., Kolodziej, S. A., … Steinbicker, A. U. (2014). The type I BMP receptor Alk3 is required for the induction of hepatic hepcidin gene expression by interleukin-6. *Blood*, *123*(14), 2261–2268. http://dx.doi.org/10.1182/blood-2013-02-480095.

McKie, A. T., Marciani, P., Rolfs, A., Brennan, K., Wehr, K., Barrow, D., … Simpson, R. J. (2000). A novel duodenal iron-regulated transporter, IREG1, implicated in the basolateral transfer of iron to the circulation. *Molecular Cell*, *5*(2), 299–309.

McMahon, S., Grondin, F., McDonald, P. P., Richard, D. E., & Dubois, C. M. (2005). Hypoxia-enhanced expression of the proprotein convertase furin is mediated by hypoxia-inducible factor-1: Impact on the bioactivation of proproteins. *The Journal of Biological Chemistry*, *280*(8), 6561–6569. http://dx.doi.org/10.1074/jbc.M413248200.

Nai, A., Rubio, A., Campanella, A., Gourbeyre, O., Artuso, I., Bordini, J., … Meynard, D. (2016). Limiting hepatic Bmp-Smad signaling by matriptase-2 is required for erythropoietin-mediated hepcidin suppression in mice. *Blood*, *127*(19), 2327–2336. http://dx.doi.org/10.1182/blood-2015-11-681494.

Nemeth, E., Valore, E. V., Territo, M., Schiller, G., Lichtenstein, A., & Ganz, T. (2003). Hepcidin, a putative mediator of anemia of inflammation, is a type II acute-phase protein. *Blood*, *101*(7), 2461–2463. http://dx.doi.org/10.1182/blood-2002-10-3235.

Nicolas, G., Bennoun, M., Devaux, I., Beaumont, C., Grandchamp, B., Kahn, A., & Vaulont, S. (2001). Lack of hepcidin gene expression and severe tissue iron overload in upstream stimulatory factor 2 (USF2) knockout mice. *Proceedings of the National Academy of Sciences of the United States of America*, *98*(15), 8780–8785. http://dx.doi.org/10.1073/pnas.151179498.

Nicolas, G., Bennoun, M., Porteu, A., Mativet, S., Beaumont, C., Grandchamp, B., … Vaulont, S. (2002). Severe iron deficiency anemia in transgenic mice expressing liver hepcidin. *Proceedings of the National Academy of Sciences of the United States of America*, *99*(7), 4596–4601. http://dx.doi.org/10.1073/pnas.072632499.

Nicolas, G., Chauvet, C., Viatte, L., Danan, J. L., Bigard, X., Devaux, I., … Vaulont, S. (2002). The gene encoding the iron regulatory peptide hepcidin is regulated by anemia, hypoxia, and inflammation. *The Journal of Clinical Investigation*, *110*(7), 1037–1044. http://dx.doi.org/10.1172/JCI15686.

Pak, M., Lopez, M. A., Gabayan, V., Ganz, T., & Rivera, S. (2006). Suppression of hepcidin during anemia requires erythropoietic activity. *Blood*, *108*(12), 3730–3735. http://dx.doi.org/10.1182/blood-2006-06-028787.

Pantopoulos, K., Porwal, S. K., Tartakoff, A., & Devireddy, L. (2012). Mechanisms of mammalian iron homeostasis. *Biochemistry*, *51*(29), 5705–5724. http://dx.doi.org/10.1021/bi300752r.

Paul, V. D., & Lill, R. (2015). Biogenesis of cytosolic and nuclear iron-sulfur proteins and their role in genome stability. *Biochimica et Biophysica Acta*, *1853*(6), 1528–1539. http://dx.doi.org/10.1016/j.bbamcr.2014.12.018.

Peyssonnaux, C., Zinkernagel, A. S., Schuepbach, R. A., Rankin, E., Vaulont, S., Haase, V. H., ... Johnson, R. S. (2007). Regulation of iron homeostasis by the hypoxia-inducible transcription factors (HIFs). *The Journal of Clinical Investigation*, *117*(7), 1926–1932. http://dx.doi.org/10.1172/JCI31370.

Pietrangelo, A. (2010). Hereditary hemochromatosis: Pathogenesis, diagnosis, and treatment. *Gastroenterology*, *139*(2), 393–408. http://dx.doi.org/10.1053/j.gastro.2010.06.013. 408.e391–392.

Pietrangelo, A. (2015). Genetics, genetic testing, and management of hemochromatosis: 15 Years since hepcidin. *Gastroenterology*, *149*(5). http://dx.doi.org/10.1053/j.gastro.2015.06.045. 1240–1251.e1244.

Pigeon, C., Ilyin, G., Courselaud, B., Leroyer, P., Turlin, B., Brissot, P., & Loreal, O. (2001). A new mouse liver-specific gene, encoding a protein homologous to human antimicrobial peptide hepcidin, is overexpressed during iron overload. *The Journal of Biological Chemistry*, *276*(11), 7811–7819. http://dx.doi.org/10.1074/jbc.M008923200.

Pinto, J. P., Ribeiro, S., Pontes, H., Thowfeequ, S., Tosh, D., Carvalho, F., & Porto, G. (2008). Erythropoietin mediates hepcidin expression in hepatocytes through EPOR signaling and regulation of C/EBPalpha. *Blood*, *111*(12), 5727–5733. http://dx.doi.org/10.1182/blood-2007-08-106195.

Piperno, A., Galimberti, S., Mariani, R., Pelucchi, S., Ravasi, G., Lombardi, C., ... Parati, G. (2011). Modulation of hepcidin production during hypoxia-induced erythropoiesis in humans in vivo: Data from the HIGHCARE project. *Blood*, *117*(10), 2953–2959. http://dx.doi.org/10.1182/blood-2010-08-299859.

Poli, M., Luscieti, S., Gandini, V., Maccarinelli, F., Finazzi, D., Silvestri, L., ... Arosio, P. (2010). Transferrin receptor 2 and HFE regulate furin expression via mitogen-activated protein kinase/extracellular signal-regulated kinase (MAPK/Erk) signaling. Implications for transferrin-dependent hepcidin regulation. *Haematologica*, *95*(11), 1832–1840. http://dx.doi.org/10.3324/haematol.2010.027003.

Pospíšilová, D., Houda, J., Holub, D., Ludíková, B., Mojzíková, R., Pospíšilová, P., ... Džubák, P. (2012). Significance of hepcidin level assessment in the diagnosis of selected types of anaemia in childhood. *Transfuze A Hematologie Dnes*, *18*(2), 58–65.

Powell, L. W., Seckington, R. C., & Deugnier, Y. (2016). Haemochromatosis. *Lancet*, *388*, 706–716. http://dx.doi.org/10.1016/S0140-6736(15)01315-X.

Recalcati, S., Gammella, E., & Cairo, G. (2015). New perspectives on the molecular basis of the interaction between oxygen homeostasis and iron metabolism. *Hypoxia (Auckland, N.Z.)*, *3*, 93–103.

Robach, P., Recalcati, S., Girelli, D., Campostrini, N., Kempf, T., Wollert, K. C., ... Cairo, G. (2013). Serum hepcidin levels and muscle iron proteins in humans injected with low- or high-dose erythropoietin. *European Journal of Haematology*, *91*(1), 74–84. http://dx.doi.org/10.1111/ejh.12122.

Robach, P., Recalcati, S., Girelli, D., Gelfi, C., Aachmann-Andersen, N. J., Thomsen, J. J., ... Cairo, G. (2009). Alterations of systemic and muscle iron metabolism in human subjects treated with low-dose recombinant erythropoietin. *Blood*, *113*(26), 6707–6715. http://dx.doi.org/10.1182/blood-2008-09-178095.

Sankaran, V. G., & Weiss, M. J. (2015). Anemia: Progress in molecular mechanisms and therapies. *Nature Medicine*, *21*(3), 221–230. http://dx.doi.org/10.1038/nm.3814.

Sasaki, Y., Noguchi-Sasaki, M., Yasuno, H., Yorozu, K., & Shimonaka, Y. (2012). Erythropoietin stimulation decreases hepcidin expression through hematopoietic activity on bone marrow cells in mice. *International Journal of Hematology*, *96*(6), 692–700. http://dx.doi.org/10.1007/s12185-012-1217-4.

Scortegagna, M., Morris, M. A., Oktay, Y., Bennett, M., & Garcia, J. A. (2003). The HIF family member EPAS1/HIF-2alpha is required for normal hematopoiesis in mice. *Blood*, *102*(5), 1634–1640. http://dx.doi.org/10.1182/blood-2003-02-0448.

Seldin, M. M., Peterson, J. M., Byerly, M. S., Wei, Z., & Wong, G. W. (2012). Myonectin (CTRP15), a novel myokine that links skeletal muscle to systemic lipid homeostasis. *The Journal of Biological Chemistry*, *287*(15), 11968–11980. http://dx.doi.org/10.1074/jbc.M111.336834.

Silvestri, L., Pagani, A., & Camaschella, C. (2008a). Furin-mediated release of soluble hemojuvelin: A new link between hypoxia and iron homeostasis. *Blood*, *111*(2), 924–931. http://dx.doi.org/10.1182/blood-2007-07-100677.

Silvestri, L., Pagani, A., Nai, A., De Domenico, I., Kaplan, J., & Camaschella, C. (2008b). The serine protease matriptase-2 (TMPRSS6) inhibits hepcidin activation by cleaving membrane hemojuvelin. *Cell Metabolism*, *8*(6), 502–511. http://dx.doi.org/10.1016/j.cmet.2008.09.012.

Sonnweber, T., Nachbaur, D., Schroll, A., Nairz, M., Seifert, M., Demetz, E., ... Weiss, G. (2014). Hypoxia induced downregulation of hepcidin is mediated by platelet derived growth factor BB. *Gut*, *63*(12), 1951–1959. http://dx.doi.org/10.1136/gutjnl-2013-305317.

Talbot, N. P., Lakhal, S., Smith, T. G., Privat, C., Nickol, A. H., Rivera-Ch, M., ... Robbins, P. A. (2012). Regulation of hepcidin expression at high altitude. *Blood*, *119*(3), 857–860. http://dx.doi.org/10.1182/blood-2011-03-341776.

Tanno, T., Bhanu, N. V., Oneal, P. A., Goh, S. H., Staker, P., Lee, Y. T., ... Miller, J. L. (2007). High levels of GDF15 in thalassemia suppress expression of the iron regulatory protein hepcidin. *Nature Medicine*, *13*(9), 1096–1101. http://dx.doi.org/10.1038/nm1629.

Tanno, T., Porayette, P., Sripichai, O., Noh, S. J., Byrnes, C., Bhupatiraju, A., ... Miller, J. L. (2009). Identification of TWSG1 as a second novel erythroid regulator of hepcidin expression in murine and human cells. *Blood*, *114*(1), 181–186. http://dx.doi.org/10.1182/blood-2008-12-195503.

Valore, E. V., & Ganz, T. (2008). Posttranslational processing of hepcidin in human hepatocytes is mediated by the prohormone convertase furin. *Blood Cells, Molecules & Diseases*, *40*(1), 132–138. http://dx.doi.org/10.1016/j.bcmd.2007.07.009.

Wang, R. H., Li, C., Xu, X., Zheng, Y., Xiao, C., Zerfas, P., ... Deng, C. X. (2005). A role of SMAD4 in iron metabolism through the positive regulation of hepcidin expression. *Cell Metabolism*, *2*(6), 399–409. http://dx.doi.org/10.1016/j.cmet.2005.10.010.

Zhang, C. (2014). Essential functions of iron-requiring proteins in DNA replication, repair and cell cycle control. *Protein & Cell*, *5*(10), 750–760. http://dx.doi.org/10.1007/s13238-014-0083-7.

Zhang, D. L., Ghosh, M. C., & Rouault, T. A. (2014). The physiological functions of iron regulatory proteins in iron homeostasis—An update. *Frontiers in Pharmacology*, *5*, 124. http://dx.doi.org/10.3389/fphar.2014.00124.

Zhang, A. S., West, A. P., Jr., Wyman, A. E., Bjorkman, P. J., & Enns, C. A. (2005). Interaction of hemojuvelin with neogenin results in iron accumulation in human embryonic kidney 293 cells. *The Journal of Biological Chemistry*, *280*(40), 33885–33894. http://dx.doi.org/10.1074/jbc.M506207200.

Zhao, N., Zhang, A. S., & Enns, C. A. (2013). Iron regulation by hepcidin. *The Journal of Clinical Investigation*, *123*(6), 2337–2343. http://dx.doi.org/10.1172/JCI67225.

CHAPTER NINE

Context-Dependent Skeletal Effects of Erythropoietin

Sahar Hiram-Bab, Drorit Neumann, Yankel Gabet[1]
Sackler Faculty of Medicine, Tel-Aviv University, Tel-Aviv, Israel
[1]Corresponding author: e-mail address: yankel@post.tau.ac.il

Contents

Abstract

Erythropoietin (Epo) is the main hormone that regulates the production of red blood cells (hematopoiesis), by stimulating their progenitors. Beyond this vital function, several emerging roles have been noted for Epo in other tissues, including neurons, heart, and retina. The skeletal system is also affected by Epo; however, its actions on bone are, as yet, controversial. Here, we review the seemingly contradicting evidence regarding Epo effects on bone remodeling. We also discuss the evidence pointing to a direct vs indirect effect of Epo on the osteoblastic and osteoclastic cell lineages. The current controversy may derive from a context-dependent mode of function of Epo, namely, opposite skeletal actions during bone regeneration and steady-state bone remodeling. Differences in conclusions deriving from the published in vitro studies may thus relate to the different experimental conditions. Taken together, the current state-of-the-art indicates definite Epo effects on bone cells and points to the complexity of the mode of function.

Vitamins and Hormones, Volume 105
ISSN 0083-6729
http://dx.doi.org/10.1016/bs.vh.2017.02.003
© 2017 Elsevier Inc.
All rights reserved.

1. INTRODUCTION

The erythropoietin (Epo) hormone stimulates red blood cell differentiation. It acts via binding to its receptor (Epo-R) on erythroid progenitor cells within the bone marrow. Epo-R downstream pathways are mediated by Janus kinase 2 (JAK2), STAT3, and STAT5 and signal through phosphatidylinositol-3 kinase (PI3K)/Akt, MAP kinase, and protein kinase C (Watowich, 2011).

The primary site of Epo production resides in the fetal liver and adult kidney, where *Epo* gene expression occurs mainly under the control of an oxygen-sensing, hypoxia-inducible factor (HIF)-dependent mechanism (Richmond, Chohan, & Barber, 2005).

When oxygen supply is disrupted by anemia, or low oxygen (hypoxic) conditions, Epo levels in the peripheral blood increase dramatically. At the transcriptional level, *Epo* gene expression is tightly regulated by HIFs, which are the master transcription factors for oxygen-dependent gene regulation (Franke, Gassmann, & Wielockx, 2013). Active HIF is a heterodimeric complex composed of an oxygen-sensitive HIFα and a constitutive HIFβ subunit. HIF-1α has been suggested to represent the response to acute hypoxia, while HIF-2α is the predominant subunit responding to chronic exposure to low oxygen, as occurring at high altitude (van Patot & Gassmann, 2011). Activity of these subunits is controlled by HIF-prolyl hydroxylases (PHD1–4). These deoxygenases need oxygen as a cofactor to initiate the inactivation of HIF (Eltzschig & Carmeliet, 2011), thus enable the cell to instantaneously sense and adapt (Jewell & Gassmann, 2001) to an inappropriately low oxygen pressure.

The expression of Epo-R in nonerythroid tissues such as the brain, retina, heart, kidney, smooth muscle, and vascular endothelium, as well as macrophages and dendrocytes, has been associated with the discovery of novel biological functions of endogenous Epo signaling, unrelated to hematopoiesis (Lifshitz, Tabak, Gassmann, Mittelman, & Neumann, 2010; Lipsic et al., 2006; Nakano et al., 2007; Pankratova et al., 2010; Rex, Wong, Kodali, & Merry, 2009; Tsai et al., 2006; Westenbrink et al., 2007). However, the expression and functional relevance of Epo-R in nonerythroid cells are still a matter of debate (Elliott et al., 2013; Singbrant et al., 2011). It is expected that the recent development of a more specific and sensitive antibody against human Epo-R will contribute to resolving these questions (Maxwell et al., 2015).

Osteoclasts are derived from the monocyte/macrophage lineage (Bar-Shavit, 2007). They differentiate in the proximity of bone surface and their main function is resorption of the mineralized matrix. Osteoclastogenesis is under the tight regulation of the TNF-related cytokine receptor activator of NFκB (RANK) ligand (RANKL) and colony-stimulating factor-1 (CSF-1, M-CSF). Activation of RANK, the receptor of RANKL, on the surface of monocyte/macrophage precursor cells induces expression of osteoclastic related genes, including those encoding tartrate-resistant acid phosphatase (TRAP), cathepsin K (CatK), calcitonin receptor, and the $\alpha v\beta_3$-integrin, leading to osteoclast maturation and resorbing activity (Florencio-Silva, Sasso, Sasso-Cerri, Simões, & Cerri, 2015).

Osteoblasts arise from a common pluripotent mesenchymal stem cell (MSC) (Fattore, Teti, & Rucci, 2012). Bone morphogenetic proteins (BMPs) and Wnt pathways are especially crucial for the early steps of osteoblastogenesis, where they promote the expression of osteogenic genes such as RUNX2 and Osterix, leading to their commitment toward an osteochondroprogenitor. RUNX2 has been shown to upregulate osteoblast-related genes such as *Col1α1*, tissue nonspecific alkaline phosphatase (TNSALP), bone sialoprotein 2 (*BSP*), and osteocalcin (OCN, a.k.a. *BGLAP*). The main role of osteoblasts is the deposition of bone matrix and mineralization, and they have also been implicated in extra-skeletal function such as energy metabolism and male fertility (Florencio-Silva et al., 2015).

A cross talk between hematopoietic and osteogenic cells has been known for decades. It has been shown that osteogenic activities are enhanced following direct marrow ablation or trauma (Bab, 1995). On the other hand, chronically bled mice show reduction in bone (Gurevitch, Khitrin, Valitov, & Slavin, 2007). The emerging functions of Epo in nonhematopoietic tissues and the close relation between the hematopoietic and osteogenic cells have thus prompted a search for a role of Epo in bone formation and homeostasis.

For over a decade, different groups have investigated the involvement of Epo in bone biology, to elucidate the physiologic roles of Epo in the regulation of bone homeostasis (Tables 1–4), and possibly as a new therapeutic approach. In this chapter, we summarize the main bone phenotypes and cellular responses and elucidate some of the controversies regarding the skeletal actions of Epo.

2. IN VITRO EFFECTS OF EPO

2.1 Expression of Epo-R in Bone Cells

The expression of Epo-R in nonerythroid cells in the bone marrow was challenged by a study based on lineage tracing of Epo-R promoter activity,

Table 1 Epo Effects on Bone Cells and Skeletal Homeostasis—A Summary of the Latest Literature

Model	Epo Doses[a]	Main Findings	Refs.
Rat alveolar bone regeneration model	10 or 20 U once, in the tooth socket	Increased BMD and bone formation and inhibited resorption of residual alveolar ridge	Li et al. (2015)
Human simultaneous fractures of the tibia and fibula	4000 U three times, injected into the fracture site	The mean union period was shorter in the EPO group Epo reduced the incidence of nonunion	Bakhshi et al. (2013)
Segmental bone defects in mice femur	500 U/kg/day for 2 weeks or for 5 weeks 5000 U/kg/6 days 500 U/kg/day for 2 and 10 weeks	Increase periosteal callus diameter and bone mineral density at 2 weeks, no differences by 5 weeks Increase vessel density in the healing zone	Garcia et al. (2011) Holstein et al. (2007) Holstein et al. (2011)
Mandibular distraction in rabbit	150 U/kg four doses, administered at 48-h intervals	The number of osteoblasts and blood vessels was significantly higher, whereas the number of osteoclasts was significantly lower	Mihmanli, Dolanmaz, Avunduk, and Erdemli (2009)
Femur fracture model in mouse	5 U/mouse, every other day, at the fracture site. From day 4 to day 12 postsurgery	Increase callus size and cartilage area. More vascularity (2 weeks). Increased bone mass and density and biomechanical properties (4 weeks)	Wan et al. (2014)
Scaffold implants with BMP2 placed onto cranial bone defect in mice Mice 5–6 weeks embedded in scaffold with/without BMP2	1000 U/mL every other day for 2 weeks, injected into the defect area 200 U/mL	Epo had no effect by itself. Mice treated with BMP2 + Epo exhibited larger calcified areas and increased vascularization	Sun, Jung, Shiozawa, Taichman, and Krebsbach (2012) Patel et al. (2015)
Rabbit posterolateral spinal fusion model	250 IU/kg/19 days	Increase in blood vessels	Rölfing et al. (2012)
Pig calvarial defect model	900 IU/mL Epo on a carrier	No significant benefit for Epo	Rölfing, Jensen,

Table 2 Epo Effects on Bone Remodeling

Model	Epo Doses	Main Findings	Refs.
Tg6	Overexpressing hEpo	Decrease in trabecular bone mass. increase in bone resorption and decrease in bone formation	Rauner et al. (2016) Hiram-Bab et al. (2015)
		Decrease in trabecular bone mass. Modest decrease in cortical thickness. Decreased osteoblast numbers and bone formation. Normal osteoclast numbers and activity	Oikonomidou et al. (2016)
Adult 8–12 weeks Adult 8–12 weeks Osmotic pump 9-week mice	1500 U/kg every other day for 4 weeks 9000 U/kg/twice a week for 3 weeks 350 U/kg/day for 7 days and 500 U/kg/day for 30 days 300 U/kg every other day for 10 days	Decreased bone density. Increase in bone resorption and decrease in bone formation. Increase in osteoclast number and decrease in osteoblast number	Hiram-Bab et al. (2015) Rauner et al. (2016) Singbrant et al. (2011)
Growing young mice 4 weeks	6000 U/kg, three times/week for 4 weeks	Decreased in trabecular bone mass	Hiram-Bab et al. (2015)
		Increase in BMD and bone volume Induces expression of BMP2 in HSCs	Shiozawa et al. (2010)
Osmotic pump	150/kg/day for 30 days	Bone loss, decrease in bone formation, no effect on bone resorption	Rauner et al. (2016)
New born mice	1500–6000 U/kg, three times/week for 4 weeks	At doses above 4500 U/mL, increase in BMD and osteoblasts number	Shiozawa et al. (2010)

using the CRE-loc technology (Singbrant et al., 2011). Based on that study, it was concluded that expression of Epo-R is essentially restricted to erythroid lineage cells. In contrast, other groups focusing on Epo-R expression at the RNA level, demonstrated its expression in macrophages, differentiating preosteoclasts and mature osteoblasts (Hiram-Bab et al., 2015; Rauner

Table 3 Epo Effects In Vitro—Osteoclasts

Model	Epo Doses	Main Findings	Refs.
RAW264.7 cells, mouse bone marrow-derived macrophages (mBMDM)	20 U/mL	Stimulates osteoclastogenesis and enhance EphrinB2-expressing osteoclasts, while reducing their functional activity	Li et al. (2015) Kim et al. (2012) Sun et al. (2012)
mBMDM	10 U/mL	Epo-R mRNA decrease upon differentiation Increase osteoclastogenesis. Resorption is not affected. Effects are mediated by PI3K and JAK signaling pathways	Hiram-Bab et al. (2015)
MMCs	10 U/mL	Epo alone induced osteoclastogenesis, in the absence of RANKL no resorption activity	Shiozawa et al. (2010)

et al., 2016; Rölfing, Jensen, et al., 2014; Shiozawa et al., 2010; Westenbrink et al., 2007). Moreover, specific Epo-R signaling and functional assays in response to Epo stimulation was shown in isolated cells from the osteoblastic and monocytic lineages (Hiram-Bab et al., 2015; Kim et al., 2012; Lifshitz et al., 2010; Rauner et al., 2016; Rölfing, Baatrup, et al., 2014). Whereas the functional relevance of Epo-R expression has been repeatedly demonstrated in vitro, the physiologic and therapeutic actions of Epo-R in bone cells remain to be validated in vivo. This can be accomplished, for instance, by using animal models, such as conditional deletion of Epo-R in specific bone cells.

2.2 Actions of Epo Signaling in Osteoblastic Cells

Epo-R expression was identified in bone cells originating from mice, humans, and other rodents, using RT-PCR, western blot, flow cytometry, and/or functional assays. In mouse osteoblasts, Epo-R expression increases normally during differentiation (Hiram-Bab et al., 2015; Rauner et al., 2016), as well as upon Epo administration in vitro (Guo et al., 2012; Shiozawa et al., 2010). In other nonhematopoietic organs, tissue-protective actions of Epo are functionally linked to the CD131, also known as βc receptor, through an interaction with the Epo-R (Brines et al., 2004).

Table 4 Epo Effects In Vitro—Osteoblasts

Model	Epo Doses	Main Findings	Refs.
Murine cell line ST2, human, rat, and mouse bone marrow mesenchymal stem cells (h/r/m BMSCs), primary mouse calvaria osteoblasts, human osteoblast cell line hFOB 1.19	>10 U/mL	Induces an osteoblastic phenotype, effect is mediated by mTOR, JAK2, and PI3K	Li et al. (2015) Rauner et al. (2016) Guo et al. (2012) Shiozawa et al. (2010) Kim et al. (2012) Rölfing, Baatrup, et al. (2014)
ST2		No effect on osteoblasts differentiation and proliferation	Kim et al. (2012)
Human osteoblast cell line hFOB 1.19, mBMSC	5 U/mL < Epo < 10 U/mL	EPO-induced osteoblast differentiation	Guo et al. (2012) and Shiozawa et al. (2010) Singbrant et al. (2011)
Mouse osteoblast cell line Kusa4b10, hBMSCs, primary mouse calvaria osteoblasts, hBMSCs, murine cell line MC3T3-E1		No effect on osteoblasts differentiation and proliferation	Hiram-Bab et al. (2015) Rauner et al. (2016) Singbrant et al. (2011) Rölfing, Baatrup, et al. (2014)
Primary mouse calvaria osteoblasts	0.5–5 mU/mL	Inhibition of osteoblasts differentiation	Rauner et al. (2016)

The tissue-protective receptor is a multisubunit complex comprising Epo-R, CD131, and in some instances, also the vascular endothelial growth factor receptor (VEGFR2), that is expressed following tissue injury or hypoxia (Brines & Cerami, 2012). In human bone marrow osteoblasts, CD131 was found to be expressed in nearly all Epo-R-positive cells, and the heterodimeric Epo-R/CD131 receptor was proposed to exert Epo signals (Rölfing, Jensen, et al., 2014). The Epo stimulation of either mouse, or human osteoblasts at supraphysiological concentration, between 20–100 U/mL and even up to 250 U/mL (Rauner et al., 2016), induces increase in mineralization (Guo et al., 2012; Kim et al., 2012; Li et al., 2015; Rölfing, Jensen, et al., 2014; Shiozawa et al., 2010), ALP activity, and expression of osteoblastic genes. At the lower limit of this dose range, the effect of Epo on the mouse bone marrow-derived stromal cell line ST2 is still obscure. One study reported that Epo increases ALP activity and mineralization in ST2 cells (Li et al., 2015), while another group found that Epo was unable to induce differentiation at this dose (Kim et al., 2012).

The osteogenic effect of Epo was shown to be mediated by the mTOR, JAK2, and PI3K signaling pathways as was demonstrated using specific inhibitors. However, none of the inhibitors completely abolished ALP activity and calcium deposition (Kim et al., 2012; Rölfing, Jensen, et al., 2014). Interestingly, rapamycin blocked Akt phosphorylation induced by Epo in hBMSC, but not in ST2 cells, and it totally blocked Epo-mediated S6 phosphorylation in both cell types (Kim et al., 2012). It is important to consider the physiologic relevance of these findings, in view of the relatively high Epo doses necessary to trigger the osteogenic response, whereas physiologically, Epo basal plasma concentrations range from 6 to 32 mU/mL (Jelkmann, 2011). Importantly, a recent article reported that at low doses between 0.5 and 5 mU/mL, osteoblast mineralization was significantly inhibited by Epo, whereas doses up to 5 U/mL had no effect (Rauner et al., 2016).

It seems, therefore, that the function of Epo in osteoblasts is dose dependent. Hence, at very high concentrations, Epo stimulates osteogenesis, while at lower, physiologically relevant doses, Epo does the opposite, i.e., attenuation of osteoblast differentiation and mineralization.

2.3 Epo Signaling in Osteoclastic Cells

Osteoclasts originate from the monocyte/macrophage lineage. Bone marrow-derived macrophages (BMDM) express Epo-R transcripts, and stimulation with Epo activates multiple signaling pathways, including signal STAT1 and 5, mitogen-activated protein kinase (MAPK), phosphatidylinositol 3-kinase (PI3K), and nuclear factor kappa B (Lifshitz et al., 2010). Fig. 1 presents immunofluorescent staining of Epo-R and CD11b in bone marrow-derived preosteoclasts. CD11b is a negative regulator of the earliest stages of osteoclast

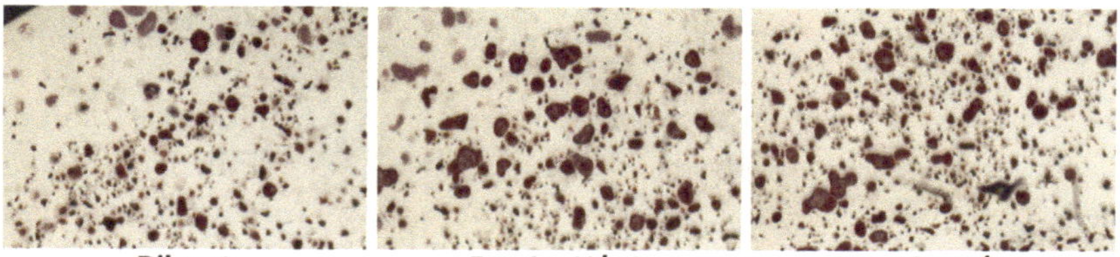

Fig. 1 Low doses of Epo increase osteoclastogenesis in vitro. Mouse BMDM were differentiated in osteoclastogenic medium (RANKL 50 ng/mL, M-CSF 20 ng/mL) with indicated Epo doses for 5 days. Cells were fixed and stained for TRAP.

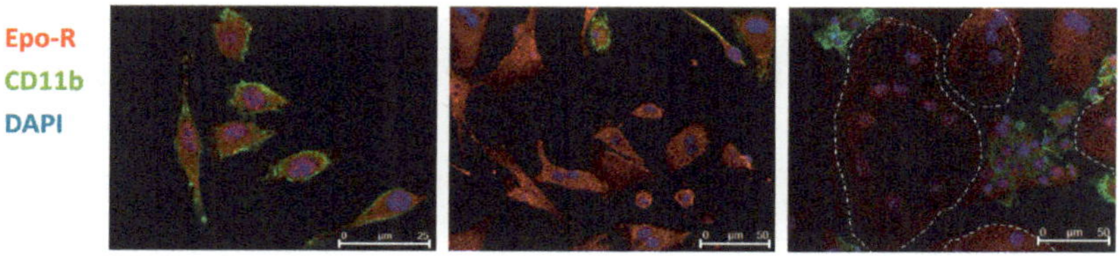

Fig. 2 Epo-R expression is high in bone marrow macrophages and decreases upon osteoclast differentiation. Immunofluorescent staining for Epo-R (*red*) and CD11b (*green*) of BMDM during osteoclastogenesis. Cells were seeded on cover slips with osteoclastogenic medium for 2–4 days before fixation and staining with anti-mEpo-R antibody (1:200) followed by PE (*red*) conjugated secondary Ab (1:500), and anti-FITC (*green*)-conjugated CD11b Ab (1:1000). Mounting medium containing DAPI (*blue*) was used to label cell nuclei.

differentiation. It acts by preventing RANKL-induced induction of the master regulator of osteoclastogenesis NFATc1 (Park-Min et al., 2014). In line with our recently published data (Hiram-Bab et al., 2015), proliferating BMDM abundantly expresses CD11b and Epo-R, whereas both CD11b and Epo-R expressions are completely absent in fully differentiated osteoclasts (Fig. 1).

Different studies found Epo to promote in vitro osteoclast formation at doses ranging from 5 to 20 U/mL (Hiram-Bab et al., 2015; Kim et al., 2012; Li et al., 2015; Shiozawa et al., 2010). In light of the dose dependency of Epo effect in osteoblasts (see earlier), we have reexamined the effect of Epo on osteoclastogenesis at doses that are a 1000-fold lower, i.e., 1–10 mU/mL. The results showed that Epo-stimulated osteoclast differentiation even at these low, "physiological" concentrations (Fig. 2). The stimulatory effect of Epo in preosteoclasts was mediated by JAK2 and PI3K, independently of MAPK (Hiram-Bab et al., 2015).

In some reports, Epo administration during RANKL-induced osteoclastogenesis in primary osteoclast precursors resulted in the formation of inactive osteoclasts (Kim et al., 2012; Li et al., 2015). However, in another study, using a different calcium-coated plate type, osteoclastogenesis was associated with a similar significant increase in pit formation, and the resorbing activity per cell was not impaired by Epo (Hiram-Bab et al., 2015). This was further supported by in vivo data (see later) showing an overall increase in the bone resorption marker TRAP5b, associated with bone resorption in Epo-overexpressing (Tg6), as well as in Epo-injected mice. Similar results were obtained on carboxy-terminal collagen crosslinks (CTX), representing another marker of bone resorption (Hiram-Bab et al., 2015; Rauner et al., 2016).

2.4 Epo Effects on Bone Formation via Osteoclastic Epo-R

An additional Epo function that deserves attention is its signaling in osteoclasts that may lead to regulation of osteoblast function. Hence, osteoclastic cells express EphrinB2, a coupling factor that positively regulates osteoblast differentiation by binding to its receptor EphB4, harbored on osteoblast precursors. The EphinB2/EphB4 interaction also negatively regulates osteoclast differentiation by reverse signaling (Zhao et al., 2006). Epo was reported to upregulate the expression of EphrinB2 in osteoclasts and that of EphB4 receptor in both bone marrow stromal cells (osteoblast progenitors), as well as in the ST2 osteoblastic cell line (Li et al., 2015). Thus, the functional occurrence of this process, namely, Epo stimulation of osteoblast differentiation via increased expression of EphrinB2 in osteoclasts, remains to be investigated by using relevant experimental strategies, such as cocultures, or conditional knockout models.

3. REGULATION OF BONE HOMEOSTASIS BY EPO

Physiologic bone remodeling is a life-lasting process, where mature bone tissue is removed (bone resorption) and replaced with new bone tissue (bone formation). These processes control the remodeling, or the replacement of bone following microdamage that occurs during normal activity. An imbalance in the regulation of bone resorption and formation results in changes in bone mineral density. The process and regulation of bone remodeling differ greatly from bone modeling (skeletal growth and development) and healing (fracture repair, bone regeneration). In contrast with bone remodeling, these latter traumatic models are accompanied with noticeable inflammatory reaction, neoangiogenesis, and hypoxia. To study the skeletal effect of Epo during steady-state bone remodeling, independently of these processes, we first review the skeletal actions of Epo in atraumatic animal models, and the effects of Epo in bone regeneration models will be discussed at the following section.

We and others have used several mouse models, namely, Epo-overexpressing mice (Tg6), intermittent injections of moderate and high Epo doses, and continuous administration of moderate and low doses (Hiram-Bab et al., 2015; Rauner et al., 2016; Singbrant et al., 2011; Wang, Vogel, Kuhn, Gassmann, & Vogel, 2013). In Tg6 mice, the increase in Epo levels is similar to the stimulation of Epo at high altitude (Robach, Fulla, Westerterp, & Richalet, 2004). This model is also relevant for the pathophysiologic increase of endogenous Epo levels, observed after bleeding/repeated phlebotomies or bone injuries, without the confounding direct effect of hypoxia on bone cells (Arnett, 2010; Chang, Jackson, Wardale, & Jones, 2014). On the other hand, the Epo-injection model is relevant for therapeutic administration of exogenous Epo (Lu et al., 2012). Noteworthy, recombinant human Epo is widely prescribed for treating anemia in patients with end-stage renal disease, and for years it has been the highest expenditure drug by Medicare in the United States.

In all the atraumatic adult mouse models, high serum levels of Epo resulted in a lower trabecular bone mass (bone volume fraction), mainly due to lower trabecular number (Dewamitta, Russell, Nandurkar, & Walkley, 2013; Hiram-Bab et al., 2015; Oikonomidou et al., 2016; Rauner et al., 2016; Singbrant et al., 2011). This low bone mass was associated with lower bone formation and higher bone resorption, as compared to controls.

In that regard, Singbrant et al. (2011) agreed with the notion on stimulatory effect of Epo on osteoclastogenesis, but they concluded that Epo also increased bone formation, based on the observed increase in osteoblast number. However, a careful analysis of their presented data reveals that in Epo-treated mice, the measure of single labeled calcein, which is a surrogate of osteoblasts' activity, was significantly decreased (Singbrant et al., 2011), in line with our reported inhibition of bone formation by Epo (Hiram-Bab et al., 2015). The attenuation of bone formation was further supported by the significant decrease in serum levels of OCN and P1NP (Rauner et al., 2016) manifested in the Tg6 mice (Hiram-Bab et al., 2015). The stimulatory doses of Epo used to induce in vitro osteoclastogenesis were ~10 times higher than those used for osteoblast differentiation (Rauner et al., 2016). The dose dependency was recently demonstrated in vivo using an osmotic pump model. A continuous administration of 3 U Epo/day for

4 weeks resulted in an osteoblast-driven bone phenotype, with no effect on osteoclasts. However, when Epo concentration was increased to 10 U/day, an induced bone resorption became evident, similar to the phenotype observed in the Tg6 mice (Rauner et al., 2016). Two mouse models of polycythemia, the CD68:cre-PHD2$^{f/f}$ and the Vav-Cre/JAK2^{V617F}, were also used to study Epo effects on bone homeostasis. The CD68:cre-PHD2$^{f/f}$ (Rauner et al., 2016) mice displayed elevated HIF signaling, and as a consequence, elevated Epo levels. The Vav-Cre/JAK2^{V617F} (Oikonomidou et al., 2016) mice are characterized by constitutively active Epo-R signaling in the broad hematopoietic lineage, but do not display Epo elevation. As expected, both mouse models had markedly elevated hemoglobin concentrations. Interestingly, both cases also displayed bone loss, reduced osteoblast number, and bone formation rate. However, osteoclast number and activity were normal (Oikonomidou et al., 2016; Rauner et al., 2016). Although the bone phenotype was similar, it is reasonable to assume that the mechanism of bone loss differed under these conditions. Because deletion of PHD2 in osteoblasts and osteoclasts only, did not result in bone loss, the authors attributed the bone phenotype of the CD68:cre-PHD2$^{f/f}$ mice to the slight elevation of serum Epo levels, which attenuated bone formation. Since the effect on osteoclasts requires higher Epo levels, bone resorption was not affected. Noteworthy, the bone phenotype in Vav-Cre/JAK2^{V617F} mice is more challenging to interpret. As the knock-in is restricted to the hematopoietic lineage, the constitutively active Epo-R should be carried by preosteoclasts, but not by the osteoblasts. However, the bone phenotype was related to suppressed osteoblasts, with no apparent effect on osteoclasts (Oikonomidou et al., 2016). Since stimulation of the osteoclastic effect requires relatively high Epo levels, lack of the effect may be due to the reduced Epo serum levels that partially attenuate Epo-R signaling in these cells. Nonetheless, attenuation of bone formation in the presence of low Epo serum levels is puzzling. One possibility is that Epo-R signaling in hematopoietic cells (including preosteoclasts) entrains the secretion of antiosteogenic signals. Another explanation may reside in the bone marrow ablation and transplantation as carried out in the animals in order to generate the Vav-Cre/JAK2^{V617F} model. This procedure may preclude the inclusion of this study along with the "atraumatic" models discussed in this section. As detailed later, traumatic models display different modes of action of Epo in the skeletal system.

In contrast to adult animals, one study on growing mice (newborn and 4–6 weeks of age) treated for 4 weeks, showed that Epo increased bone mass in an atraumatic model (Shiozawa et al., 2010). Using the same experimental conditions in growing female mice (4 weeks of age) treated for 4 weeks, we found that the effect of Epo was very similar to that observed in adult animals, in both the femur and vertebrae (Hiram-Bab, Neumann, & Gabet, 2017). This discrepancy between our findings and the earlier report on a similar experimental design (Shiozawa et al., 2010) might be attributed to genetic drift of the mouse strain, to environmental conditions and/or to certain differences in the growth rate of the animals.

4. EPO EFFECTS ON BONE HEALING AND REGENERATION

Bone healing can occur in two main processes, namely, primary/cortical or secondary/endochondral (Perren, 2008). Primary cortical healing takes place when the stability of the fracture is absolute. Healing is characterized by cortical osteonal healing, leading to direct fusion of the fracture ends by cortical bone. Secondary healing occurs in cases of nonrigid fixation and is mediated by the formation of a cartilaginous callus that rapidly forms, calcifies, and remodels into woven and trabecular bone. Ultimately, the intermediate trabecular bone will remodel to restore the original cortical bone morphology. Secondary healing is reminiscent of embryological bone development and includes both models of intramembraneous and endochondral bone formation. In both primary and secondary bone healing, the early stages include an acute inflammatory response and activation of angiogenesis (Perren, 2008; Tsiridis, Upadhyay, & Giannoudis, 2007).

Studies on the effects of Epo on bone were carried out in vivo mainly on models of bone regeneration, including mice, rats, rabbits, and pigs, as well as in clinical trials (Bakhshi et al., 2013; Betsch et al., 2014; Garcia et al., 2011; Holstein et al., 2007; Mihmanli et al., 2009; Rölfing, Baatrup, et al., 2014; Rölfing et al., 2012; Sun et al., 2012; Wan et al., 2014). Holstein et al. (2007) reported for the first time the beneficial effects of short-term Epo treatment in fracture repair. Although not all articles agree in that respect (Rölfing, Jensen, et al., 2014), the general effect of Epo in fracture healing includes increased callus volume and mineral content, accelerated fracture bridging and enhanced biomechanical strength (Garcia et al., 2011; Holstein et al., 2007, 2011; Rölfing et al., 2012; Rölfing, Jensen, et al., 2014). In some reports, the beneficial effects of Epo subsided with time progression (Garcia et al., 2011; Holstein et al., 2007, 2011; Rölfing et al., 2012).

It is interesting to note that in newborn mice treated from birth to day 28, bone growth is strongly associated with hypoxia and angiogenesis, similarly to traumatic bone healing (Gerber et al., 1999). The important stimulatory role of Epo on these two processes involved during bone growth and other traumatic bone models appears, therefore, to supersede the inhibitory effect of Epo on steady-state bone formation. It is likely to assume that the opposite action of Epo in the traumatic/skeletal growth models (Shiozawa et al., 2010) vs adult bone remodeling (Hiram-Bab et al., 2015; Singbrant et al., 2011) is attributed to the fact that hypoxia and angiogenesis play dominant roles in the former, but not in the latter process.

Interestingly, Epo's bone anabolic effect was reported on both models of primary and secondary healing. It may therefore be concluded that the effect of Epo is not dependent on chondrocytes. This assumption is supported by a report showing no beneficial effect of Epo in a model of osteochondral defect (Betsch et al., 2014). In that study, an improvement in cartilage healing was evident only when Epo was administered in addition to bone marrow aspiration concentrate containing osteoblast progenitors.

4.1 Epo Effects via Angiogenesis

Epo-induced osteogenesis in bone healing and regeneration models was repeatedly attributed to the direct stimulation of osteoblast proliferation and differentiation observed in vitro. In addition, Epo was found to act on other cells, leading to an indirect stimulation of osteoblasts (Shiozawa et al., 2010). Different studies suggested that such indirect effects were mediated by hematopoietic stem cells (HSCs) and/or osteoclasts. Epo was suggested to enhance HSCs population (Shiozawa et al., 2010; Sun et al., 2012) and to increase BMP2 expression in these cells (Shiozawa et al., 2010), which in turn, stimulates osteogenic differentiation of osteoblasts. These studies suggest that the osteogenic function of Epo is mediated, at least in part, by BMP2 expressed in HSCs. The importance of this pathway is advocated by the fact that Epo administration had no additive effect over BMP2 alone (Sun et al., 2012), and that Epo's impact on bone regeneration was blunted in the absence of BMP2 (Patel et al., 2015).

In addition to the direct activation of osteoblasts, or their neighboring HSCs and osteoclasts, the beneficial role of Epo could be attributed also to early effects on the posttraumatic inflammatory and angiogenic responses. Epo has been associated with reduced inflammation in several conditions, such as secondary burn progression (Tobalem, Harder, Rezaeian, & Wettstein, 2013), intestinal inflammation (Nakamura et al., 2015), hepatic injury (Liu et al., 2015), and suppression of TLR2/NFκB-mediated inflammation (Garcia et al., 2011; Liu et al., 2015).

Neoangiogenesis plays a critical role in the process of osteogenesis. Simultaneously to endochondral ossification, new blood vessels invade the callus. Indeed, in most of the studied models of bone repair, Epo was associated with an increased number of blood vessels (Garcia et al., 2011; Holstein et al., 2011; Mihmanli et al., 2009; Rölfing et al., 2012; Rölfing, Jensen, et al., 2014; Sun et al., 2012; Wan et al., 2014).

One of the suggested mechanisms is that Epo increases the number of endothelial progenitor cells in the peripheral blood, and their recruitment to the site of injury (Garcia et al., 2011). Epo was also shown to increase expression of VEGF (Holstein et al., 2011), an important angiogenic factor. Interestingly, Sun et al. (2012) showed that Epo-induced angiogenesis was largely blunted when osteoclastogenesis was blocked by bisphosphonates or osteoprotegerin in a metatarsal organ culture assay. It thus appears that osteoclastogenesis is essential for the angiogenesis process facilitated by Epo. This is also in line with the reported increase in osteoclast number in Epo-treated mice (Li et al., 2015; Sun et al., 2012), as discussed earlier.

In all these models of fracture healing and regeneration, bone trauma triggers a complex reaction involving a variety of cytokines and cell types. Notably, Epo seems to affect a series of different steps during this process, such as acute inflammatory response, callus formation and mineralization, as well as angiogenesis. These processes involve the mobilization of a variety of cell types, including osteoclasts, osteoblasts, endothelial, and blood cells, and their respective progenitors. During the surgical phase of these experimental models, blood loss and bone trauma often result in a significant increase in Epo production, in addition to the exogenously administered Epo. In line with the Epo dosage required to induce osteogenesis in vitro, the bone anabolic action of Epo is likely to be restricted to its very high circulating concentrations.

5. CONCLUDING REMARKS

In adult bone homeostasis, bone loss is driven mainly by the inhibition of bone formation and stimulation of bone resorption. Contradicting reports on the effect of Epo on bone during skeletal growth and bone healing may be attributed to the involvement of Epo-stimulated angiogenesis and response to hypoxia. In vitro, the stimulatory effect of Epo on osteoclastogenesis is widely accepted. Controversy remains as to the stimulation of bone resorbing activity. Regarding direct effect of Epo-R signaling in osteoblasts, the discrepancy between stimulatory (in vitro) and inhibitory (in vivo) actions of Epo may be due either to the involvement of other cell types and/or to the dosage. Based on the physiological concentrations of serum Epo, one may assume that concentrations below 100 mU/mL are more clinically relevant than the doses that induce an osteogenic response in vitro. As such, further studies are warranted to examine the direct effect of Epo on osteoblasts at low concentrations. This discrepancy may also result from the involvement of other cell types in vivo. As an example, Epo may act on preosteoclasts to modulate the expression of osteoclasts-to-osteoblasts coupling signals (Sims & Martin, 2015).

ACKNOWLEDGMENTS

This work was supported by Israel Science Foundation (ISF) Grants Nos. 1822/12 and 1367/12 to Y.G.

REFERENCES

Arnett, T. R. (2010). Acidosis, hypoxia and bone. *Archives of Biochemistry and Biophysics*, *503*, 103–109.

Bab, I. A. (1995). Postablation bone marrow regeneration: An in vivo model to study differential regulation of bone formation and resorption. *Bone*, *17*, 437S–441S.

Bakhshi, H., Kazemian, G., Emami, M., Nemati, A., Karimi Yarandi, H., & Safdari, F. (2013). Local erythropoietin injection in tibiofibular fracture healing. *Trauma Monthly*, *17*, 386–388.

Bar-Shavit, Z. (2007). The osteoclast: A multinucleated, hematopoietic-origin, bone-resorbing osteoimmune cell. *Journal of Cellular Biochemistry*, *102*, 1130–1139.

Betsch, M., Thelen, S., Santak, L., Herten, M., Jungbluth, P., Miersch, D., et al. (2014). The role of erythropoietin and bone marrow concentrate in the treatment of osteochondral defects in mini-pigs. *PloS One*, *9*, e92766.

Brines, M., & Cerami, A. (2012). The receptor that tames the innate immune response. *Molecular Medicine*, *18*, 486–496.

Brines, M., Grasso, G., Fiordaliso, F., Sfacteria, A., Ghezzi, P., Fratelli, M., et al. (2004). Erythropoietin mediates tissue protection through an erythropoietin and common β-subunit heteroreceptor. *Proceedings of the National Academy of Sciences of the United States of America*, *101*, 14907–14912.

Chang, J., Jackson, S. G., Wardale, J., & Jones, S. W. (2014). Hypoxia modulates the phenotype of osteoblasts isolated from knee osteoarthritis patients, leading to undermineralized bone nodule formation. *Arthritis & Rheumatology (Hoboken, N.J.)*, *66*, 1789–1799.

Dewamitta, S. R., Russell, M. R., Nandurkar, H., & Walkley, C. R. (2013). Darbepoietin-alfa has comparable erythropoietic stimulatory effects to recombinant erythropoietin whilst preserving the bone marrow microenvironment. *Haematologica*, *98*, 686–690.

Elliott, S., Swift, S., Busse, L., Scully, S., Van, G., Rossi, J., et al. (2013). Epo receptors are not detectable in primary human tumor tissue samples. *PloS One*, *8*, e68083.

Eltzschig, H. K., & Carmeliet, P. (2011). Hypoxia and inflammation. *The New England Journal of Medicine*, *364*, 656–665.

Fattore, A. D., Teti, A., & Rucci, N. (2012). Bone cells and the mechanisms of bone remodelling. *Frontiers in Bioscience (Elite Edition)*, *4*, 2302–2321.

Florencio-Silva, R., Sasso, G. R., Sasso-Cerri, E., Simões, M. J., & Cerri, P. S. (2015). Biology of bone tissue: Structure, function, and factors that influence bone cells. *BioMed Research International*, *2015*, 421746.

Franke, K., Gassmann, M., & Wielockx, B. (2013). Erythrocytosis: The HIF pathway in control. *Blood*, *122*, 1122–1128.

Garcia, P., Speidel, V., Scheuer, C., Laschke, M. W., Holstein, J. H., Histing, T., et al. (2011). Low dose erythropoietin stimulates bone healing in mice. *Journal of Orthopaedic Research*, *29*, 165–172.

Gerber, H. P., Vu, T. H., Ryan, A. M., Kowalski, J., Werb, Z., & Ferrara, N. (1999). VEGF couples hypertrophic cartilage remodeling, ossification and angiogenesis during endochondral bone formation. *Nature Medicine*, *5*, 623–628.

Guo, L., Luo, T., Fang, Y., Yang, L., Wang, L., Liu, J., et al. (2012). Effects of erythropoietin on osteoblast proliferation and function. *Clinical and Experimental Medicine*, *14*, 69–76.

Gurevitch, O., Khitrin, S., Valitov, A., & Slavin, S. (2007). Osteoporosis of hematologic etiology. *Experimental Hematology*, *35*, 128–136.

Hiram-Bab, S., Liron, T., Deshet-Unger, N., Mittelman, M., Gassmann, M., Rauner, M., et al. (2015). Erythropoietin directly stimulates osteoclast precursors and induces bone loss. *The FASEB Journal*, *29*, 1890–1900.

Hiram-Bab, S., Neumann, D., & Gabet, Y. (2017). Erythropoietin in bone—Controversies and consensus. *Cytokine*, *89*, 155–159.

Holstein, J. H., Menger, M. D., Scheuer, C., Meier, C., Culemann, U., Wirbel, R. J., et al. (2007). Erythropoietin (EPO): EPO-receptor signaling improves early endochondral ossification and mechanical strength in fracture healing. *Life Sciences*, *80*, 893–900.

Holstein, J. H., Orth, M., Scheuer, C., Tami, A., Becker, S. C., Garcia, P., et al. (2011). Erythropoietin stimulates bone formation, cell proliferation, and angiogenesis in a femoral segmental defect model in mice. *Bone*, *49*, 1037–1045.

Jelkmann, W. (2011). Regulation of erythropoietin production. *The Journal of Physiology*, *589*, 1251–1258.

Jewell, U. R., & Gassmann, M. (2001). Mammalian gene expression in hypoxic conditions. *Zoology*, *104*, 192–197.

Kim, J., Jung, Y., Sun, H., Joseph, J., Mishra, A., Shiozawa, Y., et al. (2012). Erythropoietin mediated bone formation is regulated by mTOR signaling. *Journal of Cellular Biochemistry*, *113*, 220–228.

Li, C., Shi, C., Kim, J., Chen, Y., Ni, S., Jiang, L., et al. (2015). Erythropoietin promotes bone formation through EphrinB2/EphB4 signaling. *Journal of Dental Research*, *94*, 455–463.

Lifshitz, L., Tabak, G., Gassmann, M., Mittelman, M., & Neumann, D. (2010). Macrophages as novel target cells for erythropoietin. *Haematologica*, *95*, 1823–1831.

Lipsic, E., Schoemaker, R. G., van der Meer, P., Voors, A. A., van Veldhuisen, D. J., & van Gilst, W. H. (2006). Protective effects of erythropoietin in cardiac ischemia: From bench to bedside. *Journal of the American College of Cardiology*, *48*, 2161–2167.

Liu, Q. S., Cheng, Z. W., Xiong, J. G., Cheng, S., He, X. F., & Li, X. C. (2015). Erythropoietin pretreatment exerts anti-inflammatory effects in hepatic ischemia/reperfusion-injured rats via suppression of the TLR2/NF-κB pathway. *Transplantation Proceedings*, *47*, 283–289.

Lu, J., Yao, Y. Y., Dai, Q. M., Ma, G. S., Zhang, S. F., Cao, L., et al. (2012). Erythropoietin attenuates cardiac dysfunction by increasing myocardial angiogenesis and inhibiting interstitial fibrosis in diabetic rats. *Cardiovascular Diabetology*, *11*, 105.

Maxwell, P., Melendez-Rodríguez, F., Matchett, K. B., Aragones, J., Ben-Califa, N., Jaekel, H., et al. (2015). Novel antibodies directed against the human erythropoietin receptor: Creating a basis for clinical implementation. *British Journal of Haematology*, *168*, 429–442.

Mihmanli, A., Dolanmaz, D., Avunduk, M. C., & Erdemli, E. (2009). Effects of recombinant human erythropoietin on mandibular distraction osteogenesis. *Journal of Oral and Maxillofacial Surgery*, *67*, 2337–2343.

Nakamura, S., Sho, M., Koyama, F., Ueda, T., Nishigori, N., Inoue, T., et al. (2015). Erythropoietin attenuates intestinal inflammation and promotes tissue regeneration. *Scandinavian Journal of Gastroenterology*, *50*, 1094–1102.

Nakano, M., Satoh, K., Fukumoto, Y., Ito, Y., Kagaya, Y., Ishii, N., et al. (2007). Important role of erythropoietin receptor to promote VEGF expression and angiogenesis in peripheral ischemia in mice. *Circulation Research*, *100*, 662–669.

Oikonomidou, P. R., Casu, C., Yang, Z., Crielaard, B., Shim, J. H., Rivella, S., et al. (2016). Polycythemia is associated with bone loss and reduced osteoblast activity in mice. *Osteoporosis International*, *27*, 1559–1568.

Pankratova, S., Kiryushko, D., Sonn, K., Soroka, V., Køhler, L. B., Rathje, M., et al. (2010). Neuroprotective properties of a novel, non-haematopoietic agonist of the erythropoietin receptor. *Brain*, *133*, 2281–2294.

Park-Min, K. H., Lim, E., Lee, M. J., Park, S. H., Giannopoulos, E., Yarilina, A., et al. (2014). Inhibition of osteoclastogenesis and inflammatory bone resorption by targeting BET proteins and epigenetic regulation. *Nature Communications*, *5*, 5418.

Patel, J. J., Modes, J. E., Flanagan, C. L., Krebsbach, P., Edwards, S. P., & Hollister, S. (2015). Dual delivery of EPO and BMP2 from a novel modular poly-ε-caprolactone construct to increase the bone formation in prefabricated bone flaps. *Tissue Engineering. Part C, Methods*, *21*, 889–897.

Perren, S. M. (2008). Fracture healing. The evolution of our understanding. *Acta Chirurgiae Orthopaedicae et Traumatologiae Cechoslovaca*, *75*, 241–246.

Rauner, M., Franke, K., Murray, M., Singh, R. P., Hiram-Bab, S., Platzbecker, U., et al. (2016). Increased EPO levels are associated with bone loss in mice lacking PHD2 in EPO-producing cells. *Journal of Bone and Mineral Research*, *31*, 1877–1887.

Rex, T. S., Wong, Y., Kodali, K., & Merry, S. (2009). Neuroprotection of photoreceptors by direct delivery of erythropoietin to the retina of the retinal degeneration slow mouse. *Experimental Eye Research*, *89*, 735–740.

Richmond, T. D., Chohan, M., & Barber, D. L. (2005). Turning cells red: Signal transduction mediated by erythropoietin. *Trends in Cell Biology*, *15*, 146–155.

Robach, P., Fulla, Y., Westerterp, K. R., & Richalet, J. P. (2004). Comparative response of EPO and soluble transferrin receptor at high altitude. *Medicine and Science in Sports and Exercise*, *36*, 1493–1498. discussion 1492.

Rölfing, J., Baatrup, A., Stiehler, M., Jensen, J., Lysdahl, H., & Bünger, C. (2014). The osteogenic effect of erythropoietin on human mesenchymal stromal cells is dose-dependent and involves non-hematopoietic receptors and multiple intracellular signaling pathways. *Stem Cell Reviews and Reports*, *10*, 69–78.

Rölfing, J. H. D., Bendtsen, M., Jensen, J., Stiehler, M., Foldager, C. B., Hellfritzsch, M. B., et al. (2012). Erythropoietin augments bone formation in a rabbit posterolateral spinal fusion model. *Journal of Orthopaedic Research*, *30*, 1083–1088.

Rölfing, J. H. D., Jensen, J., Jensen, J. N., Greve, A.-S., Lysdahl, H., Chen, M., et al. (2014). A single topical dose of erythropoietin applied on a collagen carrier enhances calvarial bone healing in pigs. *Acta Orthopaedica*, *85*, 201–209.

Shiozawa, Y., Jung, Y., Ziegler, A. M., Pedersen, E. A., Wang, J., Wang, Z., et al. (2010). Erythropoietin couples hematopoiesis with bone formation. *PloS One*, *5*, e10853.

Sims, N. A., & Martin, T. J. (2015). Coupling signals between the osteoclast and osteoblast: How are messages transmitted between these temporary visitors to the bone surface? *Frontiers in Endocrinology*, *6*, 41.

Singbrant, S., Russell, M. R., Jovic, T., Liddicoat, B., Izon, D. J., Purton, L. E., et al. (2011). Erythropoietin couples erythropoiesis, B-lymphopoiesis, and bone homeostasis within the bone marrow microenvironment. *Blood*, *117*, 5631–5642.

Sun, H., Jung, Y., Shiozawa, Y., Taichman, R., & Krebsbach, P. (2012). Erythropoietin modulates the structure of bone morphogenetic protein 2-engineered cranial bone. *Tissue Engineering. Part A*, *18*, 2095–2105.

Tobalem, M., Harder, Y., Rezaeian, F., & Wettstein, R. (2013). Secondary burn progression decreased by erythropoietin. *Critical Care Medicine*, *41*, 963–971.

Tsai, P. T., Ohab, J. J., Kertesz, N., Groszer, M., Matter, C., Gao, J., et al. (2006). A critical role of erythropoietin receptor in neurogenesis and post-stroke recovery. *The Journal of Neuroscience*, *26*, 1269–1274.

Tsiridis, E., Upadhyay, N., & Giannoudis, P. (2007). Molecular aspects of fracture healing: Which are the important molecules? *Injury*, *38*, S11–S25.

van Patot, M. C., & Gassmann, M. (2011). Hypoxia: Adapting to high altitude by mutating EPAS-1, the gene encoding HIF-2alpha. *High Altitude Medicine & Biology*, *12*, 157–167.

Wan, L., Zhang, F., He, Q., Tsang, W. P., Lu, L., Li, Q., et al. (2014). EPO promotes bone repair through enhanced cartilaginous callus formation and angiogenesis. *PloS One*, *9*, e102010.

Wang, Z., Vogel, O., Kuhn, G., Gassmann, M., & Vogel, J. (2013). Decreased stability of erythroblastic islands in integrin β3-deficient mice. *Physiological Reports*, *1*, e00018.

Watowich, S. S. (2011). The erythropoietin receptor: Molecular structure and hematopoietic signaling pathways. *Journal of Investigative Medicine*, *59*, 1067–1072.

Westenbrink, B. D., Lipsic, E., van der Meer, P., van der Harst, P., Oeseburg, H., Du Marchie Sarvaas, G. J., et al. (2007). Erythropoietin improves cardiac function through endothelial progenitor cell and vascular endothelial growth factor mediated neovascularization. *European Heart Journal*, *28*, 2018–2027.

Zhao, C., Irie, N., Takada, Y., Shimoda, K., Miyamoto, T., Nishiwaki, T., et al. (2006). Bidirectional ephrinB2-EphB4 signaling controls bone homeostasis. *Cell Metabolism*, *4*, 111–121.

CHAPTER TEN

Erythropoietin-Mediated Neuroprotection in Insects Suggests a Prevertebrate Evolution of Erythropoietin-Like Signaling

Ralf Heinrich[1], Verena Günther, Natasa Miljus
Institute for Zoology and Anthropology, Georg-August-University Goettingen, Goettingen, Germany
[1]Corresponding author: e-mail address: rheinri1@gwdg.de

Contents

Abstract

The cytokine erythropoietin (Epo) mediates protective and regenerative functions in mammalian nervous systems via activation of poorly characterized receptors that differ from the "classical" homodimeric Epo receptor expressed on erythroid progenitor cells. Epo genes have been identified in vertebrate species ranging from human to fish, suggesting that Epo signaling evolved earlier than the vertebrate lineage.

Studies on insects (*Locusta migratoria*, *Chorthippus biguttulus*, *Tribolium castaneum*) revealed Epo-mediated neuroprotection and neuroregeneration. Recombinant human Epo (rhEpo) prevents apoptosis by binding to a janus kinase-associated receptor, stimulation of STAT transcription factors, and generation of factors that prevent the activation of proapoptotic caspases. Insect neurons were also protected by a neuroprotective but nonerythropoietic Epo splice variant, suggesting similarity with mammalian neuroprotective but not with homodimeric "classical" Epo receptors. Additionally, rhEpo promotes the regeneration of neurites in primary cultured insect brain neurons and after nerve crush in an in vivo preparation. In contrast to neuroprotective and regenerative effects shared with mammalian species, no evidence for a role of Epo signaling in the regulation of neuro- or gliogenesis was found in insects.

Similar structural and functional characteristics of the Epo binding receptors, partly shared transduction pathways that prevent apoptosis and the functional implication in

Vitamins and Hormones, Volume 105
ISSN 0083-6729
http://dx.doi.org/10.1016/bs.vh.2017.02.004
© 2017 Elsevier Inc.
All rights reserved.

neuroprotective and neuroregenerative processes in both mammalian and insect species, suggest that Epo-like signaling was already established in their last common ancestor. Originally functioning as a tissue-protective response to unfavorable physiological situations, cell injury, and pathogen invasion, Epo was later adapted as a humoral regulator of erythropoiesis in the vertebrate lineage.

1. ERYTHROPOIETIN IN VERTEBRATES

Erythropoietin (Epo) is a cytokine with pleiotropic biological functions. As a circulating hormone released by liver or kidneys in vertebrates, Epo promotes the production of erythrocytes in the bone marrow by interfering with apoptotic processes in erythroid progenitor cells to enhance oxygen delivery to the entire organism (reviewed by Constantinescu, Ghaffari, & Lodish, 1999; Jelkmann & Metzen, 1996; Lundby & Olsen, 2011). In addition to its role in hematopoiesis, Epo initiates adaptive responses to physiological and pathophysiological challenges in various nonhematopoietic tissues including for instance reproductive organs, muscle myoblasts, retina, pancreatic islets, adipocytes in white adipose tissue, and the nervous system (reviewed by Ghezzi & Brines, 2004; Juul, Anderson, Li, & Christensen, 1998; Wang, Di, & Noguchi, 2014). In vertebrate nervous systems, Epo is mainly expressed by astrocytes and also by oligodendrocytes, neurons, and endothelial cells (review: Ogunshola & Bogdanova, 2013). Epo signaling is crucial for mammalian brain development (Genc, Koroglu, & Genc, 2004; Yu et al., 2002) and enhances cognitive functions in adult rodents and humans (Adamcio et al., 2008; El-Kordi, Radyushkin, & Ehrenreich, 2009). Epo acts neuroprotectively after hypoxic/ischemic insults and during glutamate excitotoxicity (Bernaudin et al., 1999; Celik et al., 2002; Morishita, Masuda, Nagao, Yasuda, & Sasaki, 1997), suppresses neuroinflammatory processes including activation of microglia (Bond & Rex, 2014; Mitkovski et al., 2015), and promotes regeneration after axonal damage (Kretz, Happold, Marticke, & Isenmann, 2005; Zhang et al., 2009).

The "classical" Epo receptor (EpoR) is a cytokine class I receptor that transduces the ligand binding signal via intracellular association with tyrosine kinases of the janus kinase family (Constantinescu et al., 1999). EpoR is expressed by erythroid progenitor cells and also in various other tissues outside the hematopoietic system. In the nervous system, neurons, astrocytes, oligodendrocytes, microglia, and endothelial cells express EpoR (Marti et al., 1996; Nagai et al., 2001; Sugawa, Sakurai, Ishikawa-Ieda, Suzuki, & Asou, 2002). However, while Epo stimulates erythroid progenitor cells via binding to a preformed EpoR homodimer (Livnah et al., 1999), its cytoprotective effects on various nonhematopoietic cell types are mediated by a different receptor. With respect to the nervous system various Epo derivatives (e.g., asialo-Epo, carbamylated Epo, Epobis), the molecularly unrelated peptide ARA 290 and the human Epo splice variant EV3 were shown to be neuroprotective without stimulating erythropoiesis (Bonnas, 2009; Brines et al., 2008; Erbayraktar et al., 2003; Leist et al., 2004; Pankratova et al., 2012). It has been suggested that the neuroprotective Epo receptor could be a heteromer consisting of EpoR and one or more different proteins (e.g., beta common chain receptor (Brines et al., 2004)) or a completely different receptor (e.g., ephrin B4 receptor (Pradeep et al., 2015)). However, the expression of EpoR and/or alternative Epo receptors may depend on cell type, developmental stage, physiological condition, and previous exposure to challenging stimuli (Brines & Cerami, 2005; Sanchez et al., 2009; Shein, Horowitz, Alexandrovich, Tsenter, & Shohami, 2005; Sinor & Greenberg, 2000; Um, Gross, & Lodish, 2007). Since different challenges require different cellular adaptations, different Epo receptors that couple to particular intracellular signaling pathways could be expressed as a flexible response.

2. ERYTHROPOIETIN-LIKE SIGNALING IN INSECTS

The last common ancestor of vertebrates (including mammals) and arthropods (including insects), the so-called urbilaterian, lived between 700 and 600 million years ago (Peterson, Cotton, Gehling, & Pisani, 2008). According to a scenario that is favored in the majority of recent publications, the urbilaterian contained more than 7000 genes (Wyder, Kriventseva, Schröder, Kadowaki, & Zdobnov, 2007) and possessed a centralized nervous system that already contained the precursors of brain centers, including neurosecretory systems, present in extant protostomes and deuterostomes (Tessmar-Raible, 2007; Tomer, Denes, Tessmar-Raible, & Arendt, 2010; Wirmer, Bradler, & Heinrich, 2012). While insects deliver oxygen via tracheae and rely on innate responses to pathogenic invaders, vertebrates evolved various types of blood cells, including erythrocytes for oxygen transport and immune cells for specific immune responses (review: Jiravanichpaisal, Lee, & Söderhäll, 2006). However, the discovery of Epo functions outside the hematopoietic system and the identification of *Epo* and *EpoR* genes in all vertebrates from fish to human led to the hypothesis of a prevertebrate evolution of Epo signaling in the context of innate immunity processes (Brines & Cerami, 2005; Ghezzi & Conklin, 2013).

2.1 Epo-Mediated Neuroprotection in Insects

Although sequence comparison revealed no orthologs of Epo and EpoR in any of the available insect genomes, neuroprotective and regenerative effects of recombinant human Epo (rhEpo), similar to those described in mammalian nervous systems, were observed in insects. Studies on primary cultures of locust (*Locusta migratoria*) brain cells found that rhEpo increased the survival of neurons but decreased the survival of glia (Fig. 1A and B), demonstrating a cell type-specific cellular protection mediated by Epo (Ostrowski, Ehrenreich, & Heinrich, 2011). The neuroprotective effect in locusts was dose dependent with a most beneficial concentration of 4 U/mL and with both higher and lower concentrations being less effective (Fig. 1C). Primary cultured brain neurons of the beetle *Tribolium castaneum* were best protected by much lower concentrations of rhEpo of only 0.1 U/mL. Higher concentrations of rhEpo, including the most protective concentration for locust neurons, even decreased neuronal survival. Similar optimum-type dose responses and opposite effects on the survival of neurons and astrocytes have previously been reported in rats (Keswani et al., 2004; Sinor & Greenberg, 2000; Weishaupt et al., 2004), while astroglial survival was supported in another study with different experimental conditions (Diaz, Assaraf, Miller, & Schipper, 2005). These examples from insects and mammals show that beneficial effects of Epo on cell survival depend on cell type, physiological state, dosage of Epo, and type of physiological challenge.

Insects typically tolerate hypoxic conditions much better than mammalian species (reviewed by Schmitz & Harrison, 2004). Only prolonged periods of hypoxia induce significant apoptotic cell death, partly mediated by activation of caspase 3 (Fig. 1D), in primary cultures of locust brain neurons. Apoptotic cell death induced by 36-h hypoxia (O_2 <2%) was completely prevented by rhEpo applied to the cultured neurons at 12 h before onset of the hypoxic stimulus (Fig. 1E). Prevention of neuronal death by rhEpo was apparently achieved by interference with apoptotic processes as indicated by the reduction of anti-cleaved caspase 3-immunoreactive cells in the presence of rhEpo (Fig. 1F). Epo-mediated protection of erythroid progenitors, neurons, and other mammalian cell types also relies on

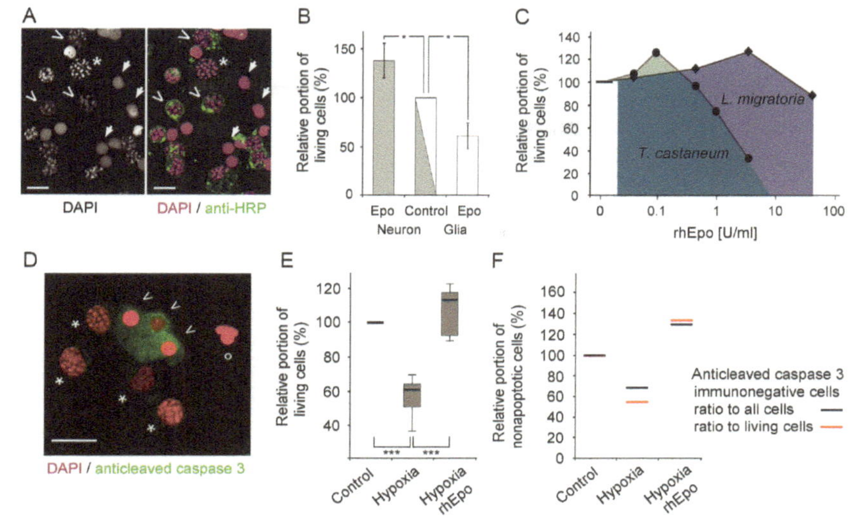

Fig. 1 Epo-mediated neuroprotection in insects. (A) DAPI nuclear stain and anti-HRP immunocytochemistry in primary cultured locust brain cells. Anti-HRP-positive intact neuron (V), anti-HRP-negative intact glia cell (*), condensed nuclei of dead cells (*arrows*). (B) rhEpo (4 U/mL) increases the survival of locust brain neurons but not of glia during 3 days in vitro. Portions of viable neurons and glia cells were normalized to portions of viable cells of the respective cell type in serum-free controls. (C) Survival of locust (*L. migratoria*) and beetle (*T. castaneum*) primary cultured brain neurons incubated with different concentrations of rhEpo. Portions of viable neurons were normalized to viable neurons in serum-free cultures. (D) Locust brain neurons in vitro labeled with DAPI and by anti-cleaved caspase 3 immunocytochemistry. Anti-cleaved caspase 3-immunonegative intact neurons (*), cleaved caspase 3-immunopositive dead or dying cells (V), nuclei of dead cells (O). (E) rhEpo (4 U/mL) prevents hypoxia-induced (36 h, $O_2 < 2\%$) apoptotic cell death in primary cultures of locust brain neurons ($N = 8$). (F) rhEpo increases the portion of intact, anti-cleaved caspase 3-immunonegative (=nonapoptotic) cells in hypoxia-treated cultures of locust brain neurons (5542 cells evaluated).

interference with apoptotic processes, including the prevention of caspase activation (Li et al., 2015; Schober et al., 2014).

Neuroprotective effects of Epo in mammals are mediated by intracellular signaling pathways that are largely similar to those activated in erythroid progenitor cells. Transduction starts with the phosphorylation of intracellular domains of the activated EpoRs by janus kinase 2 (JAK2) which enables activation of downstream signaling pathways. These include signal transducer and activator of transcription 5 (STAT5), phosphoinositol-3-kinase/protein kinase B (PI3K/Akt), mitogen-activated protein kinase, and nuclear factor kappa B (NFκB, not in erythroid cells) (Chong, Kang, & Maiese, 2003; Digicaylioglu & Lipton, 2001; Sirén et al., 2001). These pathways either activate transcription factors that promote the expression of antiapoptotic proteins or directly interfere with apoptotic processes. Orthologs of components of these transduction pathways are also present in insects (Miljus et al., 2014). Inhibitors that were shown to interfere with particular steps in insect signaling pathways were tested for their ability to prevent rhEpo-mediated protection of hypoxia-treated locust brain neurons. Antiapoptotic effects of rhEpo were abolished by coincubation of the JAK inhibitor AG490 and the STAT inhibitor sc-355979 (Fig. 2A and B), indicating that rhEpo mediates its antiapoptotic effects via activation of JAK/STAT signaling in locust neurons. In contrast, Epo-mediated neuroprotection was not affected by inhibition of PI3K and NFκB signaling (Fig. 2C and D), suggesting that these transduction pathways are not involved. Whether the exclusive requirement of JAK/STAT signaling is a particular characteristic of the majority of locust brain neurons or a general feature of insect neurons remains to be shown.

Activation of the transcription factor STAT and reduction of caspase 3 activation suggested that rhEpo stimulates expression of antiapoptotic genes in insect neurons. While inhibition of transcription provided inconclusive results, suppression of translation with anisomycin prevented the antiapoptotic effect of rhEpo on hypoxia-treated locust brain neurons (Fig. 3). The absence of additional negative effects on the survival of hypoxia-treated neurons suggested a rather specific interference with

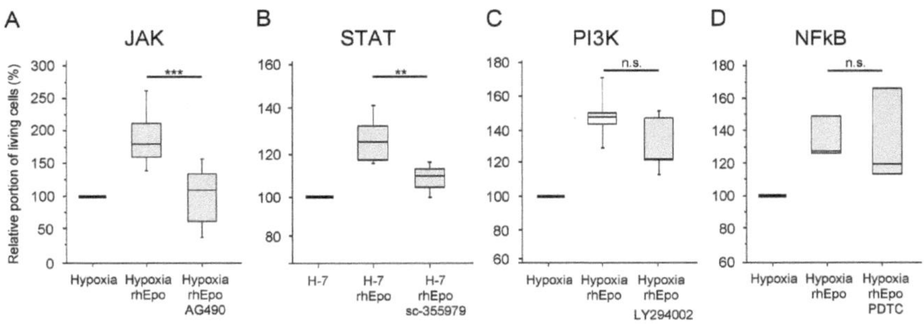

Fig. 2 rhEpo-mediated neuroprotection of primary cultured locust brain neurons involves (A) janus kinase (JAK, inhibitor: AG490, $N = 13$) and (B) signal transducer and activator of transcription (STAT, inhibitor: sc-355979, $N = 7$) but not (C) phosphoinositol-3-kinase (PI3K, inhibitor: LY294002, $N = 5$) and (D) nuclear factor kappa B (NFκB, inhibitor: PDTC, $N = 3$). Portions of viable cells were normalized to numbers of surviving neurons in cultures challenged with hypoxia (O_2 <2%, 36h) or H-7 (50μM, 18h). Nonparametric Kruskal–Wallis ANOVA with Mann–Whitney post hoc analysis, $^{***}P < 0.001$, $^{**}P < 0.01$.

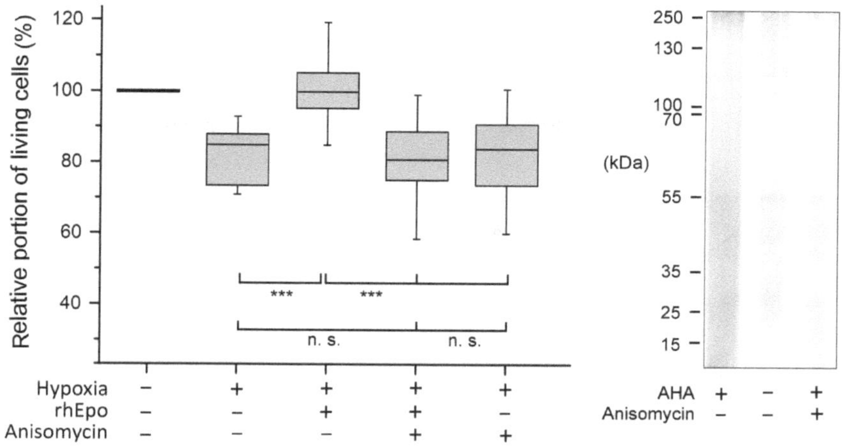

Fig. 3 *Left*: rhEpo-mediated neuroprotection of hypoxia-treated locust brain neurons depends on translation. 10^{-5} M anisomycin abolishes the antiapoptotic effect of rhEpo (4 U/mL) on hypoxia-treated neurons. *Right*: Translated proteins were labeled by the methionine analogue L-azidohomoalanine (AHA) that was incorporated into newly synthesized protein during 5 h incubation and subsequently visualized by click-chemistry reaction with fluorescent TAMRA alkyne. TAMRA-associated fluorescence revealed strong reduction of newly synthesized labeled protein in the presence of 10^{-5} M anisomycin.

Epo-induced beneficial processes instead of an anisomycin-triggered harmful effect that was not related to Epo-stimulated pathways. A summary of Epo-stimulated mechanisms involved in the protection of insect brain neurons is provided in Fig. 5.

2.2 Epo-Mediated Neuroregeneration in Insects

Studies on various in vitro and in vivo mammalian systems have demonstrated that Epo can promote the regeneration of metabolically or anatomically compromised neurons (Böcker-Meffert et al., 2002; Kretz et al., 2005; Zhong, Yao, Deng, Cheng, & Zhou, 2007). After peripheral nerve transection, Epo accelerated axonal regeneration and improved reestablishment of previous functions (Entezari, Esmaeili, & Yaseri, 2014; Zhang et al., 2015, 2009). Signaling pathways that mediate antiapoptotic and regenerative functions of Epo only partly overlap and both processes may require different optimal concentrations of externally applied Epo (Byts et al., 2008; King et al., 2007; Kretz et al., 2005).

Evidence for a potential role of Epo-like signaling in the regeneration of insect neurons has been provided by in vitro and in vivo studies (Ostrowski et al., 2011). During the process of establishing dissociated cell cultures from locust brains, neurons lose all their dendrites and axons (Fig. 4A). Fresh primary cultures initially contain round neuronal cell bodies that start to regrow neurites within a few hours in vitro (Fig. 4B). Regrowth of neurites was accelerated by incubation with rhEpo (Ostrowski et al., 2011). Both extension of the longest neurite and cumulative length of all neurites significantly increased in the presence of rhEpo compared to serum-free control cultures

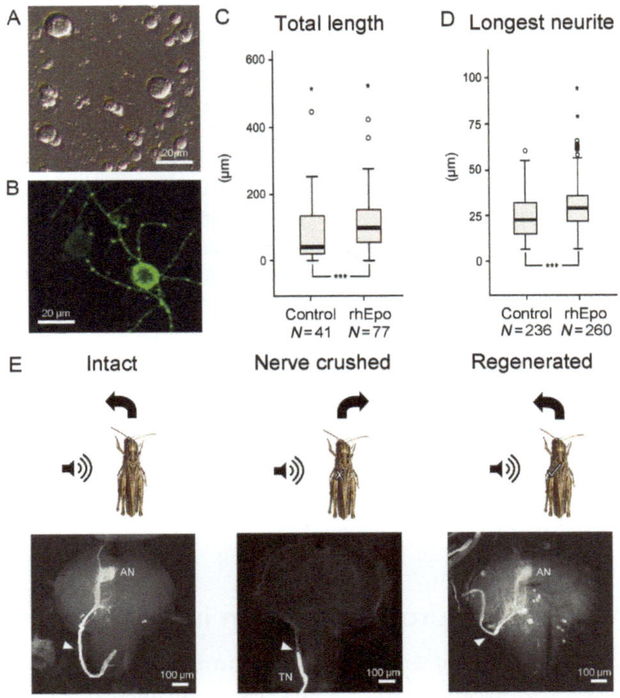

Fig. 4 Epo promotes the regeneration of neurites in insect neurons. (A) Locust brain neurons in cell culture immediately after dissociation. (B) Anti-HRP-labeled neuron with regenerated neurites after 5 days in cell culture. (C and D) rhEpo (4 U/mL) increases the total length of all neurites (C) and the extension of the longest neurite (D) after 5 days in vitro in comparison to serum-free cultured control neurons. Mann–Whitney *U* test for independent samples, $^{***}P < 0.001$. (E) Unilateral auditory nerve (*arrowheads*) crush transects centrally projecting axons of auditory receptor cells and prevents orientation toward attractive sound signals presented to the injured side. rhEpo accelerates the regeneration of axonal projections into the auditory neuropil (AN) and improves the reestablishment of correct orientation toward acoustic stimuli.

derived from the same pool of locust brain neurons (Fig. 4C), while diameters of neuronal cell bodies were not different. A neuroregenerative function of rhEpo in insects was also demonstrated in vivo (Fig. 4D). Male grasshoppers of the species *C. biguttulus* are attracted to songs of conspecific females. Males compare auditory input of the right and left ear and orient toward the side with the louder input, thereby approaching the female or a speaker that presents a recording of a female song. Crushing of the auditory nerve transects auditory receptor axons and prevents auditory input from the injured side, making the male always turn to the intact side even when sound stimuli are presented from the injured side. Previous studies have demonstrated that injured auditory receptor cells can regenerate their centrally projecting axons and reestablish correct orientation toward acoustic signals (Lakes-Harlan & Pfahlert, 1995). A single high-dose systemic application of rhEpo right after auditory nerve crush reduced the regeneration period until the first correct orientations to the injured side appeared and increased the proportion of correct turns throughout a period of 20 days with daily behavioral tests (Ostrowski et al., 2011). Backfill experiments suggested that more auditory receptor axons regenerated their central projections into the auditory neuropil, where they contact auditory interneurons that process information about sound source localization.

In mammalian nervous systems, Epo supports the reestablishment of lost functions by targeting injured neurons directly (e.g., stimulating the regeneration of transected axons (Zhang et al., 2009)) and by acting on other cell types like neural and glial progenitors, mature glia, or endothelial cells. In this respect, Epo can stimulate the proliferation of neuronal and oligodendrocyte precursors, induce differentiation of existing pools of immature cells, reduce microglial activation and inflammation, and support vascularization of damaged brain regions (Brines & Cerami, 2012; Byts & Siren, 2009; Kaneko, Kako, & Sawamoto, 2013; Mitkovski et al., 2015). Various studies reported stimulatory effects of Epo on the proliferation, migration, and differentiation of neural progenitor cells (Hassouna et al., 2016; Kaneko et al., 2013; Leconte et al., 2011; Ransome & Turnley, 2007; Shingo, Sorokan, Shimazaki, & Weiss, 2001; Wang, Zhang, Wang, Zhang, & Chopp, 2004). A potential effect of rhEpo on the proliferation of insect neurons and glia was investigated in several different species. Systemic application of rhEpo had no stimulatory effect on adult neurogenesis in cricket (*Gryllus bimaculatus*) and beetle (*T. castaneum*) mushroom bodies and on neurogenesis in the developing optic lobes of locusts (*L. migratoria*). No effect on the rate of proliferation was also observed in neuronal cell lines derived from the fruit fly *Drosophila melanogaster* and the moth *Manduca sexta*. In contrast to shared functions in neuroprotection and neuroregeneration between insects and mammals, there is so far no evidence for a role of Epo-like signaling in the regulation of insect neuro- or gliogenesis.

3. THE EVOLUTIONARY ORIGIN OF EPO-LIKE SIGNALING

Insects represent the most speciose group of animals that evolved ~480 million years ago (Misof et al., 2014; Tong, Duchene, Ho, & Lo, 2015). Together with other invertebrates they share innate immune responses (e.g., wound repair, phagocytosis, encapsulation) with vertebrates but lack adaptive immune responses mediated by antibodies and lymphocytes. They employ particular receptors (including Toll-like and Nod-like receptors) to recognize pathogens and damaged cells and can initiate complex cascades of cellular reactions to inactivate or eliminate invading cells and support tissue repair (reviews by Beschin, Bilej, Torreele, & De Baetselier, 2001; Buchmann, 2014). These cellular responses are regulated by cytokines that can interact with specific cytokine receptors and/or with carbohydrate patterns on cell surfaces via lectin-like binding domains (Ghezzi & Conklin, 2013; Zanetta et al., 1996). Various cytokines and cytokine-like factors have been identified in insects, but identification of orthologs between different species, especially in comparison with vertebrate species, was largely unsuccessful due to poor gene and protein primary sequence conservation (Foster, Parrish-Novak, Fox, & Xu, 2004; Liongue & Ward, 2007). Class I cytokine receptors (including EpoR) typically possess an extracellular cytokine receptor homology domain for ligand binding and intracellular motifs required for interaction with JAK (Boulay, O'Shea, & Paul, 2003). However, functional studies are generally complicated by pleiotropic effects of cytokines, binding of multiple cytokines to particular receptors, binding of particular cytokines to several receptors, potential variability in the formation of heteromeric cytokine receptors, and convergence of intracellular transduction pathways (Liongue & Ward, 2007). Orthologs of all typical transduction pathways for Epo signaling in vertebrates have been identified in insects and other invertebrates (Miljus et al., 2014).

Sequence comparison has not identified genes for *Epo* and *EpoR* in any insect species, though transcription of an orthologous *EpoR* gene has been reported in the nervous system of the mussel *Mytilus edulis* (Gerber et al., 2007). *Epo* genes have been identified in all major groups of vertebrates.

Their primary sequence similarity decreases with increasing phylogenetical distance from 99% between human and common chimpanzee, to 80% between human and mouse, to 36%–32% between human and various fish species (Ostrowski et al., 2011). However, the spatial organization of a few critical amino acid residues rather than similarity of the complete sequence of a cytokine is important for receptor binding. Therefore, comparison of gene and protein sequences may not lead to the identification of endogenous "Epo-like" ligands in insects and other invertebrates. Cell-protective functions in mammalian nervous systems and other tissues are (at least partly) mediated by Epo receptors that differ from the homodimeric classical EpoR involved in hematopoiesis (Brines & Cerami, 2005). Various candidates that could mediate tissue-protective effects as homomeric or part of heteromeric receptors have been identified, including βcR (Brines et al., 2004), ephrin receptor Epbh4 receptor (Pradeep et al., 2015), and leukemia inhibitory factor LIFR (Bonnas, 2009). Most likely, different cell types may express different types of EpoRs, depending on their physiological state and the type of insult to mediate appropriate adaptive responses.

Studies on insects from two different phylogenetic subgroups (orthoptera and coleoptera) revealed rhEpo-mediated neuroprotective and neuroregenerative effects similar to those described in mammalian species. On locust neurons rhEpo binds to a JAK-associated receptor which activates the transcription factor STAT that stimulates the expression of antiapoptotic proteins, including factors that interfere with proapoptotic caspase activation (Fig. 5A). In addition, rhEpo promoted neurite growth in vitro and regeneration of compromised functions after nerve crush in vivo by yet unidentified intracellular pathways. Our recent results demonstrated that Epo receptors on locust neurons can be activated by a neuroprotective but nonhematopoietic human Epo splice variant EV-3 (characterized by Bonnas, 2009) and that binding of rhEpo or EV-3 initiates internalization of the receptor/ligand complex (Miljus et al., 2017). Altogether, the described findings provide strong cumulative evidence for the presence of an Epo receptor that shares functional and structural similarities with mammalian neuroprotective Epo receptors. This supports the previously formulated hypothesis that Epo-like signaling evolved earlier than the first vertebrates as an adaptive tissue-protective response to unfavorable physiological situations, cell injury, and pathogen invasion (Brines & Cerami, 2005; Ghezzi & Conklin, 2013; Ostrowski et al., 2011). Neuroprotective functions mediated by Epo and a nonhematopoietic Epo splice variant in both mammals and insects indicate that the molecules involved in

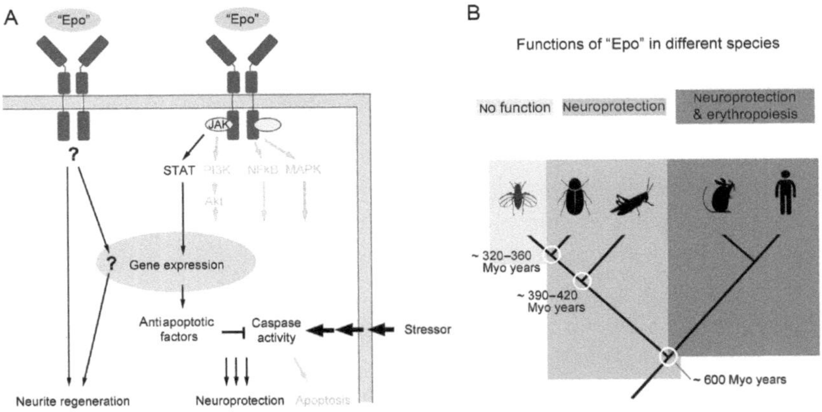

Fig. 5 (A) Mechanisms of Epo-like signaling in insect neurons. Epo or an Epo-like ligand (Epo) acts neuroprotectively via receptors that are functionally and structurally similar to mammalian neuroprotective receptors. Protective effects are mediated via JAK/STAT signaling and synthesis of antiapoptotic factors that interfere with caspase activation. Additionally, "Epo" promotes the regeneration of injured neurites by yet unidentified mechanisms. (B) Evolution of "Epo" functions. Neuroprotective functions of Epo or an Epo-like ligand were already present in the urbilaterian. In the vertebrate lineage, Epo additionally adapted its role as the major regulator of erythropoiesis. Neuroprotective effects of Epo are present in extant orthopteran (locusts and grasshoppers) and coleopteran (flour beetles) but not in dipteran (fruit flies) species.

Epo-stimulated beneficial processes already existed in common ancestors of both evolutionary lineages. However, studies on primary cultured brain neurons and a permanent neuronal cell line of *D. melanogaster* found no evidence for rhEpo-mediated neuroprotection, suggesting that this highly derived insect species has lost the respective neuroprotective functions of Epo-like signaling (Fig. 5B). In fact, *Drosophila* has lost a large number of genes for which orthologs exist in more basal insects and vertebrates (Wyder et al., 2007). Among them is the orphan cytokine type I receptor CRLF3 (cytokine receptor-like factor 3) which is absent in dipteran genomes (e.g., *Drosophila* and *Anopheles*) but expressed in the nervous systems of *T. castaneum*, mice, and humans (Liongue & Ward, 2007; Wyder et al., 2007) and therefore matches with the presence and absence of Epo-mediated neuroprotection in different species. Ongoing studies indicate that CRLF3 could be the EpoR of beetle neurons. Whether other invertebrate species besides insects also employ Epo-like signaling in neuroprotection has not been studied. It seems likely in spite of the shared ancestry of the urbilaterian who may have already possessed Epo-like signaling to adapt its cellular functionality to unfavorable conditions.

REFERENCES

Adamcio, B., Sargin, D., Stradomska, A., Medrihan, L., Gertler, C., Theis, F., et al. (2008). Erythropoietin enhances hippocampal long-term potentiation and memory. *BMC Biology, 6*, 37.

Bernaudin, M., Marti, H. H., Roussel, S., Divoux, D., Nouvelot, A., MacKenzie, E. T., et al. (1999). A potential role for erythropoietin in focal permanent cerebral ischemia in mice. *Journal of Cerebral Blood Flow and Metabolism, 19*, 643–651.

Beschin, A., Bilej, M., Torreele, E., & De Baetselier, P. (2001). On the existence of cytokines in invertebrates. *Cellular and Molecular Life Sciences, 58*, 801–814.

Böcker-Meffert, S., Rosenstiel, P., Röhl, C., Warneke, N., Held-Feindt, J., Sievers, J., et al. (2002). Erythropoietin and VEGF promote neural outgrowth from retinal explants in postnatal rats. *Investigative Ophthalmology & Visual Science, 43*, 2021–2026.

Bond, W. S., & Rex, T. S. (2014). Evidence that erythropoietin modulates neuroinflammation through differential action on neurons, astrocytes, and microglia. *Frontiers in Immunology, 5*, 523.

Bonnas, C. B. (2009). *Identification of erythropoietin isoforms and evaluation of their biological importance.* Berlin: Medical Faculty Charité. http://www.diss.fu-berlin.de/diss/receive/FUDISS_thesis_000000009488.

Boulay, J. L., O'Shea, J. J., & Paul, W. E. (2003). Molecular phylogeny within type I cytokines and their cognate receptors. *Immunity, 19*, 159–163.

Brines, M., & Cerami, A. (2005). Emerging biological roles for erythropoietin in the nervous system. *Nature Reviews Neuroscience, 6*, 484–494.

Brines, M., & Cerami, A. (2012). The receptor that tames the innate immune response. *Molecular Medicine, 18*, 486–496.

Brines, M., Grasso, G., Fiordaliso, F., Sfacteria, A., Ghezzi, P., Fratelli, M., et al. (2004). Erythropoietin mediates tissue protection through an erythropoietin and common beta-subunit heteroreceptor. *Proceedings of the National Academy of Sciences of the United States of America, 101*, 14907–14930.

Brines, M., Patel, N. S., Villa, P., Brines, C., Mennini, T., De Paola, M., et al. (2008). Non-erythropoietic, tissue-protective peptides derived from the tertiary structure of erythropoietin. *Proceedings of the National Academy of Sciences of the United States of America, 105*, 10925–10930.

Buchmann, K. (2014). Evolution of innate immunity: Clues from invertebrates via fish to mammals. *Frontiers in Immunology, 5*, 459.

Byts, N., Samoylenko, A., Fasshauer, T., Ivanisevic, M., Hennighausen, L., Ehrenreich, H., et al. (2008). Essential role for Stat5 in the neurotrophic but not in the neuroprotective effect of erythropoietin. *Cell Death and Differentiation, 15*, 783–792.

Byts, N., & Siren, A. L. (2009). Erythropoietin: A multimodal neuroprotective agent. *Experimental & Translational Stroke Medicine, 1*, 4.

Celik, M., Gökmen, N., Erbayraktar, S., Akhisaroglu, M., Konakc, S., Ulukus, C., et al. (2002). Erythropoietin prevents motor neuron apoptosis and neurologic disability in experimental spinal cord ischemic injury. *Proceedings of the National Academy of Sciences of the United States of America, 99*, 2258–2263.

Chong, Z. Z., Kang, J.-Q., & Maiese, K. (2003). Erythropoietin: Cytoprotection in vascular and neuronal cells. *Current Drug Targets. Cardiovascular & Hematological Disorders, 3*, 141–154.

Constantinescu, S. N., Ghaffari, S., & Lodish, H. F. (1999). The erythropoietin receptor: Structure, activation and intracellular signal transduction. *Trends in Endocrinology and Metabolism, 10*, 18–23.

Diaz, Z., Assaraf, M. I., Miller, W. H., Jr., & Schipper, H. M. (2005). Astroglial cytoprotection by erythropoietin pre-conditioning: Implications for ischemic and degenerative CNS disorders. *Journal of Neurochemistry, 93*, 392–402.

Digicaylioglu, M., & Lipton, S. A. (2001). Erythropoietin-mediated neuroprotection involves cross-talk between Jak2 and NF-kappaB signalling cascades. *Nature, 412*, 641–647.

El-Kordi, A., Radyushkin, K., & Ehrenreich, H. (2009). Erythropoietin improves operant conditioning and stability of cognitive performance in mice. *BMC Biology*, *7*, 37.

Entezari, M., Esmaeili, M., & Yaseri, M. (2014). A pilot study of the effect of intravenous erythropoietin on improvement of visual function in patients with recent indirect traumatic optic neuropathy. *Graefe's Archive for Clinical and Experimental Ophthalmology*, *252*, 1309–1313.

Erbayraktar, S., Grasso, G., Sfacteria, A., Xie, Q. W., Coleman, T., Kreilgaard, M., et al. (2003). Asialoerythropoietin is a nonerythropoietic cytokine with broad neuroprotective activity in vivo. *Proceedings of the National Academy of Sciences of the United States of America*, *100*, 6741–6746.

Foster, D., Parrish-Novak, J., Fox, B., & Xu, W. (2004). Cytokine receptor pairing: Accelerating discovery of cytokine function. *Nature Reviews Drug Discovery*, *3*, 160–170.

Genc, S., Koroglu, T. F., & Genc, K. (2004). Erythropoietin and the nervous system. *Brain Research*, *1000*, 19–31.

Gerber, S., Cadet, P., Sheehan, M., Stefano, G. B., & Mantione, K. J. (2007). Vertebrate interleukins originated in invertebrates? *Invertebrate Survival Journal*, *4*, 95–100.

Ghezzi, P., & Brines, M. (2004). Erythropoietin as an antiapoptotic, tissue-protective cytokine. *Cell Death and Differentiation*, *11*(Supp. 1), 37–44.

Ghezzi, P., & Conklin, D. (2013). Tissue-protective cytokines: Structure and evolution. *Methods in Molecular Biology*, *982*, 43–58.

Hassouna, I., Ott, C., Wüstefeld, L., Offen, N., Neher, R. A., Mitkovski, M., et al. (2016). Revisiting adult neurogenesis and the role of erythropoietin for neuronal and oligodendroglial differentiation in the hippocampus. *Molecular Psychiatry*, *21*, 1752–1767. http://dx.doi.org/10.1038/mp.2015.212..

Jelkmann, W., & Metzen, E. (1996). Erythropoietin in the control of red cell production. *Annals of Anatomy*, *178*, 391–403.

Jiravanichpaisal, P., Lee, B. L., & Söderhäll, K. (2006). Cell-mediated immunity in arthropods: Hematopoiesis, coagulation, melanization and opsonization. *Immunobiology*, *211*, 213–236.

Juul, S. E., Anderson, D. K., Li, Y., & Christensen, R. D. (1998). Erythropoietin and erythropoietin receptor in the developing human central nervous system. *Pediatric Research*, *43*, 40–49.

Kaneko, N., Kako, E., & Sawamoto, K. (2013). Enhancement of ventricular zone-derived nerogenesis and oligodendrogenesis by erythropoietin and its derivates. *Frontiers in Cellular Neuroscience*, *7*, 235.

Keswani, S. C., Buldanlioglu, U., Fischer, A., Reed, N., Polley, M., Liang, H., et al. (2004). A novel endogenous erythropoietin mediated pathway prevents axonal degeneration. *Annals of Neurology*, *56*, 815–826.

King, C. E., Rodger, J., Bartlett, C., Esmaili, T., Dunlop, S. A., & Beazley, L. D. (2007). Erythropoietin is both neuroprotective and neuroregenerative following optic nerve transection. *Experimental Neurology*, *205*, 48–55.

Kretz, A., Happold, C. J., Marticke, J. K., & Isenmann, S. (2005). Erythropoietin promotes regeneration of adult CNS neurons via Jak2/Stat3 and PI3K/AKT pathway activation. *Molecular and Cellular Neurosciences*, *29*, 569–579.

Lakes-Harlan, R., & Pfahlert, C. (1995). Regeneration of axotomized tympanal nerve fibres in the adult grasshopper *Chorthippus biguttulus* (L.) (Orthoptera: Acrididae) correlates with regaining the localization ability. *Journal of Comparative Physiology A*, *176*, 797–807.

Leconte, C., Bihel, E., Lepelletier, F. X., Bouet, V., Saulnier, R., Petit, E., et al. (2011). Comparison of the effects of erythropoietin and its caramylated derivative on behaviour and hippocampal neurogenesis in mice. *Neuropharmacology*, *60*, 354–365.

Leist, M., Ghezzi, P., Grasso, G., Bianchi, R., Villa, P., Fratelli, M., et al. (2004). Derivatives of erythropoietin that are tissue protective but not erythropoietic. *Science*, *305*, 239–242.

Li, K., Cao, R. J., Zhu, X. J., Liu, X. Y., Li, L. Y., & Cui, S. S. (2015). Erythropoietin attenuates the apoptosis of adult neurons after brachial plexus root avulsion by downregulating JNK phosphorylation and c-Jun expression and inhibiting c-PARP cleavage. *Journal of Molecular Neuroscience*, *56*, 917–925.

Liongue, C., & Ward, C. (2007). Evolution of class I cytokine receptors. *BMC Evolutionary Biology*, *7*, 120.

Livnah, O., Stura, E. A., Middleton, S. A., Johnson, D. L., Jolliffe, L. K., & Wilson, I. A. (1999). Crystallographic evidence for preformed dimers of erythropoietin receptor before ligand activation. *Science*, *283*, 987–990.

Lundby, C., & Olsen, N. V. (2011). Effects of recombinant human erythropoietin in normal humans. *The Journal of Physiology*, *589*, 1265–1271.

Marti, H. H., Wenger, R. H., Rivas, L. A., Straumann, U., Digicaylioglu, M., Henn, V., et al. (1996). Erythropoietin gene expression in human, monkey and murine brain. *The European Journal of Neuroscience*, *8*, 666–676.

Miljus, N., Heibeck, S., Jarrar, M., Micke, M., Ostrowski, D., Ehrenreich, H., et al. (2014). Erythropoietin-mediated protection of insect brain neurons involves JAK and STAT but not PI3K transduction pathways. *Neuroscience*, *258*, 218–227.

Miljus, N., Massih, B., Weis, M. A., Rison, J. V., Bonnas, C. B., Sillaber, I., et al. (2017). Neuroprotection and endocytosis: Erythropoietin receptors in insect nervous systems. *Journal of Neurochemistry*. http://dx.doi.org/10.1111/jnc.13967

Misof, B., et al. (2014). Phylogenomics resolves the timing and pattern of insect evolution. *Science*, *346*, 763–767.

Mitkovski, M., Dahm, L., Heinrich, R., Monnheimer, M., Gerhart, S., Stegmüller, J., et al. (2015). Erythropoietin dampens injury-induced microglial motility. *Journal of Cerebral Blood Flow and Metabolism*, *35*, 1233–1236.

Morishita, E., Masuda, S., Nagao, M., Yasuda, Y., & Sasaki, R. (1997). Erythropoietin receptor is expressed in rat hippocampal and cerebral cortical neurons, and erythropoietin prevents in vitro glutamate-induced neuronal death. *Neuroscience*, *76*, 105–116.

Nagai, A., Nakagawa, E., Choi, H. B., Hatori, K., Kobayashi, S., & Kim, S. U. (2001). Erythropoietin and erythropoietin receptors in human CNS neurons, astrocytes microglia, and oligodendrocytes grown in culture. *Journal of Neuropathology and Experimental Neurology*, *60*, 386–392.

Ogunshola, O. O., & Bogdanova, A. Y. (2013). Epo and non-hematopoietic cells: What do we know? *Methods in Molecular Biology*, *982*, 13–41.

Ostrowski, D., Ehrenreich, H., & Heinrich, R. (2011). Erythropoietin promotes survival and regeneration of insect neurons in vivo and in vitro. *Neuroscience*, *188*, 95–108.

Pankratova, S., Gu, B., Kiryushko, D., Korshunova, I., Kohler, L. B., Rathje, M., et al. (2012). A new agonist of the erythropoietin receptor, Epobis, induces neurite outgrowth and promotes neuronal survival. *Journal of Neurochemistry*, *121*, 915–923.

Peterson, K. J., Cotton, J. A., Gehling, J. G., & Pisani, D. (2008). The Ediacaran emergence of bilaterians: Congruence between the genetic and the geological fossil records. *Philosophical Transactions of the Royal Society, B: Biological Sciences*, *363*, 1435–1443.

Pradeep, S., et al. (2015). Erythropoietin stimulates tumor growth via ephb4. *Cancer Cell*, *28*, 610–622.

Ransome, M. I., & Turnley, A. M. (2007). Systemically delivered erythropoietin transiently enhances adult hippocampal neurogenesis. *Journal of Neurochemistry*, *102*, 1953–1965.

Sanchez, P. E., Fares, R. P., Risso, J. J., Bonnet, C., Bouvard, S., Le-Cavorsin, M., et al. (2009). Optimal neuroprotection by erythropoietin requires elevated expression of its receptor in neurons. *Proceedings of the National Academy of Sciences of the United States of America*, *106*, 9848–9853.

Schmitz, A., & Harrison, J. F. (2004). Hypoxic tolerance in air-breathing invertebrates. *Respiratory Physiology & Neurobiology*, *141*, 229–242.

Schober, M. E., Requena, D. F., Block, B., Davis, L. J., Rodesch, C., Casper, T. C., et al. (2014). Erythropoietin improved cognitive function and decreased hippocampal caspase activity in rat pups after traumatic brain injury. *Journal of Neurotrauma*, *31*, 358–369.

Shein, N. A., Horowitz, M., Alexandrovich, A. G., Tsenter, J., & Shohami, E. (2005). Heat acclimation increases hypoxia-inducible factor 1alpha and erythropoietin receptor

expression: Implication for neuroprotection after closed head injury in mice. *Journal of Cerebral Blood Flow and Metabolism*, *25*, 1456–1465.
Shingo, T., Sorokan, S. T., Shimazaki, T., & Weiss, S. (2001). Erythropoietin regulates the in vitro and in vivo production of neuronal progenitors by mammalian forebrain neural stem cells. *The Journal of Neuroscience*, *21*, 9733–9743.
Sinor, A. D., & Greenberg, D. A. (2000). Erythropoietin protects cultured cortical neurons, but not astroglia, from hypoxia and AMPA toxicity. *Neuroscience Letters*, *290*, 213–215.
Sirén, A. L., Fratelli, M., Brines, M., Goemans, C., Casagrande, S., et al. (2001). Erythropoietin prevents neuronal apoptosis after cerebral ischemia and metabolic stress. *Proceedings of the National Academy of Sciences of the United States of America*, *98*, 4044–4049.
Sugawa, M., Sakurai, Y., Ishikawa-Ieda, Y., Suzuki, H., & Asou, H. (2002). Effects of erythropoietin on glial cell development, oligodendrocyte maturation and astrocyte proliferation. *Neuroscience Research*, *44*, 391–403.
Tessmar-Raible, K. (2007). The evolution of neurosecretory centres in bilaterian forebrains: Insights from protostomes. *Seminars in Cell & Developmental Biology*, *18*, 492–501.
Tomer, R., Denes, A. S., Tessmar-Raible, K., & Arendt, D. (2010). Profiling by image registration reveals common origin of annelid mushroom bodies and vertebrate pallium. *Cell*, *142*, 800–809.
Tong, K. J., Duchene, S., Ho, S. Y. W., & Lo, N. (2015). Comment on "Phylogenomics resolves the timing and pattern of insect evolution". *Science*, *349*, 487–489.
Um, M., Gross, A. W., & Lodish, H. F. (2007). A "classical" homodimeric erythropoietin receptor is essential for the antiapoptotic effects of erythropoietin on differentiated neuroblastoma SH-SY5Y and pheochromocytoma PC-12 cells. *Cellular Signalling*, *19*, 634–645.
Wang, L., Di, L., & Noguchi, C. T. (2014). Erythropoietin, a novel versatile player regulating energy metabolism beyond the erythroid system. *International Journal of Biological Sciences*, *10*, 921–939.
Wang, L., Zhang, Z., Wang, Y., Zhang, R., & Chopp, M. (2004). Treatment of stroke with erythropoietin enhances neruogenesis and angiogenesis and improves neurological function in rats. *Stroke*, *35*, 1732–1737.
Weishaupt, J. H., Rohde, G., Pölking, E., Siren, A. L., Ehrenreich, H., & Bähr, M. (2004). Effect of erythropoietin on axotomy-induced apoptosis in rat retinal ganglion cells. *Investigative Ophthalmology & Visual Science*, *45*, 1514–1522.
Wirmer, A., Bradler, S., & Heinrich, R. (2012). Homology of insect corpora allata and vertebrate adenohypophysis? *Arthropod Structure & Development*, *41*, 409–417.
Wyder, S., Kriventseva, E. V., Schröder, R., Kadowaki, T., & Zdobnov, E. M. (2007). Quantification of ortholog losses in insects and vertebrates. *Genome Biology*, *8*, R242.
Yu, X., Shacka, J. J., Eells, J. B., Suarez-Quian, C., Przygodzki, R. M., Beleslin-Cokic, B., et al. (2002). Erythropoietin receptor signalling is required for normal brain development. *Development*, *129*, 505–516.
Zanetta, J. P., Scior, T., Wantyghem, J., Wermuth, C., Aubery, M., Stercker, G., et al. (1996). Lectin activities of cytokines and growth factors: Function and implications for pathology. *Histology and Histopathology*, *11*, 1101–1108.
Zhang, W., Sun, B., Yu, Z., An, J., Liu, Q., & Ren, T. (2009). High dose erythropoietin promotes functional recovery of rats following facial nerve crush. *Journal of Clinical Neuroscience*, *16*, 554–556.
Zhang, W., Gao, Y., Zhou, Y., Liu, J., Zhang, L., Long, A., et al. (2015). Localized and sustained delivery of erythropoietin from PLGA microspheres promotes functional recovery and nerve regeneration in peripheral nerve injury. *BioMed Research International*, *2015*, 478103.
Zhong, Y., Yao, H., Deng, L., Cheng, Y., & Zhou, X. (2007). Promotion of neurite outgrowth and protective effect of erythropoietin on the retinal neurons of rats. *Graefe's Archive for Clinical and Experimental Ophthalmology*, *245*, 1859–1867.

CHAPTER ELEVEN

Erythropoietin: Endogenous Protection of Ischemic Brain

Robert T. Mallet*[,1], Myoung-Gwi Ryou*[,†]
*Institute for Cardiovascular and Metabolic Diseases, University of North Texas Health Science Center, Fort Worth, TX, United States
†Tarleton State University, Fort Worth, TX, United States
[1]Corresponding author: e-mail address: robert.mallet@unthsc.edu

Contents

Vitamins and Hormones, Volume 105
ISSN 0083-6729
http://dx.doi.org/10.1016/bs.vh.2017.01.002
© 2017 Elsevier Inc.
All rights reserved.

Abstract

The human brain requires uninterrupted delivery of blood-borne oxygen and nutrients to sustain its function. Focal ischemia, particularly, ischemic stroke, and global ischemia imposed by cardiac arrest disrupt the brain's fuel supply. The resultant ATP depletion initiates a complex injury cascade encompassing intracellular Ca^{2+} overload, glutamate excitotoxicity, oxido-nitrosative stress, extracellular matrix degradation, and inflammation, culminating in neuronal and astroglial necrosis and apoptosis, neurocognitive deficits, and even death.

Unfortunately, brain ischemia has proven refractory to pharmacological intervention. Many promising treatments afforded brain protection in animal models of focal and global ischemia, but failed to improve survival and neurocognitive recovery of stroke and cardiac arrest patients in randomized clinical trials. The culprits are the blood–brain barrier (BBB) that limits transferral of medications to the brain parenchyma, and the sheer complexity of the injury cascade, which presents a daunting array of targets unlikely to respond to monotherapies.

Erythropoietin is a powerful neuroprotectant capable of interrupting multiple aspects of the brain injury cascade. Preclinical research demonstrates erythropoietin's ability to suppress glutamate excitotoxicity and intracellular Ca^{2+} overload, dampen oxidative stress and inflammation, interrupt the apoptotic cascade, and preserve BBB integrity. However, the erythropoietin dosages required to traverse the BBB and achieve therapeutically effective concentrations in the brain parenchyma impose untoward side effects. Recent discoveries that hypoxia induces erythropoietin production within the brain and that neurons, astroglia, and cerebrovascular endothelium harbor membrane erythropoietin receptors, raise the exciting prospect of harnessing endogenous erythropoietin to protect the brain from the ravages of ischemia-reperfusion.

ABBREVIATIONS

AMPA α-amino-3-hydroxy-5-methyl-4-isoazolepropionic acid
BBB blood–brain barrier
DMOG dimethyloxaloylglycine
EPOR erythropoietin receptor
ERK extracellular signal-regulated kinase
HIF hypoxia-inducible factor
MCA middle cerebral artery
MMP matrix metalloproteinase
NF-κB nuclear factor κB
NMDA *N*-methyl-D-aspartate
NOS nitric oxide synthase
PARP poly(ADP-ribose) polymerase
rhEPO recombinant human erythropoietin
rtPA recombinant tissue plasminogen activator
STAT signal transducer and activator of transcription
VEGF vascular endothelial growth factor
ΔG_{ATP} Gibbs-free energy of ATP hydrolysis

1. EPIDEMIOLOGIC IMPACT OF ACUTE BRAIN ISCHEMIA

Acute ischemic syndromes of the central nervous system rank among the leading causes of death and long-term disability in the United States. Approximately 795,000 strokes occur in the United States annually (Mozaffarian, Benjamin, Go, American Heart Association Statistics Committee, & Stroke Statistics Committee, 2016), of which 87% are ischemic strokes, the direct result of cerebral artery occlusion. Stroke is the fifth-leading cause of death in the United States, claiming 130,000 lives annually (Mozaffarian et al., 2016). Of the nearly 7 million American adults who have suffered a stroke (Roger, Go, Lloyd-Jones, American Heart Association Statistics Committee, & Stroke Statistics Committee, 2012), approximately 75% experience persistent neurocognitive impairment that substantially impacts their quality of life (Go, Mozaffarian, Roger, & American Heart Association Statistics Committee and Stroke Statistics Subcommittee, 2013; Green, 2008). Two-thirds of strokes afflict persons aged 65 and older, and the incidence of stroke and its socioeconomic impact can be expected to mount as the US population ages. Moreover, acute brain ischemia has been proposed to hasten the onset and accelerate the progression of neurodegenerative disorders including Alzheimer's disease (Kalaria, 2000; Zhou et al., 2015), a looming epidemic in the aging population.

Annually, approximately 360,000 Americans suffer cardiac arrest outside the hospital setting. Despite improvements in emergency medical services, widespread installation of automatic electrical defibrillators, public education in basic life support and other measures, mortality rates remain persistently high (Mozaffarian et al., 2016). Only 10%–15% of victims of out-of-hospital cardiac arrest achieve return of spontaneous circulation, and only 2%–5% of these victims survive to hospital discharge (Idris et al., 2005). Another 210,000 Americans suffer cardiac arrest while in the hospital, and only 25% of these patients survive to hospital discharge, despite ready access to advanced medical care.

Although cessation of cardiac rhythm imposes profound ischemia on the entire body, endangering every internal organ, most deaths are due to hypoxia-ischemia damage to the central nervous system. Brain injury was identified as the cause of death in 68% of out-of-hospital and 23% of in-hospital cardiac arrests (Laver, Farrow, Turner, & Nolan, 2004). Many patients who initially regain sinus rhythm die later due to irreversible brain injury (Girotra, Chan, & Bradley, 2015; Nolan et al., 2008). Forty percent of initial survivors of cardiac arrest enter a permanent vegetative state, 80% of whom die within 1 year (Madl & Holzer, 2004). Only 5%–14% of resuscitated cardiac arrest victims survive without significant neurocognitive impairment (Westfal, Reissman, & Doering, 1996).

2. CASCADE OF BRAIN ISCHEMIA-REPERFUSION INJURY

The adult human brain contains approximately 100 billion neurons (Begley & Brightman, 2003). Immense numbers of astrocytes and oligodendrocytes provide these neurons nutritive and structural support, and a vast capillary network delivers O_2, energy-yielding fuels, and neurotransmitter and antioxidant precursors, and removes metabolic waste products. The physiological activities of these cells are energy-intensive. Relative to its size, human brain consumes enormous amounts of ATP. The brain's limited energy reserves require that its ATP be replenished as it is consumed; ATP production requires continuous delivery of O_2 and metabolic substrates via ca. 400 miles of capillaries (Begley & Brightman, 2003). The brain comprises only 2% of lean body mass, yet it receives 15% of the resting cardiac output and consumes approximately 25% of the body's resting O_2 demand (Clark & Sokoloff, 1999). The brain's high O_2 demand makes it the least ischemia-tolerant of the internal organs. Occlusions of the cerebrovascular network, e.g., by thrombi, atherosclerosis or vasospasm, or cessation of cardiac output initiate a complex injury cascade (Fig. 1).

2.1 Ischemic ATP Depletion

ATP synthesis fails when O_2 delivery to the mitochondria is interrupted, causing an abrupt decline in Gibbs-free energy of ATP hydrolysis (ΔG_{ATP}), the thermodynamic driving force for the membrane ion transporters that repolarize neurons and prevent intracellular Ca^{2+} accumulation (Kimura et al., 2002).

2.2 Glutamate Excitotoxicity

Depolarized glutamatergic neurons release glutamate, the principal excitatory neurotransmitter in brain, producing excitotoxicity that contributes to brain injury (Hazell, 2007). Glutamate hyperexcitation of neurons via α-amino-3-hydroxy-5-methyl-4-isoazolepropionic acid (AMPA) and

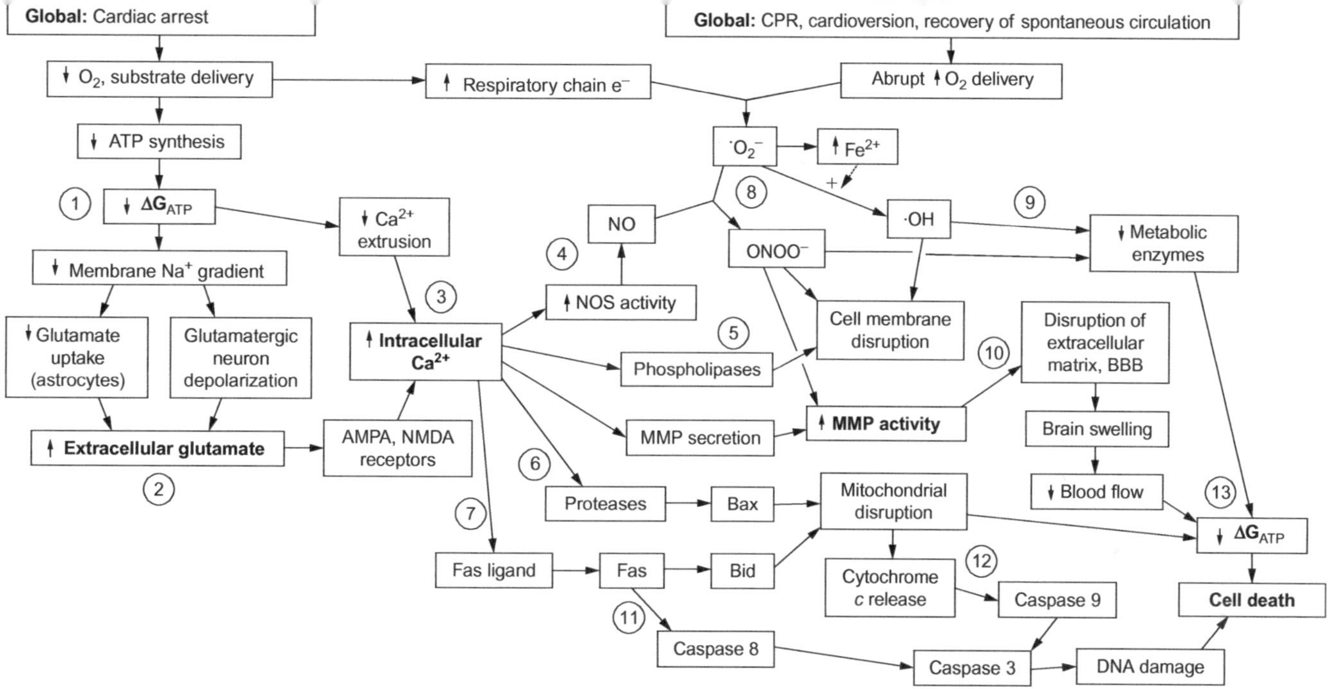

Fig. 1 Ischemia-reperfusion-induced cascade of brain injury. Ischemia initiates a cascade of mechanisms that culminate in death of the brain parenchyma, irreversible brain damage and persistent neurocognitive, and motor impairment. Interruption of blood flow causes an abrupt decline in Gibbs-free energy of ATP hydrolysis, ΔG_{ATP} (1), which provokes glutamate secretion from depolarized excitatory neurons (2). Energy depletion and glutamate stimulation impair cellular Ca^{2+} homeostasis, thereby increasing intracellular Ca^{2+} (3). Ca^{2+}, in turn, activates nitric oxide synthase to overproduce NO (4); phospholipases that degrade membrane phospholipids (5); proteases (6) that degrade cellular proteins and activate proapoptotic Bax; and secretion of Fas ligand, which initiates the extrinsic apoptosis cascade (7). Reperfusion reintroduces O_2 to the electron (e^-)-rich environment of the ischemic cells, triggering formation of superoxide (8), a precursor of hydroxyl radical and, via biradical condensation with NO, $ONOO^-$. These compounds inactivate energy-generating and antioxidant enzymes (9) and activate MMPs (10) which degrade the extracellular matrix and blood–brain barrier (BBB), causing inflammation, edema, brain swelling, and decreased cerebral blood flow. The extrinsic (11) and intrinsic (12) apoptotic cascades activate caspase 3 to degrade cellular components including DNA. This damage, and the secondary energy depletion due to underperfusion of the edematous tissue and enzyme inactivation (13), conspires to kill neurons and other cells.

N-methyl-D-aspartate (NMDA) receptors provokes intense Ca^{2+} influx; the resultant intracellular Ca^{2+} overload (Abramov & Duchen, 2008; Prentice, Modi, & Wu, 2015) activates proteases and phospholipases which degrade structural proteins and membrane components (Konstady, 2012; White et al., 2000), and the Ca^{2+}-activated neuronal and endothelial nitric oxide synthase (NOS) isoenzymes (Manukhina, Downey, & Mallet, 2006; Prentice et al., 2015). Hyperactivated NOS generates excessive nitric oxide, which condenses with $^{\bullet}O_2^-$ yielding the cytotoxic metabolite, peroxynitrite ($ONOO^-$) (Beckman & Koppenol, 1996; Ischiropoulos, 1998). Peroxynitrite provokes incorporation of the proapoptotic proteins Bax and Bad into the inner mitochondrial membrane, causing mitochondrial membrane depolarization and failure of mitochondrial ATP production (Prentice et al., 2015).

Astrocytes clear glutamate from the perineuronal space (Kim et al., 2011) by Na^+:glutamate cotransport. The driving force for glutamate uptake, the transmembrane Na^+ electrochemical gradient is maintained by the plasmalemmal Na^+,K^+ ATPase, a ΔG_{ATP}-dependent process impaired by ischemic ATP depletion. Failure of glutamate uptake increases extracellular glutamate leading to excitotoxicity (Dallas et al., 2007; Hazell, 2007), while factors that augment astrocytic glutamate transporter expression are neuroprotective (Ghosh, Yang, Rothstein, & Robinson, 2011; Kim et al., 2011). Antisense oligonucleotide suppression of glutamate transporter expression elevated extracellular glutamate concentration and produced excitotoxicity-like neurodegeneration in rat brain (Rothstein et al., 1996).

2.3 Oxyradical Formation

The brain's high metabolic rate, limited antioxidant armamentarium, and high contents of iron and unsaturated fatty acids make it highly susceptible to oxidative and nitrosative stress (Olmez & Ozyurt, 2012). Mitochondrially generated reactive oxygen species contribute to brain injury in stroke (Kalogeris, Bao, & Korthuis, 2014). Cerebral underperfusion compromises O_2 delivery to mitochondrial respiratory complex IV, causing electrons to accumulate within the respiratory chain (Selivanov et al., 2011). Univalent reduction of residual O_2 ensues, generating superoxide ($^{\bullet}O_2^-$) (Selivanov et al., 2011). Although not highly reactive, $^{\bullet}O_2^-$ is the precursor of cytotoxic compounds including hydroxyl radical ($^{\bullet}OH$) and $ONOO^-$ (Manukhina et al., 2006; Prentice et al., 2015). These reactive oxygen and nitrogen derivatives disable the mitochondrial respiratory complexes, intensifying

$^{\bullet}O_2^-$ formation (Brown & Borutaite, 2004). Fe^{2+} displaced from respiratory complex heme moieties by $^{\bullet}O_2^-$ catalyzes lipid peroxidation (White et al., 2000). Reactive oxygen and nitrogen intermediates disable crucial metabolic enzymes, including pyruvate dehydrogenase (Robertson, Scafidi, McKenna, & Fiskum, 2009), glyceraldehyde 3-phosphate dehydrogenase (Eaton, Wright, Hearse, & Shattock, 2002), aconitase (Hausladen & Fridovich, 1994), and creatine kinase (Reddy, Jones, Cross, Wong, & Van Der Vliet, 2000). $ONOO^-$ provokes membrane phospholipid peroxidation (Ferdinandy & Schulz, 2003), and nitrosylates tyrosyl and cysteinyl residues in proteins (Manukhina et al., 2006).

2.4 Reperfusion Oxyradical Burst

Defibrillation of the arrested heart and recovery of spontaneous circulation initiates global brain reperfusion following cardiac arrest. Similarly, cerebral artery revascularization effects focal reperfusion in stroke. Although reperfusion is essential to salvage brain tissue, the abrupt reintroduction of O_2 into the electron-rich environment of the ischemic tissue precipitates massive overproduction of $^{\bullet}O_2^-$ in the brain parenchyma (Kalogeris et al., 2014; Olmez & Ozyurt, 2012) and cerebrovascular endothelium (Kimura, Oike, & Ito, 2000). Sequential single-electron reductions of $^{\bullet}O_2^-$ generate first hydrogen peroxide and then the extremely aggressive, cytotoxic $^{\bullet}OH$. Also, $^{\bullet}O_2^-$ undergoes biradical condensation with NO, yielding $ONOO^-$, which attacks phospholipids and proteins and gives rise to a host of other reactive nitrogen and oxygen derivatives (Manukhina et al., 2006).

2.5 Matrix Metalloproteinases and Brain Injury

Matrix metalloproteinases (MMPs) are a class of zinc proteases that participate in normal modeling of tissue architecture by controlled degradation of extracellular matrix proteins, but when hyperactivated are complicit in brain damage and inflammation. Ischemia-reperfusion activates several MMPs including MMP-2, -3, and -9 (Jin, Yang, & Li, 2010) harbored in neurons, glia, and leukocytes (Chaturvedi & Kaczmarek, 2014). MMPs 3 and 9 are activated by reactive oxygen species and proinflammatory cytokines, whereas MMP-2 is constitutive and activated locally (Xiong et al., 2008). MMPs are harmful in the early stages of ischemic stroke, but may be beneficial during subsequent stroke recovery, when they participate in neurovascular repair (Jin et al., 2010) by remodeling the extracellular matrix and permitting neural precursor cells to take up residence in the infarcted region (Chen, Wang, Asavaritkrai, & Noguchi, 2010; Yang & Rosenberg, 2015).

The cerebrovascular endothelial tight junctions which embody the blood–brain barrier (BBB) contain integral proteins claudin, occludin, and junctional adhesion molecules, all potential MMP targets (Abbott, Patabendige, Dolman, Yusof, & Begley, 2010). By degrading claudins, occludin and other intercellular junction components, MMPs 2 and 9 disrupt the BBB following stroke (Chang, Emanuel, Mack, Tsivgoulis, & Alexandrov, 2014; Turner & Sharp, 2016). Extracellular matrix degradation and BBB disruption precipitate hemorrhagic transformation (Turner & Sharp, 2016), a devastating complication of ischemic stroke (Chaturvedi & Kaczmarek, 2014; Jickling et al., 2014), and facilitate extravasation of inflammatory neutrophils which invade the parenchyma and contribute to infarct expansion following stroke (Kogure & Kogure, 1997; Vafadari, Salamian, & Kaczmarek, 2016). Activation of MMPs by reactive oxygen species contributes to postischemic disruption of the BBB (Fraser, 2011; Gu, Dee, & Shen, 2011). MMP-9 activity in particular directly correlates with infarct volume and unfavorable outcome in animal models of stroke (Ramos-Fernandez, Bellolio, & Stead, 2011).

In preclinical stroke models, inactivation of MMPs, e.g., by pharmacological MMP inhibitors, free radical scavengers, or transgenic deletions of MMP-encoding genes preserves BBB integrity (Dejonckheere, Vandenbroucke, & Libert, 2011; Svedin, Hagberg, Sävman, Zhu, & Mallard, 2007). However, such measures have not been implemented for clinical treatment of brain ischemia.

2.6 Cell Death Mechanisms: Necrosis vs Apoptosis

In stroke, cells in the intensely ischemic core are the first to degenerate and die by necrosis (Rami, Bechmann, & Stehle, 2008). In the penumbra, ischemia is less severe and cell death more gradual (Kato & Kogure, 1999), such that prompt revascularization may salvage the tissue (Paciaroni, Caso, & Agnelli, 2009). In stroke, apoptosis is the predominant cell death mechanism in the ischemic penumbra (Zheng, Zhao, Steinberg, & Yenari, 2003), although necrosis may occur there as well (Ferrer & Planas, 2003). In stroke, the ischemic penumbra is characterized by activation of proapoptotic caspases (Ferrer & Planas, 2003), while in cardiac arrest resuscitation, the entire brain is subject to apoptosis (Martin et al., 1998).

The apoptosis cascade is orchestrated by signaling cascades that proceed at a measured pace, affording opportunities for therapeutic intervention (Maiese, 2001). There are two apoptotic mechanisms in brain (Broughton, Reutens, & Sobey, 2009; Sims & Muyderman, 2010). In the extrinsic, receptor-mediated pathway, Fas ligand secreted by neurons, astroglia, and invasive leukocytes, binds its membrane receptor followed by sequential activation of caspase 8 and caspase 3; the latter cleaves the Bcl family protein Bid, yielding truncated Bid which combines with pro-apoptotic Bad to form a channel in the mitochondrial membrane, through which mitochondrial cytochrome *c* is released (Broughton et al., 2009). In the intrinsic, i.e., mitochondrial, mechanism, translocation of proapoptotic Bax and Bid to the inner mitochondrial membrane activate cytochrome *c* release in a manner suppressed by Bcl-2 overexpression (Ferrer & Planas, 2003). In the cytosol, cytochrome *c* binds apoptotic protease-activating factor-1 and dATP, forming an "apoptosome" that recruits and activates caspase 9, which in turn activates caspase 3, the effector caspase of the intrinsic apoptotic pathway. Caspase 3 cleaves and activates poly(ADP-ribose) polymerase (PARP), transforming PARP from a DNA-repairing to an NAD^+ consuming enzyme that compromises metabolic ATP production (Amé et al., 1999; Ferrer & Planas, 2003).

3. INTERVENTIONS FOR BRAIN ISCHEMIA AND THEIR LIMITATIONS

The sheer complexity of the ischemia-reperfusion brain injury cascade (Fig. 1) challenges translation of preclinical findings into clinically effective neuroprotectants (Brouns & De Deyn, 2009; Candelario-Jalil, 2009; Schouten, 2007). Despite massive investment of resources toward developing effective treatments for stroke, most promising pharmaceuticals have failed in clinical trials (Turner, Dodson, Rosen, & Huber, 2013). Only the serine protease, recombinant tissue plasminogen activator (rtPA) (Adibhatla & Hatcher, 2008) is approved for stroke by the US Food and Drug Administration. rtPA does not address intrinsic brain injury mechanisms; rather, it dissolves thrombi, thereby restoring blood flow. rtPA's therapeutic window is limited to 4.5 h after onset of symptoms (Green, 2008; Jin et al., 2010). Later rtPA treatment could produce hemorrhagic transformation and brain edema (Derex & Nighoghossian, 2008) by activating vascular endothelial growth factor (VEGF), which increases BBB permeability as a necessary step in angiogenesis (Suzuki, Nagai, & Umemura, 2016). Also, rtPA may activate MMP-9, a mediator of hemorrhagic transformation (Adibhatla & Hatcher, 2008; Chaturvedi & Kaczmarek, 2014).

Targeted temperature management, i.e., therapeutic hypothermia, is the only treatment found to afford neuroprotection in resuscitated cardiac arrest patients, although evidence of sustained benefit is equivocal (Yu et al., 2015; Zhang et al., 2015). In one randomized trial, cooling the patients to 32–34°C for 24 h postarrest decreased 6 months mortality by 26% and increased survival with good neurological outcome by 40% vs normothermic patients (Hypothermia after Cardiac Arrest Study Group, 2002), and in another, cooling the patients to 33°C for 12 h increased by 85% the likelihood of survival with good neurological outcome (Bernard et al., 2002). Conversely, therapeutic hypothermia did not improve survival or neurological outcomes in out-of-hospital cardiac arrest victims with "nonshockable" asystole or pulseless electrical activity, while in patients with "shockable" ventricular tachycardia or fibrillation, therapeutic hypothermia improved survival, and neurological recovery (Maznyczka & Gershlick, 2015). Moreover, the specialized hypothermia equipment is not available at many hospitals.

Of the many neuroprotectants that showed promise in preclinical stroke studies, none has proven unequivocally safe and effective in randomized, phase III clinical trials (Ovbiagele, Kidwell, Starkman, & Saver, 2003). Antioxidants have not shown efficacy despite the prominent role of reactive oxygen species in brain injury (Doeppner & Hermann, 2010). For example, a phase III trial found the oxyradical scavenger NXY-059 to be no better than placebo (Shuaib et al., 2007). Glutamate receptor antagonists have proven similarly ineffective. Randomized, placebo-controlled clinical trials of the NMDA receptor antagonists aptiganel hydrochloride (Albers, Goldstein, Hall, Lesko, & Aptiganel acute stroke investigators, 2001) and selfotel (Davis et al., 2000) were suspended due to lack of clinical efficacy and higher mortality rates than placebo. The AMPA antagonist ZK200775 transiently worsened neurological function and produced glial cell toxicity in ischemic stroke patients (Elting et al., 2002). The failure of these compounds directed against major components of the injury cascade underscores the unmet need for effective neuroprotectants which traverse the BBB and address multiple mechanisms of injury.

4. ERYTHROPOIETIN: NEUROPROTECTIVE CYTOKINE

Studies in a variety of animal models of stroke, cardiac arrest, and traumatic brain injury have demonstrated the remarkable neuroprotective capabilities of the natural hematopoietic hormone erythropoietin. These studies have established erythropoietin's diverse neuroprotective mechanisms in ischemic, postischemic and contused brain, and its impact on multiple aspects of the brain injury cascade (Fig. 2). Many of these studies utilized recombinant human erythropoietin (rhEPO), already in routine clinical use for management of anemia, e.g., in end-stage renal disease (Lankhorst & Wish, 2010). This preclinical evidence supports the proposed use of rhEPO for clinical treatment of stroke, traumatic brain injury, intracerebral hemorrhage, neonatal hypoxia-ischemia, and spinal cord injury (Rangarajan & Juul, 2014).

Ischemic stroke is often modeled in rodents by temporary or permanent occlusion of the middle cerebral artery (MCA), an end-artery that supplies a substantial portion of the cerebral hemisphere without significant collateral circulation from other arteries. In a rat model of ischemic stroke produced by 2 h MCA occlusion, erythropoietin infusion into the reperfused artery decreased infarct volume and neurobehavioral deficits assessed 24 h later (Zhao et al., 2015). In neonatal rats subjected to 45 min MCA occlusion, a single ip injection of 5000 IU/kg rhEPO partially preserved volume of the affected hemisphere and increased differentiation of progenitor cells into neurons within the lesion (Gonzalez et al., 2007). In rats subjected to 60 min MCA occlusion and 24 h reperfusion, rhEPO (1000 IU/kg, iv)

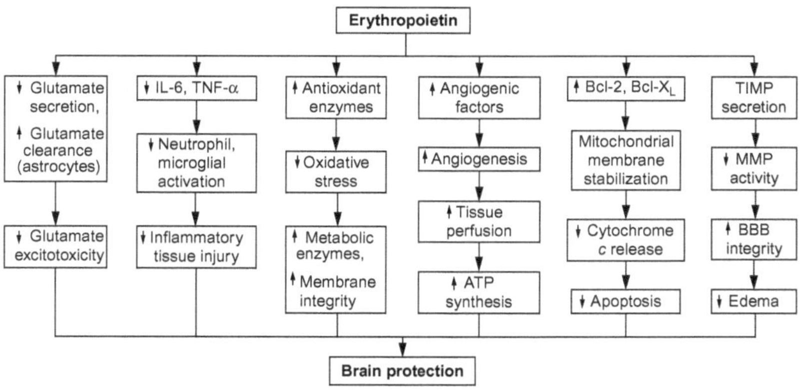

Fig. 2 Erythropoietin's protective mechanisms in ischemic and postischemic brain. Erythropoietin activates several complementary mechanisms which collectively increase the brain parenchyma's resistance to ischemia-reperfusion.

administered 10 min before ischemia decreased neurological deficit, brain edema, and circulating activity of the brain injury marker, neuron-specific enolase (Ratilal et al., 2014). In adult male rats, infusion of erythropoietin directly into the reperfused MCA decreased neurological deficit and infarct volume, and apoptosis in the ischemic core (Dang et al., 2011).

Systemically administered rhEPO traversed the BBB and decreased infarct volume by 50%–75% in a rat model of stroke, even when rhEPO administration was delayed up to 6 h after the ischemic event (Brines et al., 2000). In neonatal rats subjected to permanent MCA occlusion, ip rhEPO injections at 15 min, 1 and 2 days focal ischemia decreased infarct volume and apoptosis in the ischemic tissue (Sola, Rogido, Lee, Genetta, & Wen, 2005). In neonatal rats subjected to 90 min combined right carotid artery occlusion and hypoxia (6% O_2), erythropoietin, injected ip 1 h before and 24 and 48 h after insult, improved neurobehavioral performance, decreased death of neurons and oligodendrocytes, and suppressed microglial activation (Lan et al., 2016).

Importantly, erythropoietin was neuroprotective even when administered after the acute ischemic event. Erythropoietin injected ip in rats decreased infarct and penumbral apoptosis by 70%–75% whether given 24 h before, during, or 3 h after MCA occlusion, and was partially protective even when given as late as 6 h postocclusion (Brines, 2002). In adult rats, erythropoietin injected ip 1, 2, and 3 days after traumatic brain injury prevented hippocampal cell loss, increased angiogenesis and neurogenesis in the injured cerebral cortex and ipsilateral hippocampus, and improved sensorimotor function (Zhang et al., 2009). In neonatal rats, delayed erythropoietin administration begun 1 week after 3 h MCA occlusion afforded appreciable improvements in motor function and preserved volume of the ipsilateral cerebral hemisphere (Larpthaveesarp, Georgevits, Ferriero, & Gonzalez, 2016). Erythropoietin injected ip 6 h, 3 days, and 7 days after traumatic brain injury decreased cortical lesion volume, suppressed cell death in the dentate gyrus, and improved recovery of sensorimotor function and spatial learning in mice (Xiong et al., 2008). In 10-day-old rat pups subjected to combined 60 min hypoxia and cerebrovascular occlusion-reperfusion, rhEPO administration 60 min after reperfusion minimized cerebral injury and MMP-9 activity (Souvenir et al., 2011). Erythropoietin's effective therapeutic window for brain protection is not firmly established, and may depend on the nature of brain injury as well as the dosage and route of administration.

5. NEUROPROTECTIVE MECHANISMS OF ERYTHROPOIETIN

Unlike pharmaceuticals targeting isolated elements of the injury cascade, erythropoietin's diverse actions address several aspects including reactive oxygen and nitrogen intermediates, inflammation, MMPs, excitotoxicity, and cerebrovascular angiogenesis (Fig. 2).

5.1 Erythropoietin's Antioxidant Actions

In mice, rhEPO exerted neuroprotection against the neurotoxin 1-methyl-4-phenyl-1,2,5,6-tetrahydropyridine, in part by restoring, primarily in astroglia, activity of the pivotal antioxidant enzyme, glutathione peroxidase, following its inactivation by the toxin (Genc, Akhisaroglu, Kuralay, & Genc, 2002). In Mongolian gerbils subjected to 5 min bilateral carotid artery occlusion, rhEPO (50–100 IU) injected ip at reperfusion, attenuated hippocampal edema, neuronal death, and lipid peroxidation and suppressed formation of the NO precursor, $ONOO^-$ (Calapai et al., 2000). Thus, postischemic erythropoietin treatment can protect ischemia-sensitive brain regions, and NOS suppression may contribute to the protection.

5.2 Antiinflammatory Mechanisms of Erythropoietin

Erythropoietin exerts antiinflammatory protection on all brain cells which harbor membrane erythropoietin receptors, i.e., erythropoietin receptor (EPOR) (Merelli, Czornyj, & Lazarowski, 2015). In rats undergoing MCA occlusion, rhEPO (5000 U/kg, ip) suppressed astrocyte activation and recruitment of leukocytes and microglia into the infarct, and decreased contents of proinflammatory cytokines interleukin-6, TNF-α, and monocyte chemoattractant protein-1 by >50% (Villa et al., 2003). In a rat model of diffuse traumatic axonal injury, rhEPO improved sensorimotor and cognitive recovery, suppressed microglial activation and macrophage infiltration, minimized axonal damage, and increased EPOR expression and content (Hellewell, Yan, Alwis, Bye, & Morganti-Kossmann, 2013). rhEPO also preserved learning and memory in a rat endotoxemia model of septic encephalopathy (Zhou & Yu, 2013). In mice subjected to proinflammatory intracerebral hemorrhage, rhEPO, injected ip during the first 3 days post-hemorrhage, preserved the BBB, prevented tissue edema, restrained increases in MMP-2 content, and activated the endogenous MMP inhibitor, TIMP-2 (Li et al., 2008).

5.3 Antiapoptotic Actions

Erythropoietin suppressed apoptosis in diverse animal models of brain ischemia. Erythropoietin (5000 U/kg, ip) injected at the outset of 60 min MCA occlusion in rats, suppressed apoptosis in the ischemic penumbra, and decreased infarct size by 75% (Sirén & Ehrenreich, 2001). Erythropoietin attenuated caspase 3 activation in cerebrovascular endothelium when infused into the reperfused MCA after 2 h occlusion (Zhao et al., 2015) and suppressed apoptotic cell death in hippocampal neurons subjected to a cell culture model of hypoxia-ischemia (Wen et al., 2002).

The balance of pro- and antiapoptotic Bcl family proteins is a major determinant of cellular apoptosis vs survival (Baines, 2010). Erythropoietin attenuates brain apoptosis by increasing expression and synthesis of the anti-apoptotic Bcl family proteins Bcl-X_L and Bcl-2, while suppressing the proapoptotic Bcl proteins Bax, Bak, and Bid. Thus, erythropoietin increased the abundance of Bcl-X_L mRNA while lowering Bax mRNA in hypoxic hippocampal neurons (Wen et al., 2002), induced Bcl-X_L expression and afforded neuroprotection in permanently ischemic neonatal rat cerebrum (Sola et al., 2005), and increased Bcl-X_L mRNA abundance and protein content, preserved hippocampal pyramidal neurons, and prevented learning disability in gerbils subjected to cerebrovascular occlusion (Wen et al., 2002). In neuron-like PC12 cells, erythropoietin increased Bcl-X_L mRNA abundance by 2.5-fold, while suppressing mRNA encoding proapoptotic Bak by 75% (Renzi et al., 2002). Erythropoietin similarly increased the Bcl/Bax mRNA ratio in rat microglia in primary culture (Vairano et al., 2002). rhEPO increased Bcl-2 mRNA abundance and protein content in female rats subjected to traumatic brain injury (Liao, Zhi, Shi, & He, 2008) while suppressing Bax mRNA and protein (Liao et al., 2009). Similarly, in neonatal rats subjected to combined left carotid artery occlusion and hypoxia, rhEPO increased Bcl-2 content while lowering Bax content and caspase-3 activity (Kim et al., 2010).

Bax, Bak, and truncated Bid form channels that increase permeability of the inner mitochondrial membrane and release cytochrome *c*, while Bcl-X_L and Bcl-2 sequester Bax and Bak, preventing channel formation (Broughton et al., 2009; Sims, Clarke, Njah, Hopkins, & Sontheimer, 2010). In a rat model of traumatic brain injury, erythropoietin injected ip 30 min after injury restored coupling of mitochondrial O_2 consumption to ATP synthesis (Xiong, Chopp, & Lee, 2009), in accordance with suppression of mitochondrial membrane channels.

5.4 Erythropoietin Dampens Glutamate Excitotoxicity

In gerbils subjected to forebrain ischemia imposed by bilateral common carotid artery occlusion, infusion of rhEPO into the left lateral cerebral ventricle prevented NMDA receptor-mediated, glutamate-induced neuronal apoptosis in the hippocampal CA1 subregion, maintained the number of synapses in that region, and preserved learning ability, without attenuating the NMDA receptor mediated increase in intracellular Ca^{2+} concentration; the protection was blunted by cerebroventricular infusion of soluble EPOR, which sequesters erythropoietin (Sakanaka et al., 1998). Erythropoietin and its analog EMP1 suppressed Ca^{2+}-activated glutamate release from cultured rat cerebellar granular cells subjected to chemical ischemia produced by potassium cyanate (Kawakami, Sekiguchi, Sato, Kozaki, & Takahashi, 2001). In primary cultures of rat neurons, erythropoietin pretreatment suppressed hypoxia-induced glutamate release, NMDA receptor-mediated glutamate excitotoxicity, and cell death (Xu, Dai, & Shi, 2009). In cultured astrocytes, erythropoietin dampened glutamate-induced intracellular Ca^{2+} oscillations, thereby decreasing insertion of aquaporin-4 water channels into the plasma membrane and the resultant water entry and cell swelling (Gunnarson et al., 2009).

In cultured cortical neural stem cells and neuronally derived B104 cells, the excitatory glutamate mimetic kainate depleted the endogenous antioxidant glutathione and produced neurotoxicity. Erythropoietin preserved glutathione and prevented kainate neurotoxicity (Sims & Muyderman, 2010), in a manner dependent on induction of the neuronal membrane cysteine:glutamate exchanger which delivers cysteine required for neuronal glutathione synthesis. Erythropoietin may also activate glutamate clearance by astrocytes. rhEPO increased by 60% glutamate uptake by cultured primary rat astroglia, as well as astroglial glutamine synthase activity, the enzyme that converts glutamate to nonexcitatory glutamine (Lourhmati et al., 2013). rhEPO-induced cytoprotection and glutamine synthase were abrogated by small-interfering RNA (siRNA) against EPOR expression, implicating EPOR-induced signaling cascades in erythropoietin's anti-excitotoxic mechanisms (Lourhmati et al., 2013).

5.5 Erythropoietin Protects the BBB

Studies in a rat model of traumatic brain injury provided evidence that exogenous erythropoietin preserves BBB integrity (Grasso et al., 2007). After dural application of a precooled copper cylinder to inflict regional cryogenic injury, rhEPO (1000 IU/kg), injected ip at 8 h intervals for 2 or 14 days, minimized BBB disruption, brain edema, and lesion volume, and augmented recovery of motor function. Global brain ischemia produced by the combination of systemic hypovolemia and 20 min bilateral occlusion of the common carotid arteries disrupted the BBB and produced infarct and lipid peroxidation in rats, but rhEPO pretreatment (3000 IU/kg ip) 24 h before ischemia minimized lipid peroxidation and brain infarct and preserved the BBB (Bahcekapili, Uzüm, Gökkusu, Kuru, & Ziylan, 2007).

In a rat model of ischemic stroke, rhEPO, injected into the MCA upon reperfusion after 2 h MCA occlusion (Wang et al., 2015), decreased infarct volume and brain edema, and improved neurobehavioral recovery. This treatment suppressed degradation of the cerebrovascular endothelial intercellular junction components claudin 5 and occludin, and preserved BBB integrity assessed from extravasation of Evans blue dye. rhEPO also suppressed activities and contents of the BBB degrading metalloproteinases, MMP-2 and -9 (Wang et al., 2015).

5.6 Activation of Angiogenesis

In mice subjected to permanent MCA occlusion, rhEPO, injected ip (5000 U/kg) 30 min before occlusion and once daily thereafter for 21 days (Li, Lu, Keogh, Yu, & Wei, 2007), decreased infarct volume, increased angiogenesis at 7 and 21 days occlusion, and increased synthesis of angiogenic factors VEGF, Tie-2, and angiopoietin-2. rhEPO normalized local blood flow in the penumbra at 14 days occlusion. Thus, rhEPO induced angiogenesis and restored cerebral blood flow in the ischemic penumbra in the setting of permanent cerebroarterial occlusion (Li et al., 2007).

6. ERYTHROPOIETIN-ACTIVATED NEUROPROTECTIVE SIGNALING MECHANISMS

Erythropoietin activates cytoprotective signaling cascades in neurons and other cells harboring EPOR (Kilic et al., 2005; Ruscher et al., 2002). Erythropoietin signaling begins with its interaction with its membrane receptors. Erythropoietin binding provokes EPOR homodimerization and autophosphorylation. The tyrosine kinase JAK-2 binds and is phosphorylated and activated by the EPOR dimer. Phosphorylated JAK-2 activates multiple signaling kinases, including the phosphatidylinositol-3-kinase

(PI3K)—Akt (protein kinase B) cascade, extracellular signal-regulated kinase (ERK)-1/2, FOXO3a, and signal transducer and activator of transcription (STAT)5, each of which mediate distinct neuroprotective mechanisms (Kilic et al., 2005; Maiese, Li, & Chong, 2004).

6.1 PI3K-Akt Pathway

In rat primary cortical neurons, rhEPO suppressed NMDA-induced increases in intracellular Ca^{2+} concentration. This cytoprotective effect was blunted by the PI3K inhibitors LY294002 and wortmannin, and by iberiotoxin, an inhibitor of Ca^{2+}-activated K^{+} channels (Andoh et al., 2011). In a rat model of ischemic preconditioning neuroprotection, three 10 min MCA occlusions, 3 days before 2 h MCA occlusion, decreased brain infarct volume by 55%, improved neurobehavioral recovery, decreased apoptosis in the peri-infarct penumbra, and increased erythropoietin and EPOR content in the ipsilateral hemisphere. The protection was blunted by soluble EPOR infusion into the ipsilateral cerebral ventricle to sequester endogenous erythropoietin, and by the PI3K inhibitor, wortmannin, implicating PI3K in the neuroprotection (Malhotra, Savitz, Ocava, & Rosenbaum, 2006).

Erythropoietin suppressed ketamine-induced apoptosis of cultured rat cortical neurons via PI3K and Akt activation (Shang et al., 2007). Phosphorylated Akt inactivated GSK3β, an inducer of caspase 3. The PI3K inhibitor LY294002 blocked erythropoietin-enhanced cell survival and abrogated caspase 3 suppression. Thus, erythropoietin protected these neurons via inactivation of GSK3β, thereby preventing caspase 3 activation and apoptosis (Shang et al., 2007).

Transgenic mice constitutively expressing human erythropoietin in brain demonstrated decreased brain injury following 30 or 90 min MCA occlusions (Kilic et al., 2005). JAK2, ERK-1/2, and Akt were constitutively phosphorylated and activated in these mice, while pharmacological inhibitors of ERK-1/2 and Akt abolished the brain's ischemic resistance. Also in mice, heat acclimatization for 30 days at 34°C produced appreciable protection against subsequent traumatic brain injury-induced cognitive and motor deficits and brain edema (Shein, Tsenter, Alexandrovich, Horowitz, & Shohami, 2007). Increases in Akt phosphorylation and EPOR content paralleled the neuroprotection. The Akt inhibitor triciribine blunted heat acclimation-induced neuroprotection, implicating Akt in the neuroprotective mechanism (Shein et al., 2007).

In a rat model of Parkinson's disease produced by 6-hydroxydopamine injection, erythropoietin decreased apoptosis of striatal neurons and improved neurobehavioral function (Jia et al., 2014). Erythropoietin increased phosphorylation of Akt and its target FoxO3a. The PI3K inhibitor LY294002 blunted erythropoietin-induced neuroprotection, implicating the PI3K/Akt/FoxO3a pathway in the protection.

6.2 STAT5 Signaling

STAT5 was shown to be a mediator of erythropoietin-induced neuroprotection in a rat model of global cerebral ischemia (Zhang, Wang, Cao, Gao, & Chen, 2007). Rats were subjected to 12 min bilateral occlusion of the vertebral and common carotid arteries, and then 50 IU rhEPO was injected into the right cerebral ventricle 20 min after ischemia. rhEPO augmented STAT5 phosphorylation, increased expression of the STAT5-induced, antiapoptotic genes Bcl-X_L and XIAP, and prevented cell death in the hippocampal CA1 subregion. The JAK2 inhibitor tyrphostin AG490 blunted STAT5 phosphorylation and rhEPO-induced cytoprotection of CA1 neurons. These results show STAT5 to be a mediator of erythropoietin's neuroprotective mechanisms, but do not exclude the contributions of other signaling mechanisms downstream of JAK2. Indeed, in the same model, the PI3K inhibitor LY294002 similarly blunted rhEPO-induced neuroprotection and phosphorylation of PI3K's effectors Akt and GSK-3β (Zhang et al., 2006). In neonatal rats subjected to permanent MCA occlusion, ip rhEPO injections 15 min, 1 days, and 2 days after focal ischemia decreased infarct volume and apoptosis, increased phosphorylation of JAK2 and STAT5, and augmented antiapoptotic Bcl-X_L content in the ischemic tissue (Sola et al., 2005).

Erythropoietin pretreatment prevented apoptosis of neuron-like PC12 cells exposed to the β-amyloid peptide, $A\beta_{25-35}$ (Ma et al., 2014). In parallel with its antiapoptotic effect, erythropoietin increased JAK2 and STAT5 phosphor activation and P-STAT5 translocation to the nucleus, increased cellular content of antiapoptotic Bcl-X_L, and decreased caspase-3 content and activity. Erythropoietin's effects on Bcl-X_L, caspase-3, and apoptosis were blunted by the JAK2 inhibitor AG490 or by siRNA suppression of STAT5 expression.

On the other hand, erythropoietin prevented glutamate-induced cell death in primary cultured hippocampal neurons from both wild-type and STAT5-null mice, indicating that STAT5 was not required for erythropoietin's cytoprotective effect. Inhibitors of PI3K/Akt signaling blocked the protection, but STAT5 inhibitors did not (Byts et al., 2008).

6.3 Extracellular Signal-Related Kinase-1/2

The role of yet another kinase, ERK-1/2, in erythropoietin-induced cerebroprotection is uncertain. It is known that phosphorylated ERK-1/2 phosphorylates proapoptotic Bad (Jin, Mao, Zhu, & Greenberg, 2002), Bim (Harada, Quearry, Ruiz-Vela, & Korsmeyer, 2004), and caspase-9 (Allan et al., 2003), thereby suppressing the proapoptotic activity of these proteins. In rats subjected to MCA occlusion-reperfusion, postischemic MCA infusion of erythropoietin activated ERK and increased brain VEGF content (Dang et al., 2011). On the other hand, pyruvate-induced, erythropoietin-mediated reductions in brain infarct and apoptosis following MCA occlusion in rats were accompanied by robust activation of Akt but not ERK-1/2 (Ryou et al., 2012). Activation of the two kinases by erythropoietin may follow different time courses.

6.4 Nuclear Factor-κB

The transcription factor, nuclear factor κB (NF-κB), may contribute to erythropoietin's antiapoptotic effects. In a cell culture model of Alzheimer's disease, exposure of primary hippocampal neurons to amyloid-β provoked externalization of plasma membrane phosphatidylserine (an early event in the apoptotic cascade), genomic DNA fragmentation, and neuronal injury (Chong, Li, & Maiese, 2005). Addition of erythropoietin to the culture medium concentration dependently triggered NF-κB translocation to the nucleus and prevented neuronal apoptosis in a manner abrogated by siRNA suppression of NF-κB expression. Thus, erythropoietin protection of hippocampal neurons from amyloid-β exposure required NF-κB as an essential mediator (Chong et al., 2005). JAK2 phosphorylates NF-κB's inhibitory subunit, IκB, releasing the transcription factor, and allowing its translocation to the nucleus to activate expression of neuroprotective genes (Digicaylioglu & Lipton, 2001).

7. CLINICAL TRIALS OF ERYTHROPOIETIN FOR NEUROPROTECTION

The mounting preclinical evidence of erythropoietin's robust neuroprotective capabilities has prompted a limited number of randomized, placebo-controlled clinical trials of erythropoietin therapy for cerebral ischemia. In very preterm infants, rhEPO proved safe and effective as a neuroprotectant when intravenous treatment was initiated 72 h after birth (Song et al., 2016). rhEPO lowered the combined incidence of mortality and moderate or severe neurological disability by 50%. In another study, rhEPO decreased neurodevelopmental disability in preterm infants (Zhang et al., 2014). In newborns undergoing hypothermia therapy for hypoxic-ischemic encephalopathy, erythropoietin (1000 U/kg, iv), given at 1, 2, 3, 5, and 7 days after birth, sharply lowered brain injury score at 5 days and improved motor development at 1 year (Wu et al., 2016). Collectively, these reports demonstrate significant clinical benefit of erythropoietin therapy in infants.

In a small pilot trial in stroke patients (Ehrenreich et al., 2002), daily iv injections of 33,000 IU erythropoietin for the first 3 days after ischemic stroke improved recovery of neurocognitive function and decreased the persistent neurological deficit evident 18–30 days after stroke. The first erythropoietin injection was administered up to 8 h after onset of stroke symptoms, indicating that erythropoietin was effective even when given after several hours of brain ischemia. However, a subsequent, larger trial showed erythropoietin to be ineffective and even harmful (Ehrenreich et al., 2009), especially when combined with rtPA. Bolus 40,000 IU dosages of erythropoietin were given iv at 6, 24, and 48 h of onset of stroke symptoms. Erythropoietin produced no benefit, but increased mortality by 82%. This untoward outcome was ascribed to an adverse interaction of erythropoietin with rtPA, which was administered to a large fraction of the patients (Ehrenreich et al., 2009). This trial raised significant safety concerns for combining erythropoietin with rtPA thrombolytic therapy.

Not all erythropoietin trials in adult stroke victims have been negative. In a randomized trial in patients with neuroimaging-confirmed ischemic stroke, erythropoietin, given in 6 iv boluses totaling 56,000 IU over 60 h, improved neurological function at 14 and 28 days (Asadi et al., 2013). Moreover, erythropoietin's favorable effects persisted for several weeks, suggesting that the improvement may be permanent.

In a clinical trial of erythropoietin therapy for victims of out-of-hospital cardiac arrest, erythropoietin (1000–1500 IU/kg) was injected iv within 1–2 min of start of cardiopulmonary resuscitation by emergency medical service personnel. Erythropoietin increased survival to ICU admission, recovery of spontaneous circulation, 24 h survival, and survival to hospital discharge, but failed to improve neurological outcome (Grmec, Strnad, Kupnik, Sinkovič, & Gazmuri, 2009). A recent meta-analysis of 26 clinical trials of rhEPO for tissue protection (not limited to brain) revealed that rhEPO failed to afford a mortality benefit in any of the trials, and even increased mortality or other adverse events in 5 of the 26 trials (Lund, Lundby, & Olsen, 2014).

Erythropoietin produced promising results in a preliminary clinical trial in patients undergoing coronary artery revascularization on cardiopulmonary bypass (Lakic et al., 2010). Cardiopulmonary bypass can produce neurologic complications as a result of the systemic inflammatory response ignited by anesthesia, surgical trauma, and especially the passage of blood through the artificial components of the extracorporeal circuit, which activates proinflammatory neutrophils (Olivencia-Yurvati & Mallet, 2012). Four of 10 patients in the control group had evidence of new ischemic lesions within 5 days after surgery, but none of the 10 patients receiving postsurgical intravenous erythropoietin therapy had evidence of brain lesions.

7.1 Surmounting the BBB

High systemic doses of erythropoietin, a 30.4-kDa glycoprotein, are required for sufficient amounts to enter the brain parenchyma (Rabie & Marti, 2008). It is estimated that the transfer of systemically administered erythropoietin across the BBB is less than 1% efficient (Brines et al., 2000; Jones, Novak, & Elliott, 2013; Juul et al., 2004). In a patient undergoing brain surgery for tumor resection, an intravenous injection of 6000 IU rhEPO increased serum erythropoietin activity from ca. 13 to >6500 IU/L for at least 60 min, but there was no increase in erythropoietin activity in the cerebrospinal fluid (Buemi et al., 2000). Circulating erythropoietin can be transported across the cerebrovascular endothelium and, thus, surmount the BBB, in a process that begins with erythropoietin binding EPOR on the luminal surface of the endothelium (Brines, 2002). However, this transport process has a much lower capacity than free diffusion, so high dosages of exogenous erythropoietin are required to achieve neuroprotective concentrations in the brain parenchyma. These high dosages may produce untoward complications, e.g., polycythemia and thrombotic secondary stroke (Zhang et al., 2010).

BBB disruption may increase erythropoietin entry. In patients with traumatic brain injury, the appearance of circulating erythropoietin in the ventricular cerebrospinal fluid correlated with the extent of compromise of the BBB (Marti et al., 1997). Thus, circulating erythropoietin does not freely cross the intact BBB, but may enter the parenchyma if the barrier is disrupted.

The high dosages of exogenous erythropoietin required to surmount the intact BBB and achieve neuroprotection raise concerns about undesirable side effects, e.g., increased risks of thrombotic events and even death (McPherson & Juul, 2008). High erythropoietin dosages were associated with increased incidence of hemorrhagic transformation, especially when combined with rtPA (Ehrenreich et al., 2009).

8. ENDOGENOUS ERYTHROPOIETIN PROTECTS THE BRAIN

Erythropoietin's erythropoietic actions have been ascribed to its production primarily in O_2-sensing, pertitubular fibroblasts at the renal corticomedullary border (Wenger & Hoogewijs, 2010). Over the past 15 years, it has become increasingly recognized that other tissues, including brain, can synthesize erythropoietin, which exerts effects distinct from its hematopoietic actions and that erythropoietin produced within the brain is remarkably neuroprotective (Nguyen, Cherry, Scott, Ryou, & Mallet, 2014). Substantial erythropoietin expression was detected in several regions of brains from preterm human fetuses; indeed, in cerebellum, pituitary gland, and cerebral cortex, erythropoietin mRNA abundance rivaled that of the conventionally erythropoietin-expressing liver and kidneys (Dame, Juul, & Christensen, 2001). In patients undergoing carotid endarterectomy, jugular venous concentrations of erythropoietin increased during carotid cross-clamp (Carelli et al., 2016).

Astrocytes are the major source of brain erythropoietin and can upregulate erythropoietin production in response to hypoxia (Swanson, Ying, & Kauppinen, 2004). Exposure to severe hypoxia (2% O_2) increased erythropoietin expression and synthesis only in mixed astrocyte–neuron cultures, while EPOR expression was detected only in the neuron-enriched cultures (Liu, Suzuki, Guo, Mizuno, & Urabe, 2006). Thus, hypoxia induces erythropoietin expression in astrocytes which protects neurons from hypoxic/ischemic insult by interacting with neuronal EPOR. Neurons, astrocytes, and microglia all harbor EPOR and may respond to astrocyte-secreted erythropoietin in a paracrine or, in astrocytes, autocrine manner (Brines et al., 2000; Ruscher et al., 2002).

Induction of endogenous erythropoietin by hypobaric hypoxia was found to be neuroprotective against kainate-induced excitotoxicity in rats (Costa et al., 2013). In nonhypoxic rats, kainate induced lipid peroxidation and NO overproduction, and activated caspase-3 and proinflammatory genes. All of these harmful effects of kainate were attenuated by antecedent hypobaric exposures (380 mm Hg) 3 h/day for 6 days, which induced endogenous erythropoietin expression in brain (Costa et al., 2013).

8.1 Regulation of Erythropoietin Gene Expression

The discovery that brain produces erythropoietin suggests novel strategies to harness the hormone's neuroprotective mechanisms based on induction of erythropoietin expression and synthesis. The erythropoietin gene is expressed under the influence of the transcription factor, hypoxia-inducible factor-1 (HIF-1), which mediates hypoxia activation of erythropoietin production within the brain (Fandrey, 2004; Marti, 2004). Although erythropoietin was among the first identified products of HIF-1 gene program, HIF-1 is now known to promote expression of >100 genes, including genes encoding VEGF and the entire glycolytic sequence (Loor & Schumacker, 2008). Collectively, this gene program imparts robust resistance to hypoxia/ischemia in brain, heart, and other organs.

The active HIF-1 transcription factor is a heterodimer consisting of an O_2-regulated α subunit and a constitutively expressed β subunit. In the presence of normal O_2 concentrations and the essential cofactors α-ketoglutarate and Fe^{2+}, two proline residues in the α subunit are hydroxylated by prolyl hydroxylase, which increases the subunit's affinity for the von Hippel Lindau tumor E3 ubiquitin ligase complex, allowing the latter to ubiquitinate HIF-1α, targeting it for proteasomal degradation (Pugh, 2016).

Hypoxia stabilizes HIF-1α by two mechanisms (Fan, Heijnen, van der Kooij, Groenendaal, & van Bel, 2009). First, hypoxia deprives prolyl hydroxylase of O_2, which is required for prolyl hydroxylation; this mechanism is the basis of renal erythropoietin production in response to arterial hypoxemia (Wenger & Hoogewijs, 2010). Second, hypoxia intensifies $^{\bullet}O_2^-$ formation by the electron transport chain. $^{\bullet}O_2^-$ is the immediate precursor of H_2O_2, which via Fenton chemistry converts Fe^{2+} to Fe^{3+}, removing the source of electrons for prolyl hydroxylase. Depriving prolyl hydroxylase of Fe^{2+} or O_2 stabilizes HIF-1α, allowing it to translocate to the nucleus and combine with the β-submit forming the active transcription factor, which interacts with hypoxia-response elements in the promoters of hypoxia-responsive genes, thereby activating expression of genes that increase the brain's resistance to ischemic insults (Siddiq, Aminova, & Ratan, 2007).

In astrocytes, hypoxia stabilizes HIF-1α and activates HIF-1-driven gene expression, including erythropoietin. Astrocytes release erythropoietin, which acts in a paracrine fashion on EPOR on adjacent neurons, thereby initiating JAK2/PI3K/Akt signaling which phosphorylates and inactivates Bad and, thus, suppresses neuronal apoptosis (Ruscher et al., 2002). In mice, 1–6 h exposure to 8% O_2 24 h before permanent MCA occlusion decreased infarct volume by 30% (Bernaudin et al., 2002). Hypoxia induced HIF-1α translocation to the nucleus and activated HIF-1-driven expression of erythropoietin and VEGF.

Nitric oxide also has been identified as a mediator of hypoxic induction of erythropoietin and EPOR in brain. Hypoxia activates NO formation in brain, which in turn activates EPOR expression in neurons, making the neurons more responsive to neuroprotective erythropoietin signaling (Chen et al., 2010). Accordingly, NO donors recapitulated EPOR induction in the absence of hypoxia.

Methylene blue, a HIF-1α stabilizer, increased survival of neuronally derived HT22 cells subjected to oxygen-glucose deprivation (Ryou et al., 2015). In addition, methylene blue increased HIF-1α content and nuclear translocation, activities of the HIF-1 gene program products hexokinase and glucose 6-phosphate dehydrogenase, and activation of Akt signaling.

Paradoxically, hyperbaric oxygen pretreatment of adult rats increased HIF-1α content, HIF-1 transcriptional activity, and erythropoietin expression and content in brains of adult rats. Moreover, the hyperbaric oxygen decreased brain infarct and improved neurobehavioral recovery after focal cerebral ischemia (Gu et al., 2008). These findings were concordant with the reactive oxygen species mechanism of HIF-1α activation, since both hypoxia and hyperoxia can increase oxyradical formation.

8.2 Prolyl Hydroxylase Inhibitors as Neuroprotectants

Several preclinical studies have demonstrated neuroprotective actions of prolyl hydroxylase inhibitors. In cultured rat hippocampal slices subjected to oxygen-glucose deprivation, the prolyl hydroxylase inhibitors $CoCl_2$ and deferoxamine increased erythropoietin secretion by astrocytes, in parallel with neuronal protection by these inhibitors (Jones et al., 2013). In neonatal mice, ip injection of the prolyl hydroxylase inhibitor FG-4497 before 6 h hypoxia (8% O_2) exposure increased brain contents of both HIF-1α and HIF-2α and increased expression of erythropoietin, VEGF, and other cerebroprotective genes while suppressing proinflammatory inducible

NOS and minimizing neuronal apoptosis (Trollmann, Richter, Jung, Walkinshaw, & Brackmann, 2014). In mice subjected to ischemic stroke, FG-4497 minimized brain injury and edema and preserved structural integrity of the BBB, in parallel with increased contents of HIF-1α, erythropoietin, and VEGF (Reischl et al., 2014).

In a mouse traumatic brain injury model, the prolyl hydroxylase inhibitor dimethyloxaloylglycine (DMOG) decreased cell death and lesion volume (Sen & Sen, 2016). DMOG increased HIF-1α content and mRNA abundances of the HIF-1 gene program products VEGF, pyruvate dehydrogenase kinase, and erythropoietin. Also in mice, DMOG, injected ip 30–60 min after permanent MCA ligation, increased cerebrocortical HIF-1α content and expression of erythropoietin, VEGF, endothelial NOS, and pyruvate dehydrogenase kinase-1; suppressed caspase-3; decreased infarct volume and behavioral deficits; and partially preserved blood flow in the ischemic territory (Ogle, Gu, Espinera, & Wei, 2012).

Another α-ketoglutarate analog, the natural intermediary metabolite pyruvate, decreased brain infarct volume and penumbral apoptosis by 85% in rats subjected to 2 h MCA occlusion and 24 h reperfusion (Ryou et al., 2012). Intravenous pyruvate infusion from 60 min MCA occlusion until 30 min reperfusion sharply increased erythropoietin contents in ipsi- and contralateral cerebral cortex. In cultured HT-22 neuronal cells, pyruvate induced HIF-1α, erythropoietin, and Akt phosphorylation, and prevented oxygen-glucose deprivation induced cell death, while soluble EPOR and siRNA against HIF-1α abolished pyruvate cytoprotection.

Collectively, these preclinical studies demonstrate HIF-1 stabilization, whether by hypoxia or prolyl hydroxylase inhibitors, to be neuroprotective. Although induction of HIF and its gene program remains to be evaluated by clinical trials, these results raise the exciting prospect that HIF-1 stabilization might overcome the limitations of less successful interventions for stroke and cardiac arrest.

8.3 Potential Limitations of Erythropoietin-Based Therapy

There are several limitations to hypoxia induction of endogenous erythropoietin for brain ischemia. De novo gene expression and protein synthesis are inherently slow processes, and several hours may be required to generate neuroprotective amounts of erythropoietin in the at-risk tissue. Thus, hypoxia may not suitable for acute treatment during the first few hours after the ischemic event. Superimposing hypoxia before resolving the cerebrovascular lesion may intensify the ischemic insult. A nonselective stimulus affecting the entire body, hypoxia could induce cardiac arrhythmias in patients with severe coronary artery stenosis, or pulmonary vasoconstriction increasing right ventricular afterload. Long-term hypoxia treatment could increase hematocrit and, thus, blood viscosity, which not only increases cardiac workload but could compromise blood flow in the cerebral microcirculation. On the other hand, moderate, normobaric hypoxia was found safe, nonarrhythmogenic, and clinically efficacious in patients with coronary artery disease (Burtscher et al., 2004) or chronic obstructive pulmonary disease (Haider et al., 2009).

9. SUMMARY

The brain is extremely vulnerable to interruptions of its supply of blood-borne oxygen and nutrients. Ischemic syndromes of the central nervous system rank among the leading causes of death and disability and are major risk factors for neurodegenerative disorders. Cerebral ischemia-reperfusion ignites a dauntingly complex injury cascade which has defied pharmacological intervention. Erythropoietin addresses multiple components of this cascade and has proven powerfully neuroprotective in many preclinical brain ischemia models, yet clinical trials of exogenous erythropoietin failed to demonstrate consistent benefit except in preterm infants. An alternative strategy, induction of HIF-driven erythropoietin expression within the brain parenchyma, could overcome the barriers to effective neuroprotection for focal and global ischemia, and awaits stringent evaluation in clinical trials.

ACKNOWLEDGMENTS

This work was funded by research grant NS076975 from the U.S. National Institute of Neurological Disorders and Stroke, and by a collaborative research grant from the Institute for Cardiovascular and Metabolic Diseases, University of North Texas Health Science Center.

REFERENCES

Abbott, N. J., Patabendige, A. A., Dolman, D. E., Yusof, S. R., & Begley, D. J. (2010). Structure and function of the blood-brain barrier. *Neurobiology of Disease*, *37*, 13–25.

Abramov, A. Y., & Duchen, M. R. (2008). Mechanisms underlying the loss of mitochondrial membrane potential in glutamate excitotoxicity. *Biochimica et Biophysica Acta*, *1777*, 953–964.

Adibhatla, R. M., & Hatcher, J. F. (2008). Tissue plasminogen activator (tPA) and matrix metalloproteinases in the pathogenesis of stroke: Therapeutic strategies. *CNS and Neurological Disorders Drug Targets*, 7, 243–253.

Albers, G. W., Goldstein, L. B., Hall, D., Lesko, L. M., & Aptiganel acute stroke investigators. (2001). Aptigel hydrochloride in acute stroke: A randomized controlled trial. *JAMA, 286*, 2673–2682.
Allan, L. A., Morrice, N., Brady, S., Magee, G., Pathak, S., & Clarke, P. R. (2003). Inhibition of caspase-9 through phosphorylation at Thr 125 by ERK MAPK. *Nature Cell Biology, 5*, 647–654.
Amé, J. C., Rolli, V., Schreiber, V., Niedergang, C., Apiou, F., Decker, P., et al. (1999). PARP-2, a novel mammalian DNA damage-dependent poly(ADP-ribose) polymerase. *The Journal of Biological Chemistry, 274*, 17860–17868.
Andoh, T., Echigo, N., Kamiya, Y., Hayashi, M., Kudoh, I., & Goto, T. (2011). Effects of erythropoietin on intracellular calcium concentration of rat primary cortical neurons. *Brain Research, 1387*, 8–18.
Asadi, B., Askari, G. R., Khorvash, F., Bagherpur, A., Mehrabi, F., Karimi, M., et al. (2013). Neuroprotective effects of erythropoietin in acute ischemic stroke. *International Journal of Preventive Medicine, 4*(Suppl. 2), S306–S312.
Bahcekapili, N., Uzüm, G., Gökkusu, C., Kuru, A., & Ziylan, Y. Z. (2007). The relationship between erythropoietin pretreatment with blood-brain barrier and lipid peroxidation after ischemia/reperfusion in rats. *Life Sciences, 80*, 1245–1251.
Baines, C. P. (2010). The cardiac mitochondrion: Nexus of stress. *Annual Review of Physiology, 72*, 61–80.
Beckman, J. S., & Koppenol, W. H. (1996). Nitric oxide, superoxide, and peroxynitrite: The good, the bad, and the ugly. *American Journal of Physiology. Cell Physiology, 271*, C1424–C1437.
Begley, D. J., & Brightman, D. W. (2003). Structural and functional aspects of the blood-brain barrier. *Progress in Drug Research, 61*, 39–78.
Bernard, S. A., Gray, T. W., Buist, M. D., Jones, B. M., Silvester, W., Gutteridge, G., et al. (2002). Treatment of comatose survivors of out-of-hospital cardiac arrest with induced hypothermia. *The New England Journal of Medicine, 346*, 557–563.
Bernaudin, M., Nedelec, A. S., Divoux, D., MacKenzie, E. T., Petit, E., & Schumann-Bard, P. (2002). Normobaric hypoxia induces tolerance to focal permanent cerebral ischemia in association with an increased expression of hypoxia-inducible factor-1 and its target genes, erythropoietin and VEGF, in the adult mouse brain. *Journal of Cerebral Blood Flow and Metabolism, 22*, 393–403.
Brines, M. (2002). What evidence supports use of erythropoietin as a novel neurotherapeutic? *Oncology, 16*, 79–89.
Brines, M. L., Ghezzi, P., Keenan, S., Agnello, D., de Lanerolle, N. C., Cerami, C., et al. (2000). Erythropoietin crosses the blood-brain barrier to protect against experimental brain injury. *Proceedings of the National Academy of Sciences of the United States of America, 97*, 10526–10531.
Broughton, B. R., Reutens, D. C., & Sobey, C. G. (2009). Apoptotic mechanisms after cerebral ischemia. *Stroke, 40*, e331–e339.
Brouns, R., & De Deyn, P. P. (2009). The complexity of neurobiological processes in acute ischemic stroke. *Clinical Neurology and Neurosurgery, 111*, 483–495.
Brown, G. C., & Borutaite, V. (2004). Inhibition of mitochondrial respiratory complex I by nitric oxide, peroxynitrite and S-nitrosothiols. *Biochimica et Biophysica Acta, 1658*, 44–49.
Buemi, M., Allegra, A., Corica, F., Floccari, F., D'Avella, D., Aloisi, C., et al. (2000). Intravenous recombinant erythropoietin does not lead to an increase in cerebrospinal fluid erythropoietin concentration. *Nephrology, Dialysis, Transplantation, 15*, 422–423.
Burtscher, M., Pachinger, O., Ehrenbourg, I., Mitterbauer, G., Faulhaber, M., Pühringer, R., et al. (2004). Intermittent hypoxia increases exercise tolerance in elderly men with and without coronary artery disease. *International Journal of Cardiology, 96*, 247–254.

Byts, N., Samoylenko, A., Fasshauer, T., Ivanisevic, M., Hennighausen, L., Ehrenreich, H., et al. (2008). Essential role for Stat5 in the neurotrophic but not in the neuroprotective effect of erythropoietin. *Cell Death and Differentiation*, *15*, 783–792.

Calapai, G., Marciano, M. C., Corica, F., Allegra, A., Parisi, A., Frisina, N., et al. (2000). Erythropoietin protects against brain ischemic injury by inhibition of nitric oxide formation. *European Journal of Pharmacology*, *401*, 349–356.

Candelario-Jalil, E. (2009). Injury and repair mechanisms in ischemic stroke: Considerations for the development of novel neurotherapeutics. *Current Opinion in Investigational Drugs*, *10*, 644–654.

Carelli, S., Ghilardi, G., Bianciardi, P., Latorre, E., Rubino, F., Bissi, M., et al. (2016). Enhanced brain release of erythropoietin, cytokines and NO during carotid clamping. *Neurological Sciences*, *37*, 243–252.

Chang, J. J., Emanuel, B. A., Mack, W. J., Tsivgoulis, G., & Alexandrov, A. V. (2014). Matrix metalloproteinase-9: Dual role and temporal profile in intracerebral hemorrhage. *Journal of Stroke and Cerebrovascular Diseases*, *23*, 2498–2505.

Chaturvedi, M., & Kaczmarek, L. (2014). Mmp-9 inhibition: A therapeutic strategy in ischemic stroke. *Molecular Neurobiology*, *49*, 563–573.

Chen, Z. Y., Wang, L., Asavaritkrai, P., & Noguchi, C. T. (2010). Up-regulation of erythropoietin receptor by nitric oxide mediates hypoxia preconditioning. *Journal of Neuroscience Research*, *88*, 3180–3188.

Chong, Z. Z., Li, F., & Maiese, K. (2005). Stress in the brain: Novel cellular mechanisms of injury linked to Alzheimer's disease. *Brain Research. Brain Research Reviews*, *49*, 1–21.

Clark, D. D., & Sokoloff, L. (1999). Circulation and energy metabolism of the brain. In G. J. Siegel, B. W. Agranoff, R. W. Albers, S. K. Fisher, & M. D. Uhler (Eds.), *Basic neurochemistry: Molecular, cellular and medical aspects* (pp. 637–670). Philadelphia: Lippincott.

Costa, D. C., Alva, N., Trigueros, L., Gamez, A., Carbonell, T., & Rama, R. (2013). Intermittent hypobaric hypoxia induces neuroprotection in kainate-induced oxidative stress in rats. *Journal of Molecular Neuroscience*, *50*, 402–410.

Dallas, M., Boycott, H. E., Atkinson, L., Miller, A., Boyle, J. P., Pearson, H. A., et al. (2007). Hypoxia suppresses glutamate transport in astrocytes. *The Journal of Neuroscience*, *27*, 3946–3955.

Dame, C., Juul, S. E., & Christensen, R. D. (2001). The biology of erythropoietin in the central nervous system and its neurotrophic and neuroprotective potential. *Biology of the Neonate*, *79*, 228–235.

Dang, S., Liu, X., Fu, P., Gong, W., Yan, F., Han, P., et al. (2011). Neuroprotection by local intra-arterial infusion of erythropoietin after focal cerebral ischemia in rats. *Neurological Research*, *33*, 520–528.

Davis, S. M., Lees, K. R., Albers, G. W., Diener, H. C., Markabi, S., Karlsson, G., et al. (2000). Selfotel in acute ischemic stroke: Possible neurotoxic effects of an NMDA antagonist. *Stroke*, *31*, 347–354.

Dejonckheere, E., Vandenbroucke, R. E., & Libert, C. (2011). Matrix metalloproteinases as drug targets in ischemia/reperfusion injury. *Drug Discovery Today*, *16*, 762–778.

Derex, L., & Nighoghossian, N. (2008). Intracerebral haemorrhage after thrombolysis for acute ischaemic stroke: An update. *Journal of Neurology, Neurosurgery, and Psychiatry*, *79*, 1093–1099.

Digicaylioglu, M., & Lipton, S. A. (2001). Erythropoietin-mediated neuroprotection involves cross-talk between Jak2 and NF-kappaB signalling cascades. *Nature*, *412*, 641–647.

Doeppner, T. R., & Hermann, D. M. (2010). Free radical scavengers and spin traps—Therapeutic implications for ischemic stroke. *Best Practice & Research. Clinical Anaesthesiology*, *24*, 511–520.

Eaton, P., Wright, N., Hearse, D. J., & Shattock, M. J. (2002). Glyceraldehyde phosphate dehydrogenase oxidation during cardiac ischemia and reperfusion. *Journal of Molecular and Cellular Cardiology*, *34*, 1549–1560.

Ehrenreich, H., Hasselblatt, M., Dembowski, C., Cepek, L., Lewczuk, P., Stiefel, M., et al. (2002). Erythropoietin therapy for acute stroke is both safe and beneficial. *Molecular Medicine*, *8*, 495–505.

Ehrenreich, H., Weissenborn, K., Prange, H., Schneider, D., Weimar, C., Wartenberg, K., et al. (2009). Recombinant human erythropoietin in the treatment of acute ischemic stroke. *Stroke*, *40*, e647–e656.

Elting, J. W., Sulter, G. A., Kaste, M., Lees, K. R., Diener, H. C., Hommel, M., et al. (2002). AMPA antagonist ZK200775 in patients with acute ischemic stroke: Possible glial cell toxicity detected by monitoring of S-100B serum levels. *Stroke*, *33*, 2813–2818.

Fan, X., Heijnen, C. J., van der Kooij, M. A., Groenendaal, F., & van Bel, F. (2009). The role and regulation of hypoxia-inducible factor-1a expression in brain development and neonatal hypoxic-ischemic brain injury. *Brain Research Reviews*, *62*, 99–108.

Fandrey, J. (2004). Oxygen-dependent and tissue-specific regulation of erythropoietin gene expression. *American Journal of Physiology. Regulatory, Integrative and Comparative Physiology*, *286*, R977–R988.

Ferdinandy, P., & Schulz, R. (2003). Nitric oxide, superoxide and peroxynitrite in myocardial ischemia-reperfusion injury and preconditioning. *British Journal of Pharmacology*, *138*, 532–543.

Ferrer, I., & Planas, A. M. (2003). Signaling of cell death and cell survival following focal cerebral ischemia: Life and death struggle in the penumbra. *Journal of Neuropathology and Experimental Neurology*, *62*, 329–339.

Fraser, P. A. (2011). The role of free radical generation in increasing cerebrovascular permeability. *Free Radical Biology and Medicine*, *51*, 967–977.

Genc, S., Akhisaroglu, M., Kuralay, F., & Genc, K. (2002). Erythropoietin restores glutathione peroxidase activity in 1-methyl-4-phenyl-1,2,5,6-tetrahydropyridine-induced neurotoxicity in C57BL mice and stimulates murine astroglial glutathione peroxidase production in vitro. *Neuroscience Letters*, *321*, 73–76.

Ghosh, M., Yang, Y., Rothstein, J. D., & Robinson, M. B. (2011). Nuclear factor-κB contributes to neuron-dependent induction of glutamate transporter-1 expression in astrocytes. *The Journal of Neuroscience*, *31*, 9159–9169.

Girotra, S., Chan, P. S., & Bradley, S. M. (2015). Post-resuscitation care following out-of-hospital and in-hospital cardiac arrest. *Heart*, *101*, 1943–1949.

Go, A. S., Mozaffarian, D., Roger, V. L., & American Heart Association Statistics Committee and Stroke Statistics Subcommittee. (2013). Heart disease and stroke statistics—2013 update. *Circulation*, *127*, e6–e245.

Gonzalez, F. F., McQuillen, P., Mu, D., Chang, Y., Wendland, M., Vexler, Z., et al. (2007). Erythropoietin enhances long-term neuroprotection and neurogenesis in neonatal stroke. *Developmental Neuroscience*, *29*, 321–330.

Grasso, G., Sfacteria, A., Meli, F., Fodale, V., Buemi, M., & Iacopino, D. G. (2007). Neuroprotection by erythropoietin administration after experimental traumatic brain injury. *Brain Research*, *1182*, 99–105.

Green, A. R. (2008). Pharmacological approaches to acute ischaemic stroke: Reperfusion certainly, neuroprotection possibly. *British Journal of Pharmacology*, *153*(Suppl. 1), S325–S338.

Grmec, Š., Strnad, M., Kupnik, D., Sinkovič, A., & Gazmuri, R. J. (2009). Erythropoietin facilitates the return of spontaneous circulation and survival in victims of out-of-hospital cardiac arrest. *Resuscitation*, *80*, 631–637.

Gu, Y., Dee, C. M., & Shen, J. (2011). Interaction of free radicals, matrix metalloproteinases and caveolin-1 impacts blood-brain barrier permeability. *Frontiers in Bioscience*, *3*, 1216–1231.
Gu, G. J., Li, Y. P., Peng, Z. Y., Xu, J. J., Kang, Z. M., Xu, W. G., et al. (2008). Mechanism of ischemic tolerance induced by hyperbaric oxygen preconditioning involves upregulation of hypoxia-inducible factor-1alpha and erythropoietin in rats. *Journal of Applied Physiology*, *104*, 1185–1191.
Gunnarson, E., Song, Y., Kowalewski, J. M., Brismar, H., Brines, M., Cerami, A., et al. (2009). Erythropoietin modulation of astrocyte water permeability as a component of neuroprotection. *Proceedings of the National Academy of Sciences of the United States of America*, *106*, 1602–1607.
Haider, T., Casucci, G., Linser, T., Faulhaber, M., Gatterer, H., Ott, G., et al. (2009). Interval hypoxic training improves autonomic cardiovascular and respiratory control in patients with mild chronic obstructive pulmonary disease. *Journal of Hypertension*, *27*, 1648–1654.
Harada, H., Quearry, B., Ruiz-Vela, A., & Korsmeyer, S. J. (2004). Survival factor-induced extracellular signal-regulated kinase phosphorylates BIM, inhibiting its association with BAX and proapoptotic activity. *Proceedings of the National Academy of Sciences of the United States of America*, *101*, 15313–15317.
Hausladen, A., & Fridovich, I. (1994). Superoxide and peroxynitrite inactivate aconitases, but nitric oxide does not. *The Journal of Biological Chemistry*, *269*, 29405–29408.
Hazell, A. S. (2007). Excitotoxic mechanisms in stroke: An update of concepts and treatment strategies. *Neurochemistry International*, *50*, 941–953.
Hellewell, S. C., Yan, E. B., Alwis, D. S., Bye, N., & Morganti-Kossmann, M. C. (2013). Erythropoietin improves motor and cognitive deficit, axonal pathology, and neuroinflammation in a combined model of diffuse traumatic brain injury and hypoxia, in association with upregulation of the erythropoietin receptor. *Journal of Neuroinflammation*, *10*, 156.
Hypothermia after Cardiac Arrest Study Group. (2002). Mild therapeutic hypothermia to improve the neurologic outcome after cardiac arrest. *The New England Journal of Medicine*, *346*, 549–556.
Idris, A. H., Roberts, L. J., II, Caruso, L., Showstark, M., Layon, A. J., Becker, L. B., et al. (2005). Oxidant injury occurs rapidly after cardiac arrest, cardiopulmonary resuscitation, and reperfusion. *Critical Care Medicine*, *33*, 2043–2048.
Ischiropoulos, H. (1998). Biological tyrosine nitration: A pathophysiological function of nitric oxide and reactive oxygen species. *Archives of Biochemistry and Biophysics*, *356*, 1–11.
Jia, Y., Mo, S. J., Feng, Q. Q., Zhan, M. L., OuYang, L. S., Chen, J. C., et al. (2014). EPO-dependent activation of PI3K/Akt/FoxO3a signalling mediates neuroprotection in in vitro and in vivo models of Parkinson's disease. *Journal of Molecular Neuroscience*, *53*, 117–124.
Jickling, G. C., Liu, D., Stamova, B., Ander, B. P., Zhan, X., Lu, A., et al. (2014). Hemorrhagic transformation after ischemic stroke in animals and humans. *Journal of Cerebral Blood Flow and Metabolism*, *34*, 185–199.
Jin, K., Mao, X. O., Zhu, Y., & Greenberg, D. A. (2002). MEK and ERK protect hypoxic cortical neurons via phosphorylation of Bad. *Journal of Neurochemistry*, *80*, 119–125.
Jin, R., Yang, G., & Li, G. (2010). Molecular insights and therapeutic targets for blood-brain barrier disruption in ischemic stroke: Critical role of matrix metalloproteinases and tissue-type plasminogen activator. *Neurobiology of Disease*, *38*, 376–385.
Jones, S. M., Novak, A. E., & Elliott, J. P. (2013). The role of HIF in cobalt-induced ischemic tolerance. *Neuroscience*, *252*, 420–430.
Juul, S. E., McPherson, R. J., Farrell, F. X., Jolliffe, L., Ness, D. J., & Gleason, C. A. (2004). Erythropoietin concentrations in cerebrospinal fluid of nonhuman primates and fetal

sheep following high-dose recombinant erythropoietin. *Biology of the Neonate, 85*, 138–144.

Kalaria, R. N. (2000). The role of cerebral ischemia in Alzheimer's disease. *Neurobiology of Aging, 21*, 321–330.

Kalogeris, T., Bao, Y., & Korthuis, R. J. (2014). Mitochondrial reactive oxygen species: A double edged sword in ischemia/reperfusion vs preconditioning. *Redox Biology, 2*, 702–714.

Kato, H., & Kogure, K. (1999). Biochemical and molecular characteristics of the brain with developing cerebral infarction. *Cellular and Molecular Neurobiology, 19*, 93–108.

Kawakami, M., Sekiguchi, M., Sato, K., Kozaki, S., & Takahashi, M. (2001). Erythropoietin receptor-mediated inhibition of exocytoxic glutamate release confers neuroprotection during chemical ischemia. *The Journal of Biological Chemistry, 276*, 39469–39475.

Kilic, U., Kilic, E., Soliz, J., Bassetti, C. I., Gassmann, M., & Hermann, D. M. (2005). Erythropoietin protects from axotomy-induced degeneration of retinal ganglion cells by activating ERK-1/-2. *The FASEB Journal, 19*, 249–251.

Kim, K., Lee, S. G., Kegelman, T. P., Su, Z. Z., Das, S. K., Dash, R., et al. (2011). Role of excitatory amino acid transporter-2 (EAAT2) and glutamate in neurodegeneration: Opportunities for developing novel therapeutics. *Journal of Cellular Physiology, 226*, 2484–2493.

Kim, M. S., Seo, Y. K., Park, H. J., Lee, K. H., Lee, K. H., Choi, E. J., et al. (2010). The neuroprotective effect of recombinant human erythropoietin via an antiapoptotic mechanism on hypoxic-ischemic brain injury in neonatal rats. *Korean Journal of Pediatrics, 53*, 898–908.

Kimura, C., Oike, M., & Ito, Y. (2000). Hypoxia-induced alterations in Ca2+ mobilization in brain microvascular endothelial cells. *American Journal of Physiology. Heart and Circulatory Physiology, 279*, H2310–H2318.

Kimura, T., Sako, K., Tanaka, K., Kusakabe, M., Tanaka, T., & Nakada, T. (2002). Effect of mild hypothermia on energy state recovery following transient forebrain ischemia in the gerbil. *Experimental Brain Research, 145*, 83–90.

Kogure, T., & Kogure, K. (1997). Molecular and biochemical events within the brain subjected to cerebral ischemia (targets for therapeutical intervention). *Clinical Neuroscience, 4*, 179–183.

Konstady, B. B. (2012). The role of glutamate in neuronal ischemic injury: The role of spark in fire. *Neurological Sciences, 33*, 223–237.

Lakic, N., Surlan, K., Jerin, A., Meglic, B., Curk, N., & Bunc, M. (2010). Importance of erythropoietin in brain protection after cardiac surgery: A pilot study. *The Heart Surgery Forum, 13*, E185–E189.

Lan, K. M., Tien, L. T., Cai, Z., Lin, S., Pang, Y., Tanaka, S., et al. (2016). Erythropoietin ameliorates neonatal hypoxia-ischemia-induced neurobehavioral deficits, neuroinflammation, and hippocampal injury in the juvenile rat. *International Journal of Molecular Sciences, 17*, E289.

Lankhorst, C. E., & Wish, J. B. (2010). Anemia in renal disease: Diagnosis and management. *Blood Reviews, 24*, 39–47.

Larpthaveesarp, A., Georgevits, M., Ferriero, D. M., & Gonzalez, F. F. (2016). Delayed erythropoietin therapy improves histological and behavioral outcomes after transient neonatal stroke. *Neurobiology of Disease, 93*, 57–63.

Laver, S., Farrow, C., Turner, D., & Nolan, J. (2004). Mode of death after admission to an intensive care unit following cardiac arrest. *Intensive Care Medicine, 30*, 2126–2128.

Li, Y., Lu, Z., Keogh, C. L., Yu, S. P., & Wei, L. (2007). Erythropoietin-induced neurovascular protection, angiogenesis, and cerebral blood flow restoration after focal ischemia in mice. *Journal of Cerebral Blood Flow and Metabolism, 27*, 1043–1054.

Li, G., Ma, R., Huang, C., Tang, Q., Fu, Q., Liu, H., et al. (2008). Protective effect of erythropoietin on beta-amyloid-induced PC12 cell death through antioxidant mechanisms. *Neuroscience Letters*, *442*, 143–147.

Liao, Z. B., Jiang, G. Y., Tang, Z. H., Zhi, X. G., Sun, X. C., Tang, W. Y., et al. (2009). Erythropoietin can promote survival of cerebral cells by downregulating Bax gene after traumatic brain injury in rats. *Neurology India*, *57*, 722–728.

Liao, Z. B., Zhi, X. G., Shi, Q. H., & He, Z. H. (2008). Recombinant human erythropoietin administration protects cortical neurons from traumatic brain injury in rats. *European Journal of Neurology*, *15*, 140–149.

Liu, R., Suzuki, A., Guo, Z., Mizuno, Y., & Urabe, T. (2006). Intrinsic and extrinsic erythropoietin enhances neuroprtection against ischemia and reperfusion injury in vitro. *Journal of Neurochemistry*, *96*, 1101–1110.

Loor, G., & Schumacker, P. T. (2008). Role of hypoxia-inducible factor in cell survival during myocardial ischemia-reperfusion. *Cell Death and Differentiation*, *15*, 686–690.

Lourhmati, A., Buniatian, G. H., Paul, C., Verleysdonk, S., Buecheler, R., Buadze, M., et al. (2013). Age-dependent astroglial vulnerability to hypoxia and glutamate: The role for erythropoietin. *PloS One*, *8*, e77182.

Lund, A., Lundby, C., & Olsen, N. V. (2014). High-dose erythropoietin for tissue protection. *European Journal of Clinical Investigation*, *44*, 1230–1238.

Ma, R., Hu, J., Huang, C., Wang, M., Xiang, J., & Li, G. (2014). JAK2/STAT5/Bcl-xL signalling is essential for erythropoietin-mediated protection against apoptosis induced in PC12 cells by the amyloid β-peptide Aβ25-35. *British Journal of Pharmacology*, *171*, 3234–3245.

Madl, C., & Holzer, M. (2004). Brain function after resuscitation from cardiac arrest. *Current Opinion in Critical Care*, *10*, 213–217.

Maiese, K. (2001). The dynamics of cellular injury: Transformation into neuronal and vascular protection. *Histology and Histopathology*, *16*, 633–644.

Maiese, K., Li, F., & Chong, Z. Z. (2004). Erythropoietin in the brain: Can the promise to protect be fulfilled? *Trends in Pharmacological Sciences*, *25*, 577–583.

Malhotra, S., Savitz, S. I., Ocava, L., & Rosenbaum, D. M. (2006). Ischemic preconditioning is mediated by erythropoietin through PI-3 kinase signaling in an animal model of transient ischemic attack. *Journal of Neuroscience Research*, *83*, 19–27.

Manukhina, E. B., Downey, H. F., & Mallet, R. T. (2006). Role of nitric oxide in cardiovascular adaptation to intermittent hypoxia. *Experimental Biology and Medicine*, *231*, 343–365.

Marti, H. H. (2004). Erythropoietin and the hypoxic brain. *The Journal of Experimental Biology*, *207*, 3233–3242.

Marti, H. H., Gassmann, M., Wenger, R. H., Kvietikova, I., Morganti-Kossmann, M. C., Kossmann, T., et al. (1997). Detection of erythropoietin in human liquor: Intrinsic erythropoietin production in the brain. *Kidney International*, *51*, 416–418.

Martin, L. J., Al-Abdulla, N. A., Brambrink, A. M., Kirsch, J. R., Sieber, F. E., & Portera-Cailliau, C. (1998). Neurodegeneration in excitotoxicity, global cerebral ischemia, and target deprivation: A perspective on the contributions of apoptosis and necrosis. *Brain Research Bulletin*, *46*, 281–309.

Maznyczka, A. M., & Gershlick, A. H. (2015). Therapeutic hypothermia in patients with out-of-hospital arrest. *Heart*, *101*, 1265–1271.

McPherson, R. J., & Juul, S. E. (2008). Recent trends in erythropoietin-mediated neuroprotection. *International Journal of Developmental Neuroscience*, *26*, 103–111.

Merelli, A., Czornyj, L., & Lazarowski, A. (2015). Erythropoietin as a new therapeutic opportunity in brain inflammation and neurodegenerative diseases. *The International Journal of Neuroscience*, *125*, 793–797.

Mozaffarian, D., Benjamin, E. J., Go, A. S., American Heart Association Statistics Committee, & Stroke Statistics Committee. (2016). Heart disease and stroke statistics—2016 update: A report from the American Heart Association. *Circulation, 133*, e38–e360.

Nguyen, A. Q., Cherry, B. H., Scott, G. F., Ryou, M. G., & Mallet, R. T. (2014). Erythropoietin: Powerful protection of ischemic and post-ischemic brain. *Experimental Biology and Medicine, 239*, 1461–1475.

Nolan, J. P., Neumar, R. W., Adrie, C., Aibiki, M., Berg, R. A., Böttiger, B. W., et al. (2008). Post-cardiac arrest syndrome: Epidemiology, pathophysiology, treatment, and prognostication. *Resuscitation, 79*, 350–379.

Ogle, M. E., Gu, X., Espinera, A. R., & Wei, L. (2012). Inhibition of prolyl hydroxylases by dimethyloxaloylglycine after stroke reduces ischemic brain injury and requires hypoxia inducible factor-1a. *Neurobiology of Disease, 45*, 733–742.

Olivencia-Yurvati, A. H., & Mallet, R. T. (2012). Inflammatory response and miniaturized cardiopulmonary bypass. In T. Gourlay & S. Gunaydin (Eds.), *Minimized cardiopulmonary bypass techniques and technologies* (pp. 86–112). Cambridge: Woodhead Publishing Ltd.

Olmez, I., & Ozyurt, H. (2012). Reactive oxygen species and ischemic cerebrovascular disease. *Neurochemistry International, 60*, 208–212.

Ovbiagele, B., Kidwell, C. S., Starkman, S., & Saver, J. L. (2003). Neuroprotective agents for the treatment of acute ischemic stroke. *Current Neurology and Neuroscience Reports, 3*, 9–20.

Paciaroni, M., Caso, V., & Agnelli, G. (2009). The concept of ischemic penumbra in acute stroke and therapeutic opportunities. *European Neurology, 61*, 321–330.

Prentice, H., Modi, J. P., & Wu, J. Y. (2015). Mechanisms of neuronal protection against excitotoxicity, endoplasmic reticulum stress, and mitochondrial dysfunction in stroke and neurodegenerative diseases. *Oxidative Medicine and Cellular Longevity, 2015*, 964518.

Pugh, C. W. (2016). Modulation of the hypoxic response. *Advances in Experimental Medicine and Biology, 903*, 259–271.

Rabie, T., & Marti, H. H. (2008). Brain protection by erythropoietin: A manifold task. *Physiology, 23*, 263–274.

Rami, A., Bechmann, I., & Stehle, J. H. (2008). Exploiting endogenous anti-apoptotic proteins for novel therapeutic strategies in cerebral ischemia. *Progress in Neurobiology, 85*, 273–296.

Ramos-Fernandez, M., Bellolio, M. F., & Stead, L. G. (2011). Matrix metalloproteinase-9 as a marker for acute ischemic stroke: A systematic review. *Journal of Stroke and Cerebrovascular Diseases, 20*, 47–54.

Rangarajan, V., & Juul, S. E. (2014). Erythropoietin: Emerging role of erythropoietin in neonatal neuroprotection. *Pediatric Neurology, 51*, 481–488.

Ratilal, B. O., Arroja, M. M., Rocha, J. P., Fernandes, A. M., Barateiro, A. P., Brites, D. M., et al. (2014). Neuroprotective effects of erythropoietin pretreatment in a rodent model of transient middle cerebral artery occlusion. *Journal of Neurosurgery, 121*, 55–62.

Reddy, S., Jones, A. D., Cross, C. E., Wong, P. S., & Van Der Vliet, A. (2000). Inactivation of creatine kinase by S-glutathionylation of the active-site cysteinyl residue. *The Biochemical Journal, 347*, 821–827.

Reischl, S., Li, L., Walkinshaw, G., Flippin, L. A., Marti, H. H., & Kunze, R. (2014). Inhibition of HIF prolyl-4-hydroxylases by FG-4497 reduces brain tissue injury and edema formation during ischemic stroke. *PloS One, 9*, e84767.

Renzi, M. J., Farrell, F. X., Bittner, A., Galindo, J. E., Morton, M., Trinh, H., et al. (2002). Erythropoietin induces changes in gene expression in PC-12 cells. *Brain Research. Molecular Brain Research, 104*, 86–95.

Robertson, C. L., Scafidi, S., McKenna, M. C., & Fiskum, G. (2009). Mitochondrial mechanisms of cell death and neuroprotection in pediatric ischemic and traumatic brain injury. *Experimental Neurology, 218*, 371–380.

Roger, V. L., Go, A. S., Lloyd-Jones, D. M., American Heart Association Statistics Committee, & Stroke Statistics Committee. (2012). Heart disease and stroke statistics–2012 update. *Circulation, 125*, e2–e220.

Rothstein, J. D., Dykes-Hoberg, M., Pardo, C. A., Bristol, L. A., Jin, L., Kuncl, R. W., et al. (1996). Knockout of glutamate transporters reveals a major role for astroglial transport in excitotoxicity and clearance of glutamate. *Neuron, 16*, 675–686.

Ruscher, K., Freyer, D., Karsch, M., Isaev, N., Megow, D., Sawitzki, B., et al. (2002). Erythropoietin is a paracrine mediator of ischemic tolerance in the brain: Evidence from an in vitro model. *The Journal of Neuroscience, 22*, 10291–10301.

Ryou, M. G., Choudhury, G. R., Li, W., Winters, A., Yuan, F., Liu, R., et al. (2015). Methylene blue-induced neuronal protective mechanism against hypoxia-reoxygenation stress. *Neuroscience, 301*, 193–203.

Ryou, M. G., Liu, R., Ren, M., Sun, J., Mallet, R. T., & Yang, S.-H. (2012). Pyruvate protects brain against ischemia-reperfusion injury by activating erythropoietin signaling. *Stroke, 43*, 1101–1107.

Sakanaka, M., Wen, T. C., Matsuda, S., Masuda, S., Morishita, E., Nagao, M., et al. (1998). In vivo evidence that erythropoietin protects neurons from ischemic damage. *Proceedings of the National Academy of Sciences of the United States of America, 95*, 4635–4640.

Schouten, J. W. (2007). Neuroprotection in traumatic brain injury: A complex struggle against the biology of nature. *Current Opinion in Critical Care, 13*, 134–142.

Selivanov, V. A., Votyakova, T. V., Pivtoraiko, V. N., Zeak, J., Sukhomlin, T., Trucco, M., et al. (2011). Reactive oxygen species production by forward and reverse electron fluxes in the mitochondrial respiratory chain. *PLoS Computational Biology, 7*, e1001115.

Sen, T., & Sen, N. (2016). Treatment with an activator of hypoxia-inducible factor 1, DMOG provides neuroprotection after traumatic brain injury. *Neuropharmacology, 107*, 79–88.

Shang, Y., Wu, Y., Yao, S., Wang, X., Feng, D., & Yang, W. (2007). Protective effect of erythropoietin against ketamine-induced apoptosis in cultured rat cortical neurons: Involvement of PI3K/Akt and GSK-3 beta pathway. *Apoptosis, 12*, 2187–2195.

Shein, N. A., Tsenter, J., Alexandrovich, A. G., Horowitz, M., & Shohami, E. (2007). Akt phosphorylation is required for heat acclimation-induced neuroprotection. *Journal of Neurochemistry, 103*, 1523–1529.

Shuaib, A., Lees, K. R., Lyden, P., Grotta, J., Davalos, A., Davis, S. M., et al. (2007). NXY-059 for the treatment of acute ischemic stroke. *The New England Journal of Medicine, 357*, 562–571.

Siddiq, A., Aminova, L. R., & Ratan, R. R. (2007). Hypoxia inducible factor prolyl 4-hydroxylase enzymes: Center stage in the battle against hypoxia, metabolic compromise and oxidative stress. *Neurochemical Research, 32*, 931–946.

Sims, B., Clarke, M., Njah, W., Hopkins, E. S., & Sontheimer, H. (2010). Erythropoietin-induced neuroprotection requires cysteine glutamate exchanger activity. *Brain Research, 1321*, 88–95.

Sims, N. R., & Muyderman, H. (2010). Mitochondria, oxidative metabolism and cell death in stroke. *Biochimica et Biophysica Acta, 1802*, 80–91.

Sirén, A. L., & Ehrenreich, H. (2001). Erythropoietin—A novel concept for neuroprotection. *European Archives of Psychiatry and Clinical Neuroscience, 251*, 179–184.

Sola, A., Rogido, M., Lee, B. H., Genetta, T., & Wen, T. C. (2005). Erythropoietin after focal cerebral ischemia activates the Janus kinase-signal transducer and activator of transcription signaling pathway and improves brain injury in postnatal day 7 rats. *Pediatric Research, 57*, 481–487.

Song, J., Sun, H., Xu, F., Kang, W., Gao, L., Guo, J., et al. (2016). Recombinant human erythropoietin improves neurological outcomes in very preterm infants. *Annals of Neurology, 80*, 24–34.

Souvenir, R., Fathali, N., Ostrowski, R. P., Lekic, T., Zhang, J. H., & Tang, J. (2011). Tissue inhibitor of matrix metalloproteinase-1 mediates erythropoietin-induced neuroprotection in hypoxia ischemia. *Neurobiology of Disease, 44*, 28–37.

Suzuki, Y., Nagai, N., & Umemura, K. (2016). A review of the mechanisms of blood-brain barrier permeability by tissue-type plasminogen activator treatment for cerebral ischemia. *Frontiers in Cellular Neuroscience, 10*, 2.

Svedin, P., Hagberg, H., Sävman, K., Zhu, C., & Mallard, C. (2007). Matrix metalloproteinase-9 gene knock-out protects the immature brain after cerebral hypoxia-ischemia. *The Journal of Neuroscience, 27*, 1511–1518.

Swanson, R. A., Ying, W., & Kauppinen, T. M. (2004). Astrocyte influences on ischemic neuronal death. *Current Molecular Medicine, 4*, 193–205.

Trollmann, R., Richter, M., Jung, S., Walkinshaw, G., & Brackmann, F. (2014). Pharmacologic stabilization of hypoxia-inducible transcription factors protects developing mouse brain from hypoxia-induced apoptotic cell death. *Neuroscience, 278*, 327–342.

Turner, R. C., Dodson, S. C., Rosen, C. L., & Huber, J. D. (2013). The science of cerebral ischemia and the quest for neuroprotection: Navigating past failure to future success. *Journal of Neurosurgery, 118*, 1072–1085.

Turner, R. J., & Sharp, F. R. (2016). Implications of MMP9 for blood brain barrier disruption and hemorrhagic transformation following ischemic stroke. *Frontiers in Cellular Neuroscience, 10*, 1–13.

Vafadari, B., Salamian, A., & Kaczmarek, L. (2016). MMP-9 in translation: From molecule to brain physiology, pathology and therapy. *Journal of Neurochemistry, 139*(Suppl. 2), 91–114.

Vairano, M., Dello Russo, C., Pozzoli, G., Battaglia, A., Scambia, G., Tringali, G., et al. (2002). Erythropoietin exerts anti-apoptotic effects on rat microglial cells in vitro. *The European Journal of Neuroscience, 16*, 584–592.

Villa, P., Bigini, P., Mennini, T., Agnello, D., Laragione, T., Cagnotto, A., et al. (2003). Erythropoietin selectively attenuates cytokine production and inflammation in cerebral ischemia by targeting neuronal apoptosis. *The Journal of Experimental Medicine, 198*, 971–975.

Wang, R., Wu, X., Liang, J., Qi, Z., Liu, X., Min, L., et al. (2015). Intra-artery infusion of recombinant human erythropoietin reduces blood-brain barrier disruption in rats following cerebral ischemia and reperfusion. *The International Journal of Neuroscience, 125*, 693–702.

Wen, T. C., Sadamoto, Y., Tanaka, J., Zhu, P. X., Nakata, K., Ma, Y. J., et al. (2002). Erythropoietin protects neurons against chemical hypoxia and cerebral ischemic injury by up-regulating Bcl-xL expression. *Journal of Neuroscience Research, 67*, 795–803.

Wenger, R. H., & Hoogewijs, D. (2010). Regulated oxygen sensing by protein hydroxylation in renal erythropoietin-producing cells. *American Journal of Physiology. Renal Physiology, 298*, F1287–F1296.

Westfal, R. E., Reissman, S., & Doering, G. (1996). Out-of-hospital cardiac arrests: An 8-year New York City experience. *The American Journal of Emergency Medicine, 14*, 364–368.

White, B. C., Sullivan, J. M., DeGracia, D. J., O'Neil, B. J., Neumar, R. W., Grossman, L. I., et al. (2000). Brain ischemia and reperfusion: Molecular mechanisms of neuronal injury. *Journal of the Neurological Sciences, 179*, 1–33.

Wu, Y. W., Mathur, A. M., Chang, T., McKinstry, R. C., Mulkey, S. B., Mayock, D. E., et al. (2016). High-dose erythropoietin and hypothermia for hypoxic-ischemic encephalopathy: A phase II trial. *Pediatrics, 137*, e20160191.

Xiong, Y., Chopp, M., & Lee, C. P. (2009). Erythropoietin improves brain mitochondrial function in rats after traumatic brain injury. *Neurological Research, 31*, 496–502.

Xiong, Y., Lu, D., Qu, C., Goussev, A., Schallert, T., Mahmood, A., et al. (2008). Effects of erythropoietin on reducing brain damage and improving functional outcome after traumatic brain injury in mice. *Journal of Neurosurgery*, *109*, 510–521.

Xu, X., Dai, H., & Shi, Y. (2009). Erythropoietin protects primary cultures of rat cortical neurons from hypoxia-induced toxicity through attenuating both glutamate release and NMDA receptor evoked neurotoxicity pathway. *Pharmazie*, *64*, 210–213.

Yang, Y., & Rosenberg, G. A. (2015). Matrix metalloproteinases as therapeutic targets for stroke. *Brain Research*, *1623*, 30–38.

Yu, T., Longhini, F., Wu, R., Yao, W., Lu, W., & Jin, X. (2015). The role of the induction of mild hypothermia in adult patient outcomes after cardiac arrest: Systematic review and meta-analysis of randomized controlled studies. *The Journal of International Medical Research*, *43*, 471–482.

Zhang, F., Signore, A. P., Zhou, Z., Wang, S., Cao, G., & Chen, J. (2006). Erythropoietin protects CA1 neurons against global cerebral ischemia in rat: Potential signaling mechanisms. *Journal of Neuroscience Research*, *83*, 1241–1251.

Zhang, F., Wang, S., Cao, G., Gao, Y., & Chen, J. (2007). Signal transducers and activators of transcription 5 contributes to erythropoietin-mediated neuroprotection against hippocampal neuronal death after transient global cerebral ischemia. *Neurobiology of Disease*, *25*, 45–53.

Zhang, J., Wang, Q., Xiang, H., Xin, Y., Chang, M., & Lu, H. (2014). Neuroprotection with erythropoietin in preterm and/or low birth weight infants. *Journal of Clinical Neuroscience*, *21*, 1283–1287.

Zhang, X. W., Xie, J. F., Chen, J. X., Huang, Y. Z., Guo, F. M., Yang, Y., et al. (2015). The effect of mild induced hypothermia on outcomes of patients after cardiac arrest: A systematic review and meta-analysis of randomised controlled trials. *Critical Care*, *19*, 417.

Zhang, F., Xing, J., Liou, A. K., Wang, S., Gan, Y., Luo, Y., et al. (2010). Enhanced delivery of erythropoietin across the blood-brain barrier for neuroprotection against ischemic neuronal injury. *Translational Stroke Research*, *1*, 113–121.

Zhang, Y., Xiong, Y., Mahmood, A., Meng, Y., Qu, C., Schallert, T., et al. (2009). Therapeutic effects of erythropoietin on histological and functional outcomes following traumatic brain injury in rats are independent of hematocrit. *Brain Research*, *1294*, 153–164.

Zhao, H., Wang, R., Wu, X., Liang, J., Qi, Z., Liu, X., et al. (2015). Erythropoietin delivered via intra-arterial infusion reduces endoplasmic reticulum stress in brain microvessels of rats following cerebral ischemia and reperfusion. *Journal of Neuroimmune Pharmacology*, *10*, 153–161.

Zheng, Z., Zhao, H., Steinberg, G. K., & Yenari, M. A. (2003). Cellular and molecular events underlying ischemia-induced neuronal apoptosis. *Drug News & Perspectives*, *16*, 497–503.

Zhou, T. F., & Yu, J. G. (2013). Recombinant human erythropoietin attenuates neuronal apoptosis and cognitive defects via JAK2/STAT3 signaling in experimental endotoxemia. *The Journal of Surgical Research*, *183*, 304–312.

Zhou, J., Yu, J. T., Wang, H. F., Meng, X. F., Tan, C. C., Wang, J., et al. (2015). Association between stroke and Alzheimer's disease: Systematic review and meta-analysis. *Journal of Alzheimer's Disease*, *43*, 479–489.

CHAPTER TWELVE

The Erythropoietin System Protects the Heart Upon Injury by Cardiac Progenitor Cell Activation

Maria P. Zafeiriou[1]
Institute of Pharmacology and Toxicology, Heart Research Center Goettingen, University Medical Center Goettingen, Georg-August-University Goettingen, Goettingen, Germany
[1]Corresponding author: e-mail address: patapia.zafeiriou@med.uni-goettingen.de

Contents

Abstract

Erythropoietin (EPO) is a growth hormone, widely known for its role in erythropoiesis. The broad expression of erythropoietin receptor (EPOR) in adult organs suggested that EPO may also affect other cells besides late erythroid progenitors. In the embryonic heart, EPOR is expressed in all cells including the immature proliferating cardiomyocytes. In contrast to the embryonic heart in adulthood, EPOR expression is decreased and mainly detected in immature proliferating cells (i.e., resident cardiac progenitor cells) rather than in terminally differentiated cells (i.e., cardiomyocytes). Since cardiac progenitor cells are considered a regenerative cell source upon cardiac injury, the protective action of the EPO system was tested by creating an erythroid-rescued EPOR knockout mouse model. Although these mice appear to have less immature proliferating myocytes during embryogenesis, they reach adulthood without apparent morphological defects. However, upon ischemia reperfusion, these animals show a greater infarct size, suggesting that the EPO/EPOR protects the heart upon injury. Indeed preclinical studies showed that EPO administration postinfarction improves cardiac function via neoangiogenesis, antiapoptotic mechanisms, and/or CPC activation. Despite the promising preclinical data, large cohort clinical studies in humans failed to show a significant amelioration in cardiac function upon systemic injection of EPO

Vitamins and Hormones, Volume 105
ISSN 0083-6729
http://dx.doi.org/10.1016/bs.vh.2017.04.001
© 2017 Elsevier Inc.
All rights reserved.

in patients with myocardial infarctions. The discrepancy between preclinical and clinical trials may be due to differences between the doses, the way of delivery, the homogeneity of the cohorts, and last but not least the species differences. These data pinpoint the importance of carrying out preclinical studies in human models of disease as engineered human cardiac tissue that will provide a better understanding of the expression pattern of EPOR and the role of its ligand in human cardiac cells. Such studies may be able to bridge the gap between preclinical rodent data and human clinical trials and thus lead to the design of more successful clinical studies.

1. INTRODUCTION

Cardiovascular diseases are the main cause of morbidity and mortality in the world (WHO, Fact sheet No. 317, September 2011), partly due to the limited regenerative capacity of the heart. Recent evidence suggests significant myocardial regeneration postinjury in the early neonatal mammalian heart (Porrello et al., 2011) that later is confined to a low postnatal cardiomyocyte turnover as shown in the human heart (Bergmann et al., 2009). In the absence of regeneration potential a compensatory mechanism is activated in order to sustain cardiac output. This process also known as cardiac remodeling leads to cardiomyocyte hypertrophy and connective tissue deposition. A common feature of the remodeling heart is the activation of the so-called fetal gene program (Dirkx, da Costa Martins, & De Windt, 2013; Oka, Xu, & Molkentin, 2007; Zelarayan et al., 2008).

Two different mechanisms of cardiomyocyte regeneration have been suggested including proliferating dedifferentiated cardiomyocytes (Bersell, Arab, Haring, & Kuhn, 2009; Kuhn et al., 2007; Senyo et al., 2013) and endogenous progenitor cell differentiation (Hsieh, Segers, et al., 2007). Unfortunately, neither of these mechanisms is under physiological conditions able to replace damaged myocardium in the adult heart. To exploit the under normal circumstances insufficient regenerative capacity of the heart it is necessary to delineate the components of signaling pathways, which govern the activity of regenerative cell niches within the heart. One of these signaling pathways is the erythropoietin (EPO)/erythropoietin receptor (EPOR) system.

2. THE EPO/EPOR SYSTEM

EPO is a 34-kDa growth hormone that exerts its action mainly by binding with high affinity (Yamamura, Noda, & Ikawa, 1994) to its receptor. EPOR is 55 kDa single-pass type I membrane protein (extracellular N-terminus, intracellular C-terminus). At the endoplasmic reticulum, during EPOR maturation, the C-terminus of the receptor is associated with the tyrosine kinase Janus kinase 2 (JAK2) (Huang, Constantinescu, & Lodish, 2001). At the membrane, in the absence of EPO, EPOR forms inactive homodimers (Fig. 1, inactive state) (Constantinescu et al., 2001). Once EPO binds to its receptor, it induces conformational changes that allow the two JAK2 molecules at the C-terminus to transphosphorylate and thereby transactivate each other (Remy, Wilson, & Michnick, 1999; Fig. 1). Subsequently, the activated JAK2 phosphorylates the tyrosine residues that are present on the C-terminus of EPOR which then recruit and activate molecules of various signaling cascades as STAT3/5 (Asaumi et al., 2007; Grebien et al., 2008), PI3K/AKT (Hirata et al., 2005; Zafiriou et al., 2014),

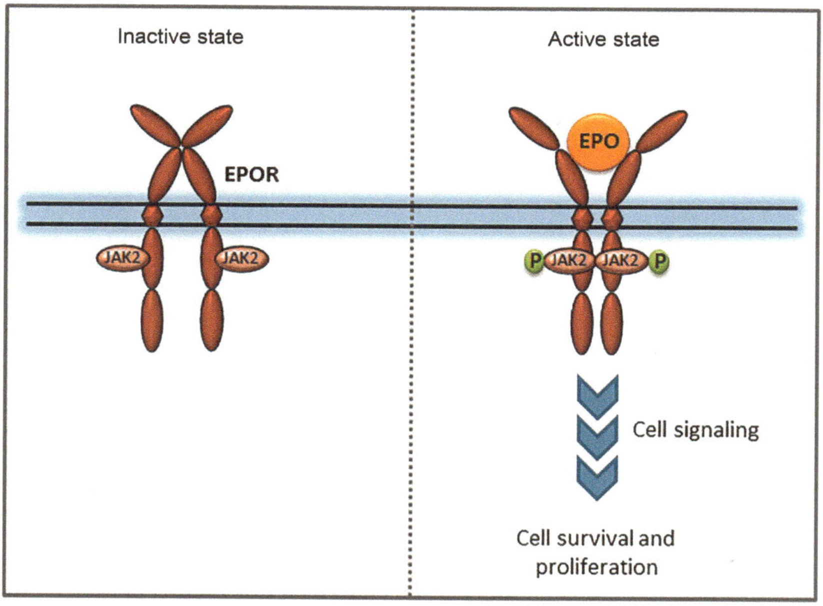

Fig. 1 Schema depicting signaling pathway activation upon EPO binding to its receptor. *Inactive state*: In the absence of EPO, EPOR forms inactive homodimers. *Active state*: Once EPO binds to its receptor, it induces conformational changes that allow the two JAK2 molecules at the C-terminus of EPOR to transphosphorylate and thereby transactivate each other. The activated JAK2 phosphorylates the tyrosine residues that are present on the C-terminus of EPOR which then recruit and activate molecules of various signaling cascades as STAT3/5, PI3K/AKT, MAPK, and NF-κb. Activation of these signaling pathways upon EPO stimulation results in transcriptional changes that promote cell survival and cell proliferation.

MAPK(Miura, Miura, Ihle, & Aoki, 1994), and NF-κb (Ferrario et al., 2011). Activation of these signaling pathways results in transcriptional regulation. The transcriptional changes occurring upon EPO stimulation promote cell survival and cell proliferation (Fig. 1).

Besides the classical activation EPO can exert an antiapoptotic action in several tissues by binding to a heterodimer complex of EPOR and the common beta-chain receptor (Brines et al., 2004). Short-term in vitro studies on adult mouse cardiomyocytes have shown that EPO can protect from staurosporine-induced apoptosis via this heterodimer complex (Brines et al., 2004). However, in vivo studies subjecting common beta-chain KO mice in myocardial infarction (MI) and treating them with EPO could show that the long-term effects of EPO as cell proliferation and neoangiogenesis are not attenuated in the absence of the common beta chain (Kanellakis et al., 2010).

3. EXPRESSION AND ROLE OF THE EPO/EPOR SYSTEM IN THE HEART

In the developing human fetus EPO is produced in the liver and the kidneys (Dame et al., 1998), whereas postnatally EPO production persists mainly in the kidney (Jacobson, Goldwasser, Fried, & Plzak, 1957). Despite the developmental stage or the organ it is being produced in, EPO expression is enhanced under hypoxic conditions since its transcription is induced by hypoxia-induced factor 1 alpha (HIF-1alpha) (Hsieh, Linde, et al., 2007). Then EPO enters circulation thereby reaching and affecting different organs. In these organs EPO will exert an effect only in cells on which EPOR is being expressed. Thus, to study the effect of EPO in a tissue, it is important to identify the cell types that express its receptor.

3.1 EPOR Expression in the Mouse Heart

EPOR expression is broad and besides bone marrow it has been found in brain (Morishita, Masuda, Nagao, Yasuda, & Sasaki, 1997; Ott et al., 2015), retina (Fu et al., 2008), vascular endothelium (Anagnostou et al., 1994), and heart (Wu, Lee, Gao, Liu, & Iruela-Arispe, 1999; Zafiriou et al., 2014). In the developing mouse heart, EPOR expression is first detected around E7.5 with a transcriptional peek around E10 (Wu et al., 1999; Zafiriou et al., 2014). Postnatally, EPOR expression is significantly reduced and in adult hearts it persists only in interstitial cells (Hoch et al., 2011; Zafiriou et al., 2014). These data are in line with data about EPOR expression in other tissues. EPOR is highly expressed in the early developing nervous system (Liu, Shen, Liu, & Noguchi, 1997), while the remaining low EPOR expression in the brain seems to be confined to adult neural progenitor cells (Tsai et al., 2006).

The expression of EPOR in late-stage erythroid progenitor cells but not differentiated erythroid cells in combination with the levels and localization of EPOR during development and maturation of the heart lead to the hypothesis that EPOR may be expressed in cardiac progenitor cells (CPCs). Indeed isolated CPCs show a stronger transcript expression of EPOR compared to the total heart. Costaining of EPOR with CPC markers showed that different progenitor cell types express EPOR including endothelial and cardiomyogenic cell types. Interestingly all of these populations seem to lose EPOR expression upon cell cycle exit and differentiation to mature cells (Zafiriou et al., 2014) a phenomenon observed also in erythroid progenitor cells (Wickrema, Krantz, Winkelmann, & Bondurant, 1992). These data support the notion that EPO/EPOR system has mainly a role in undifferentiated cells.

3.2 Role of EPO in the Developing and Adult Mouse Heart

In order to evaluate the role of EPO in the mouse heart two global KO models were initially employed, an EPO KO and an EPOR KO model. In both models, the mice die in utero around E13.5 due to severe anemia (Wu et al., 1999). Despite that fact, in comparison to wild-type embryonic hearts, KO hearts showed severe hypoplasia, epicardium detachment, and abnormal vascularization. These data are in line with the expression of EPOR found later on in cardiomyogenic, epicardial, and endothelial progenitors in the heart (Zafiriou et al., 2014). Under the same notion, exogenous EPO treatment enhanced cardiac cell proliferation in EPO KO mice, while it had no effect in cardiac cells of EPOR KO mice indicating mitogenic properties. It should be noted though that the insufficient erythropoiesis of these mouse models leads to two important limitations. On the one hand, due to the early lethality, the role of the EPO/EPOR system in the adult mouse heart could not be studied, and on the other hand, it could not be excluded that the observed morphological defects were not due to limited erythropoiesis.

Therefore, a novel erythroid-rescued EPOR KO mouse model was generated, in which only the erythropoietic cells could express EPOR. The resEPOR KO mice appeared to have less immature proliferating cardiomyocytes during embryogenesis, but they could reach adulthood without apparent cardiac morphological defects (Suzuki et al., 2002). Thus, the EPO/EPOR system is important for immature myocyte proliferation but it is not essential.

3.3 Role of EPO in the Diseased Mouse Heart

Upon injury, tissues are subjected to ischemia, resulting in the upregulation of HIF-1alpha and therefore its downstream target EPO (Makita, Duncan, & Sucov, 2005). This fact, in combination with the mitogenic role of EPO in different progenitors in developing and adult organs such as heart, bone marrow, and brain, rendered the EPOR/EPOR system into an attractive target for tissue regeneration.

To test this hypothesis the resEPOR KO mice that show no cardiac morphological defects (Suzuki et al., 2002) were subjected to ischemia reperfusion injury. Interestingly, these mice show a greater infarct size in comparison to the respective control, suggesting that a deficient EPO/EPOR system in the nonhematopoietic lineage may exacerbate left ventricular remodeling (Tada et al., 2006).

A number of preclinical studies explored the protective role of recombinant human EPO (rhEPO) administration postinfarction and could show that cardiac function amelioration is a result of neoangiogenesis, anti-apoptotic mechanisms, and/or CPC activation (Calvillo et al., 2003; Hirata et al., 2005; Moon et al., 2003; Xu et al., 2005; Ye, Du, Xia, & Ping, 2005; Zafiriou et al., 2014).

3.3.1 EPO and Antiapoptosis

In a canine model of ischemia reperfusion a single intravenous administration of rhEPO just before reperfusion significantly reduced infarct size in a dose-dependent manner (100–1000 U/kg, i.v.). rhEPO decreased the number of apoptotic cells in the ischemic myocardium of dogs, a process mediated by AKT activation (Hirata et al., 2005). Similarly in a rat model of MI, rhEPO (3000 U/kg, i.p.) administered immediately after the coronary artery ligation reduced border zone apoptosis by 50% (Moon et al., 2003). Although these studies could clearly show that EPO can partially preserve cardiac function upon ischemia by attenuating early occurring apoptosis, they provided no information on the cell types affected by EPO administration.

3.3.2 EPO and Neovascularization

In a rat model of MI intramyocardial rhEPO injection (3000 U/kg) resulted in a better cardiac function in comparison to control MI mice, 6 weeks upon ischemia. To elucidate the molecular mechanism underlying this observation the number of proliferating Ki-67+ cells was evaluated 48 h after injection. EPO increased the number of proliferating intracardiac cells in the periinfarction zone by 45% (Gabel et al., 2009). Furthermore, cyclin D1 and cyclin-dependent kinase 1, genes responsible for cell cycle progression which are typically activated by mitogenic factors were activated 24 h upon EPO treatment (Gabel et al., 2009). In an attempt to characterize the proliferating cells that were enhanced by EPO stimulation the authors costained with the endothelial marker CD31. Indeed some of the Ki-67+ cells were CD31 positive resulting in an increase of capillary density up to 17% compared to control after 7 days. These data suggested that endothelial cell proliferation and formation of capillaries underlie the protective effect of EPO (Gabel et al., 2009). However, the amount of Ki-67+ cells was higher as compared to CD31+/Ki-67+ cells, indicating that other populations of proliferating cells were influenced by EPO.

Indeed, Hoch et al. elegantly showed that resident Sca-1+ CPCs express EPOR and have the ability to differentiate into endothelial cells, adipocytes, fibroblasts, and pericytes (Hoch et al., 2011). Instead of rhEPO in this study mice were treated with continuous EPO receptor activator (CERA). CERA contains methoxy polyethylene glycol chains linked to amino groups via amide bonds and when compared to EPO it has a lower affinity to EPOR but a longer half-life (Macdougall et al., 2006). Despite its low affinity to EPOR, CERA results in a greater in vivo erythropoiesis due to its slow elimination (El-Komy, Schmidt, Widness, & Veng-Pedersen, 2011). Administration of CERA (3 days × 3 μg/kg, i.p.) in a doxorubicin-induced cardiomyopathy model could activate CPCs, restore endothelial differentiation, and preserve the cardiac microvasculature and cardiac function (Hoch et al., 2011).

Moreover, in a mouse model of MI, rhEPO (5000 U/kg, i.p.) improved survival, attenuated remodeling, enhanced neovascularization, and diminished apoptotic cells in the periinfarct area. In this model upon EPO stimulation BMC subpopulations (CD31+, c-kit+, and Sca-1+ cells) were mobilized, and homing of Sca-1+ and CXCR4+ BMCs toward an SDF-1 gradient into the ischemic myocardium was enhanced. In contrast to the observations of Hoch et al., EPO did not seem to have a beneficial effect on resident cardiac stem cells (Brunner et al., 2009). This discrepancy can be partially explained by the different disease models used in the two studies.

Finally, an alternative molecular mechanism was suggested by Teng et al. by using a mouse model in which EPOR expression is restricted to hematopoietic and endothelial cells (ΔEPOR). EPO stimulation of coronary artery endothelial cells upregulated endothelial nitric oxide synthase activity in vitro and in vivo, and enhanced nitric oxide production, in a PI3K-dependent manner. When ΔEPOR mice were subjected to MI and pretreated with rhEPO (4000 U/kg, i.p.) presented a significant but smaller EPO protective effect in comparison to WT EPO-treated mice (Teng et al., 2011). These data support the hypothesis that activation of endothelial cells is one of the cellular mechanisms involved in the cardioprotective effects of EPO but not the only one.

In conclusion, EPO can mobilize BMCs or activate different resident cardiac cell populations as Sca-1 +, CD31 + and enhance neovascularization and therefore partially protect the heart. Nevertheless, the ΔEPOR mouse model in which EPOR expression is restricted to hematopoietic and endothelial cells showed that EPO protective effects are not only the result of neovascularization. This conclusion is in line with the expression of EPOR in stem, mesenchymal, endothelial, and cardiomyogenic cells in the adult mouse heart (Zafiriou et al., 2014).

3.3.3 EPO and Cardiomyogenic Cells

Among the most prominent EPOR-expressing progenitor cell populations in the adult mouse heart are cardiomyogenic cells (Zafiriou et al., 2014). This cell population coexpresses early (TBX5, NKX2.5) and definitive (myosin heavy chain (MHC), cardiac troponin T (cTNT)) cardiomyocyte markers. Since rhEPO increased their number in vitro and in vivo by activating PI3K, this population was designated as EPO-responsive MHC-expressing cells (EMCs). In vitro, EMCs can proliferate and partially differentiate toward beating cardiomyocyte-like cells (Zafiriou et al., 2014). In a mouse model of MI rhEPO (4000 IU/kg, i.p.) resulted in a significant preservation of cardiac function and an increase of cardiac EMCs and cTNT-positive cells in the infarcted area. These data show that besides endothelial progenitors, EPO can activate different cell types including cardiomyogenic cells and contribute to cardioprotection.

3.3.4 EPO and Epicardial Cells

It is generally accepted that epicardial cells contribute to cardiac regeneration upon injury (Duim, Kurakula, Goumans, & Kruithof, 2015; Seeger et al., 2016; Wei et al., 2015). Interestingly, EPO treatment of MI mice hearts induced a pronounced epicardial thickening (Zafiriou et al., 2014), suggesting a possible role for the EPO system in epicardial cell activation.

In a heart slice primary culture, in which mechanically separated distinct cardiac cell populations were tested for their mitogenic potential on embryonic chick (E10) cardiomyocytes showed that the epicardium is required for CM proliferation and survival. Interestingly, block of either retinoic acid (RA) or EPO signaling from the epicardium inhibited both CM proliferation and survival, a phenotype rescued by exogenous RA or EPO administration, respectively. Supernatants of epicardial cultures treated with EPO were able to enhance CM proliferation, suggesting that EPO induced the expression of paracrine mitogenic factors (Stuckmann, Evans, & Lassar, 2003).

Insulin-like growth factor 2 (IGF2) was identified as one of these secreted mitogenic factors in an epicardial cell line stimulated by EPO. By the use of Raldh2(−/−) and Rxra(−/−) knockout hearts the authors showed that Igf2 mRNA reduction in the epicardium and the concomitant reduction of ventricular CM proliferation were rescued by exogenous EPO treatment (Brade et al., 2011). These data suggest that EPO can bind to epicardial cells and can induce expression of mitogenic factors that could potentially enhance proliferation of immature proliferating cardiomyocytes in the embryonic heart.

Although IGF2 upregulation mediated by EPO-treated epicardial cells was not studied in disease models, it represents an alternative mechanism of action that would be interesting to test. It should be though noted that in an open-chest porcine model, pharmacological conditioning at the time of reperfusion with low dose of IGF2 did not show a reduction in infarct size or apoptosis (Salminen et al., 2014) pinpointing that in disease EPO cardioprotective action in the epicardium may not be exclusively mediated by IGF2.

In conclusion, in the adult mouse heart, EPOR is expressed mainly in different immature proliferating cells. EPO can activate these progenitors and enhance their numbers. Depending on their nature, these cells differentiate and contribute to formation of new vessels or muscle. Thereby EPO attenuates cardiac remodeling and partially preserves cardiac function.

3.4 Expression and Role of the EPO/EPOR System in the Human Heart

The expression of EPOR in the adult human heart is not as well investigated as in the brain. To our knowledge only one publication exists where the authors could show expression of EPOR by immunohistochemistry in cardiomyocytes and endothelial cells but not in fibrocytes (Depping et al., 2005).

Despite the lack of knowledge on the expression of EPOR, the promising preclinical data, along with the approved availability of rhEPO for anemia therapy, lead to a fast transition to the clinic where the protective role of EPO was tested post-MI.

In the first pilot study, darbepoetin (single bolus injection 60,000 U, i.v.) was able to induce endothelial progenitor cell mobilization but had no significant effect in cardiac function (Lipsic et al., 2006). The amount of rhEPO in this study was based in a prior clinical trial of EPO administration on stroke patients (Ehrenreich et al., 2002). In this study, 100,000 U, i.v. proved safe as well as beneficial. Subsequent studies with larger number of patients also failed to show a significant amelioration in cardiac function by rhEPO (30,000–60,000 U, i.v.) but showed significantly less adverse cardiovascular events and a favorable clinical safety profile (Binbrek, Rao, Al Khaja, Assaqqaf, & Sobel, 2009; Voors et al., 2010). Other studies, where mixed cohorts of patients were administered 60,000–100,000 U of rhEPO shortly after percutaneous coronary intervention, showed that EPO not only did not protect the heart but also induced adverse cardiovascular events (Ludman et al., 2011; Ott et al., 2010). Interestingly, pilot studies where small amounts of rhEPO were administered (s.c. or i.v.) repetitively for a longer time period demonstrated that EPO had a beneficial effect on cardiac remodeling and left ventricular function (Bergmann et al., 2011; Taniguchi et al., 2010). In summary, low repetitive doses of rhEPO for a long time period may protect the heart in contrast to single high doses of EPO that do not exert a protective action and many times lead to adverse cardiovascular events (Table 1).

The discrepancy between the promising preclinical results of EPO and the variable outcome in the clinical trials may be due to differences in dosing, mode of delivery (intravenously vs subcutaneously), and endpoint. The high affinity of EPO to its receptor (K_d 160 pM) actually indicates that low doses of EPO would be enough to activate EPOR effectively (Syed et al., 1998). This is in agreement with the fact that studies using repetitive low EPO doses showed cardioprotective effects (Bergmann et al., 2011; Taniguchi et al.,

Cohort				Delivery			Endpoint	Safety		Effect		
Size	Sex	Cardiac Disease	Age	EPO Analogue	Way	Amount (U)	Months After Treatment	Adverse Cardiovascular Events	Favorable Clinical Safety Profile	LV Function Amelioration	Enzymatic Infarct Size Reduction	References
35	Males	Acute STEMI	62 ± 12	Rhepoetin β	i.v.	3 × 6000	6	–	+	+	+	Taniguchi et al. (2010)
24	Males	No STEMI	58 ± 2	Rhepoetin β	s.c.	24 × 2500	6	–	+	+		Bergmann et al. (2011)
22	Males	Acute STEMI	59 ± 2	Darbepoetin α	i.v.	60,000	4	–	+	–		Lipsic et al. (2006)
254	Males	Acute STEMI	48 ± 1	Rhepoetin β	i.v.	30,000	1	–	+	–	–	Binbrek et al. (2009)
529	Males	Acute STEMI	61 ± 11	Rhepoetin α	i.v.	60,000	1.5	–	+	–		Voors et al. (2010)
131	Mixed	Acute STEMI	55 ± 2	Rhepoetin α	i.v.	60,000	4	+	–	–	–	Najjar et al. (2011)
51	Mixed	Acute STEMI	60 ± 10	Rhepoetin β	i.v.	2 × 50,000	4	+	–	–		Bergmann et al. (2011)
138	Mixed	Acute STEMI	60 ± 12	Rhepoetin β	i.v.	3 × 33,000	6	+	–	–	–	Ott et al. (2010)

2010) and no cardiovascular adverse effects. Moreover, the high single bolus injection of 100,000 U may have proved safe for stroke patients (Ehrenreich et al., 2002), but it does not seem to be safe for patients with cardiovascular risks (Ludman et al., 2011; Ott et al., 2010). Furthermore, since the heart is not separated from circulation as the brain (blood–brain barrier), higher amounts of systemically administered rhEPO can reach the heart (6000 U/L detected in the plasma) (Lipsic et al., 2006) rather than the ones reaching the brain (17 U/L detected in the cerebrospinal fluid) (Ehrenreich et al., 2002). Altogether these data suggest that repetitive small doses of EPO may be more beneficial than a single bolus injection.

4. CONCLUSION AND FUTURE DIRECTIONS

The EPO/EPOR system was extensively studied in preclinical models of cardiovascular disease. In all of these models it was shown that EPOR is strongly expressed in progenitor or immature proliferating cells. EPO could activate these cells and protect the heart mainly by inducing neovascularization. Nevertheless, other cell types as cardiomyogenic cells or epicardial cells were affected by EPO and contributed to cardioprotection upon injury by mechanisms which are currently not well understood. The broad expression of EPOR in resident cardiac progenitors suggests a comprehensive role for EPO in tissue regeneration.

Unfortunately the promising preclinical data were not translated into successful clinical trials. In the majority of the trials no significant amelioration in the patient cardiac function was observed. In some of the studies where high bolus injections of rhEPO were administered to the patients, adverse cardiovascular effects were even observed. In contrast to these studies, repetitive low EPO doses administration to patient resulted in cardioprotection and no adverse events. These data in combination with the high affinity of EPO and its short half-life suggest that small repetitive slow-release doses of EPO may be more beneficial for cardiac function and safer.

Finally, very little is known about the expression of EPOR in *human* CPCs as well as in postmitotic cardiomyocytes. Human CMs deriving from human embryonic stem cells can be used to decipher the expression levels and possible role of EPO in these cells. Moreover, human bioengineered muscle provides an attractive model to study the role of EPO in healthy and diseased cardiac muscle in vitro. Such studies may provide some insight in the expression and role of the EPO/EPOR system in human CPCs as well as postmitotic cardiac cells. These data may prove useful for the design of future more successful clinical trials.

ACKNOWLEDGMENTS

I certainly failed to include all outstanding contributions to the field of EPO and cardiac regeneration and wish to apologize to the authors that I have undeliberately missed. The author is supported by the German Center for Cardiovascular Research (DZHK) as well as the Medical University of Goettingen.

REFERENCES

Anagnostou, A., Liu, Z., Steiner, M., Chin, K., Lee, E. S., Kessimian, N., et al. (1994). Erythropoietin receptor mRNA expression in human endothelial cells. *Proceedings of the National Academy of Sciences of the United States of America*, *91*(9), 3974–3978.

Asaumi, Y., Kagaya, Y., Takeda, M., Yamaguchi, N., Tada, H., Ito, K., et al. (2007). Protective role of endogenous erythropoietin system in nonhematopoietic cells against pressure overload-induced left ventricular dysfunction in mice. *Circulation*, *115*(15), 2022–2032.

Bergmann, O., Bhardwaj, R. D., Bernard, S., Zdunek, S., Barnabe-Heider, F., Walsh, S., et al. (2009). Evidence for cardiomyocyte renewal in humans. *Science*, *324*(5923), 98–102.

Bergmann, M. W., Haufe, S., von Knobelsdorff-Brenkenhoff, F., Mehling, H., Wassmuth, R., Munch, I., et al. (2011). A pilot study of chronic, low-dose epoetin-{beta} following percutaneous coronary intervention suggests safety, feasibility, and efficacy in patients with symptomatic ischaemic heart failure. *European Journal of Heart Failure*, *13*(5), 560–568.

Bersell, K., Arab, S., Haring, B., & Kuhn, B. (2009). Neuregulin1/ErbB4 signaling induces cardiomyocyte proliferation and repair of heart injury. *Cell*, *138*(2), 257–270.

Binbrek, A. S., Rao, N. S., Al Khaja, N., Assaqqaf, J., & Sobel, B. E. (2009). Erythropoietin to augment myocardial salvage induced by coronary thrombolysis in patients with ST segment elevation acute myocardial infarction. *The American Journal of Cardiology*, *104*(8), 1035–1040.

Brade, T., Kumar, S., Cunningham, T. J., Chatzi, C., Zhao, X., Cavallero, S., et al. (2011). Retinoic acid stimulates myocardial expansion by induction of hepatic erythropoietin which activates epicardial Igf2. *Development*, *138*(1), 139–148.

Brines, M., Grasso, G., Fiordaliso, F., Sfacteria, A., Ghezzi, P., Fratelli, M., et al. (2004). Erythropoietin mediates tissue protection through an erythropoietin and common beta-subunit heteroreceptor. *Proceedings of the National Academy of Sciences of the United States of America*, *101*(41), 14907–14912.

Brunner, S., Winogradow, J., Huber, B. C., Zaruba, M. M., Fischer, R., David, R., et al. (2009). Erythropoietin administration after myocardial infarction in mice attenuates ischemic cardiomyopathy associated with enhanced homing of bone marrow-derived progenitor cells via the CXCR-4/SDF-1 axis. *The FASEB Journal*, *23*(2), 351–361.

Calvillo, L., Latini, R., Kajstura, J., Leri, A., Anversa, P., Ghezzi, P., et al. (2003). Recombinant human erythropoietin protects the myocardium from ischemia-reperfusion injury and promotes beneficial remodeling. *Proceedings of the National Academy of Sciences of the United States of America*, *100*(8), 4802–4806.

Constantinescu, S. N., Keren, T., Socolovsky, M., Nam, H., Henis, Y. I., & Lodish, H. F. (2001). Ligand-independent oligomerization of cell-surface erythropoietin receptor is mediated by the transmembrane domain. *Proceedings of the National Academy of Sciences of the United States of America*, *98*(8), 4379–4384.

Dame, C., Fahnenstich, H., Freitag, P., Hofmann, D., Abdul-Nour, T., Bartmann, P., et al. (1998). Erythropoietin mRNA expression in human fetal and neonatal tissue. *Blood*, *92*(9), 3218–3225.

Depping, R., Kawakami, K., Ocker, H., Wagner, J. M., Heringlake, M., Noetzold, A., et al. (2005). Expression of the erythropoietin receptor in human heart. *The Journal of Thoracic and Cardiovascular Surgery, 130*(3), 877–878.
Dirkx, E., da Costa Martins, P. A., & De Windt, L. J. (2013). Regulation of fetal gene expression in heart failure. *Biochimica et Biophysica Acta, 1832*(12), 2414–2424.
Duim, S. N., Kurakula, K., Goumans, M. J., & Kruithof, B. P. (2015). Cardiac endothelial cells express Wilms' tumor-1: Wt1 expression in the developing, adult and infarcted heart. *Journal of Molecular and Cellular Cardiology, 81*, 127–135.
Ehrenreich, H., Hasselblatt, M., Dembowski, C., Cepek, L., Lewczuk, P., Stiefel, M., et al. (2002). Erythropoietin therapy for acute stroke is both safe and beneficial. *Molecular Medicine, 8*(8), 495–505.
El-Komy, M. H., Schmidt, R. L., Widness, J. A., & Veng-Pedersen, P. (2011). Differential pharmacokinetic analysis of in vivo erythropoietin receptor interaction with erythropoietin and continuous erythropoietin receptor activator in sheep. *Biopharmaceutics & Drug Disposition, 32*(5), 276–288.
Ferrario, M., Arbustini, E., Massa, M., Rosti, V., Marziliano, N., Raineri, C., et al. (2011). High-dose erythropoietin in patients with acute myocardial infarction: A pilot, randomised, placebo-controlled study. *International Journal of Cardiology, 147*(1), 124–131.
Fu, Q. L., Wu, W., Wang, H., Li, X., Lee, V. W., & So, K. F. (2008). Up-regulated endogenous erythropoietin/erythropoietin receptor system and exogenous erythropoietin rescue retinal ganglion cells after chronic ocular hypertension. *Cellular and Molecular Neurobiology, 28*(2), 317–329.
Gabel, R., Klopsch, C., Furlani, D., Yerebakan, C., Li, W., Ugurlucan, M., et al. (2009). Single high-dose intramyocardial administration of erythropoietin promotes early intracardiac proliferation, proves safety and restores cardiac performance after myocardial infarction in rats. *Interactive Cardiovascular and Thoracic Surgery, 9*(1), 20–25. discussion 25.
Grebien, F., Kerenyi, M. A., Kovacic, B., Kolbe, T., Becker, V., Dolznig, H., et al. (2008). Stat5 activation enables erythropoiesis in the absence of EpoR and Jak2. *Blood, 111*(9), 4511–4522.
Hirata, A., Minamino, T., Asanuma, H., Sanada, S., Fujita, M., Tsukamoto, O., et al. (2005). Erythropoietin just before reperfusion reduces both lethal arrhythmias and infarct size via the phosphatidylinositol-3 kinase-dependent pathway in canine hearts. *Cardiovascular Drugs and Therapy, 19*(1), 33–40.
Hoch, M., Fischer, P., Stapel, B., Missol-Kolka, E., Sekkali, B., Scherr, M., et al. (2011). Erythropoietin preserves the endothelial differentiation capacity of cardiac progenitor cells and reduces heart failure during anticancer therapies. *Cell Stem Cell, 9*(2), 131–143.
Hsieh, M. M., Linde, N. S., Wynter, A., Metzger, M., Wong, C., Langsetmo, I., et al. (2007a). HIF prolyl hydroxylase inhibition results in endogenous erythropoietin induction, erythrocytosis, and modest fetal hemoglobin expression in rhesus macaques. *Blood, 110*(6), 2140–2147.
Hsieh, P. C., Segers, V. F., Davis, M. E., MacGillivray, C., Gannon, J., Molkentin, J. D., et al. (2007b). Evidence from a genetic fate-mapping study that stem cells refresh adult mammalian cardiomyocytes after injury. *Nature Medicine, 13*(8), 970–974.
Huang, L. J., Constantinescu, S. N., & Lodish, H. F. (2001). The N-terminal domain of Janus kinase 2 is required for Golgi processing and cell surface expression of erythropoietin receptor. *Molecular Cell, 8*(6), 1327–1338.
Jacobson, L. O., Goldwasser, E., Fried, W., & Plzak, L. (1957). Role of the kidney in erythropoiesis. *Nature, 179*(4560), 633–634.
Kanellakis, P., Pomilio, G., Agrotis, A., Gao, X., Du, X. J., Curtis, D., et al. (2010). Darbepoetin-mediated cardioprotection after myocardial infarction involves multiple mechanisms independent of erythropoietin receptor-common beta-chain heteroreceptor. *British Journal of Pharmacology, 160*(8), 2085–2096.

Kuhn, B., del Monte, F., Hajjar, R. J., Chang, Y. S., Lebeche, D., Arab, S., et al. (2007). Periostin induces proliferation of differentiated cardiomyocytes and promotes cardiac repair. *Nature Medicine*, *13*(8), 962–969.

Lipsic, E., van der Meer, P., Voors, A. A., Westenbrink, B. D., van den Heuvel, A. F., de Boer, H. C., et al. (2006). A single bolus of a long-acting erythropoietin analogue darbepoetin alfa in patients with acute myocardial infarction: A randomized feasibility and safety study. *Cardiovascular Drugs and Therapy*, *20*(2), 135–141.

Liu, C., Shen, K., Liu, Z., & Noguchi, CT. (1997). Regulated human erythropoietin receptor expression in mouse brain. *The Journal of Biological Chemistry*, *272*(51), 32395–32400.

Ludman, A. J., Yellon, D. M., Hasleton, J., Ariti, C., Babu, G. G., Boston-Griffiths, E., et al. (2011). Effect of erythropoietin as an adjunct to primary percutaneous coronary intervention: A randomised controlled clinical trial. *Heart*, *97*(19), 1560–1565.

Macdougall, I. C., Robson, R., Opatrna, S., Liogier, X., Pannier, A., Jordan, P., et al. (2006). Pharmacokinetics and pharmacodynamics of intravenous and subcutaneous continuous erythropoietin receptor activator (C.E.R.A.) in patients with chronic kidney disease. *Clinical Journal of the American Society of Nephrology*, *1*(6), 1211–1215.

Makita, T., Duncan, S. A., & Sucov, H. M. (2005). Retinoic acid, hypoxia, and GATA factors cooperatively control the onset of fetal liver erythropoietin expression and erythropoietic differentiation. *Developmental Biology*, *280*(1), 59–72.

Miura, Y., Miura, O., Ihle, J. N., & Aoki, N. (1994). Activation of the mitogen-activated protein kinase pathway by the erythropoietin receptor. *The Journal of Biological Chemistry*, *269*(47), 29962–29969.

Moon, C., Krawczyk, M., Ahn, D., Ahmet, I., Paik, D., Lakatta, E. G., et al. (2003). Erythropoietin reduces myocardial infarction and left ventricular functional decline after coronary artery ligation in rats. *Proceedings of the National Academy of Sciences of the United States of America*, *100*(20), 11612–11617.

Morishita, E., Masuda, S., Nagao, M., Yasuda, Y., & Sasaki, R. (1997). Erythropoietin receptor is expressed in rat hippocampal and cerebral cortical neurons, and erythropoietin prevents in vitro glutamate-induced neuronal death. *Neuroscience*, *76*(1), 105–116.

Najjar, S. S., Rao, S. V., Melloni, C., Raman, S. V., Povsic, T. J., Melton, L., et al. (2011). Intravenous erythropoietin in patients with ST-segment elevation myocardial infarction: REVEAL: A randomized controlled trial. *JAMA*, *305*(18), 1863–1872. http://dx.doi.org/10.1001/jama.2011.592.

Oka, T., Xu, J., & Molkentin, J. D. (2007). Re-employment of developmental transcription factors in adult heart disease. *Seminars in Cell & Developmental Biology*, *18*(1), 117–131.

Ott, C., Martens, H., Hassouna, I., Oliveira, B., Erck, C., Zafeiriou, M. P., et al. (2015). Widespread expression of erythropoietin receptor in brain and its induction by injury. *Molecular Medicine*, *21*(1), 803–815.

Ott, I., Schulz, S., Mehilli, J., Fichtner, S., Hadamitzky, M., Hoppe, K., et al. (2010). Erythropoietin in patients with acute ST-segment elevation myocardial infarction undergoing primary percutaneous coronary intervention: A randomized, double-blind trial. *Circulation Cardiovascular Interventions*, *3*(5), 408–413.

Porrello, E. R., Mahmoud, A. I., Simpson, E., Hill, J. A., Richardson, J. A., Olson, E. N., et al. (2011). Transient regenerative potential of the neonatal mouse heart. *Science*, *331*(6020), 1078–1080.

Remy, I., Wilson, I. A., & Michnick, S. W. (1999). Erythropoietin receptor activation by a ligand-induced conformation change. *Science*, *283*(5404), 990–993.

Salminen, P. R., Dahle, G. O., Moen, C. A., Wergeland, A., Jonassen, A. K., Haaverstad, R., et al. (2014). Reperfusion therapy with low-dose insulin or insulin-like growth factor 2; myocardial function and infarct size in a porcine model of ischaemia and reperfusion. *Basic & Clinical Pharmacology & Toxicology*, *115*(5), 438–447.

Seeger, T., Xu, Q. F., Muhly-Reinholz, M., Fischer, A., Kremp, E. M., Zeiher, A. M., et al. (2016). Inhibition of let-7 augments the recruitment of epicardial cells and improves cardiac function after myocardial infarction. *Journal of Molecular and Cellular Cardiology*, *94*, 145–152.

Senyo, S. E., Steinhauser, M. L., Pizzimenti, C. L., Yang, V. K., Cai, L., Wang, M., et al. (2013). Mammalian heart renewal by pre-existing cardiomyocytes. *Nature*, *493*(7432), 433–436.

Stuckmann, I., Evans, S., & Lassar, A. B. (2003). Erythropoietin and retinoic acid, secreted from the epicardium, are required for cardiac myocyte proliferation. *Developmental Biology*, *255*(2), 334–349.

Suzuki, N., Ohneda, O., Takahashi, S., Higuchi, M., Mukai, H. Y., Nakahata, T., et al. (2002). Erythroid-specific expression of the erythropoietin receptor rescued its null mutant mice from lethality. *Blood*, *100*(7), 2279–2288.

Syed, R. S., Reid, S. W., Li, C., Cheetham, J. C., Aoki, K. H., Liu, B., et al. (1998). Efficiency of signalling through cytokine receptors depends critically on receptor orientation. *Nature*, *395*(6701), 511–516.

Tada, H., Kagaya, Y., Takeda, M., Ohta, J., Asaumi, Y., Satoh, K., et al. (2006). Endogenous erythropoietin system in non-hematopoietic lineage cells plays a protective role in myocardial ischemia/reperfusion. *Cardiovascular Research*, *71*(3), 466–477.

Taniguchi, N., Nakamura, T., Sawada, T., Matsubara, K., Furukawa, K., Hadase, M., et al. (2010). Erythropoietin prevention trial of coronary restenosis and cardiac remodeling after ST-elevated acute myocardial infarction (EPOC-AMI): A pilot, randomized, placebo-controlled study. *Circulation Journal*, *74*(11), 2365–2371.

Teng, R., Calvert, J. W., Sibmooh, N., Piknova, B., Suzuki, N., Sun, J., et al. (2011). Acute erythropoietin cardioprotection is mediated by endothelial response. *Basic Research in Cardiology*, *106*(3), 343–354.

Tsai, P. T., Ohab, J. J., Kertesz, N., Groszer, M., Matter, C., Gao, J., et al. (2006). A critical role of erythropoietin receptor in neurogenesis and post-stroke recovery. *The Journal of Neuroscience*, *26*(4), 1269–1274.

Voors, A. A., Belonje, A. M., Zijlstra, F., Hillege, H. L., Anker, S. D., Slart, R. H., et al. (2010). A single dose of erythropoietin in ST-elevation myocardial infarction. *European Heart Journal*, *31*(21), 2593–2600.

Wei, K., Serpooshan, V., Hurtado, C., Diez-Cunado, M., Zhao, M., Maruyama, S., et al. (2015). Epicardial FSTL1 reconstitution regenerates the adult mammalian heart. *Nature*, *525*(7570), 479–485.

Wickrema, A., Krantz, S. B., Winkelmann, J. C., & Bondurant, M. C. (1992). Differentiation and erythropoietin receptor gene expression in human erythroid progenitor cells. *Blood*, *80*(8), 1940–1949.

Wu, H., Lee, S. H., Gao, J., Liu, X., & Iruela-Arispe, M. L. (1999). Inactivation of erythropoietin leads to defects in cardiac morphogenesis. *Development*, *126*(16), 3597–3605.

Xu, B., Dong, G.-H., Liu, H., Wang, Y.-Q., Wu, H.-W., & Jing, H. (2005). Recombinant human erythropoietin pretreatment attenuates myocardial infarct size: A possible mechanism involves heat shock protein 70 and attenuation of nuclear factor-kappaB. *Annals of Clinical and Laboratory Science*, *35*(2), 161–168.

Yamamura, Y., Noda, M., & Ikawa, Y. (1994). Activated Ki-Ras complements erythropoietin signaling in CTLL-2 cells, inducing tyrosine phosphorylation of a 160-kDa protein. *Proceedings of the National Academy of Sciences of the United States of America*, *91*(19), 8866–8870.

Ye, L., Du, X., Xia, J., & Ping, J. (2005). Effects of rHu-EPO on myocyte apoptosis and cardiac function following acute myocardial infarction in rats. *Journal of Huazhong University of Science and Technology Medical Sciences*, *25*(1), 55–58.

Zafiriou, M. P., Noack, C., Unsold, B., Didie, M., Pavlova, E., Fischer, H. J., et al. (2014). Erythropoietin responsive cardiomyogenic cells contribute to heart repair post myocardial infarction. *Stem Cells*, *32*(9), 2480–2491.

Zelarayan, L. C., Noack, C., Sekkali, B., Kmecova, J., Gehrke, C., Renger, A., et al. (2008). Beta-catenin downregulation attenuates ischemic cardiac remodeling through enhanced resident precursor cell differentiation. *Proceedings of the National Academy of Sciences of the United States of America*, *105*(50), 19762–19767.

CHAPTER THIRTEEN

Erythropoietin Promotes Glioblastoma via miR-451 Suppression

Begum Alural*,†, Zeynep O. Ayyildiz*,†, Kemal U. Tufekci*,†, Sermin Genc*,†, Kursad Genc†,1

*Izmir International Biomedicine and Genome Institute, Dokuz Eylul University, Izmir, Turkey
†Health Science Institute, Dokuz Eylul University, Izmir, Turkey
[1]Corresponding author: e-mail addresses: kkursadgenc@hotmail.com; kemal.genc@deu.edu.tr

Contents

Vitamins and Hormones, Volume 105
ISSN 0083-6729
http://dx.doi.org/10.1016/bs.vh.2017.03.002

© 2017 Elsevier Inc.
All rights reserved.

Abstract

Erythropoietin (EPO) is an erythropoiesis stimulating growth factor and hormone. EPO has been widely used in the treatment of chronic renal failure, cancer, and chemotherapy-related anemia for three decades. However, many clinical trials showed that EPO treatment may be associated with tumorigenesis and cancer progression. EPO is able to cross blood–brain barriers, and this may lead to an increased possibility of central nervous system tumors such as glioblastoma. Indeed, EPO promotes glioblastoma growth and invasion in animal studies. Additionally, EPO increases glioblastoma cell survival, proliferation, migration, invasion, and chemoresistancy in vitro. However, the exact mechanisms of cancer progression induced by EPO treatment are not fully understood. Posttranscriptional gene regulation through microRNAs may contribute to EPO's cellular and biological effects in tumor progression. Here, we aimed to study whether tumor suppressive microRNA, miR-451, counteracts the positive effects of EPO on U87 human glioblastoma cell line. Migration and invasion were evaluated by scratch assay and transwell invasion assay, respectively. We found that EPO decreased basal miR-451 expression and increased cell proliferation, migration, invasion, and cisplatin chemoresistancy in vitro. miR-451 overexpression by transfection of its mimic significantly reversed these effects. Furthermore, ectopic expression of miR-451 inhibited expression of its own target genes, such as metalloproteinases-2 and -9, which are stimulated by EPO treatment and involved in carcinogenesis processes, especially invasion. These findings suggest that miR-451 mimic delivery may be useful as adjuvant therapy in addition to chemotherapy and anemia treatment by EPO and should be tested in experimental glioblastoma models.

1. INTRODUCTION

Erythropoietin (EPO) is a 34-kDa glycoprotein growth factor and hormone, which controls erythropoiesis through the promotion of proliferation, differentiation, and survival of erythrocytes progenitor cells and survival of mature erythrocytes (Jelkmann, 2013). EPO is initially synthesized in the liver during fetal development, but shortly after birth, production site of EPO subsequently shifts to the kidney. Peritubular fibroblast-like cells in the renal cortex are the major site of EPO production (Suzuki & Yamamoto, 2016). Expression of the EPO and EPO receptor (EPOR) is low in normal adult tissues and mainly induced by hypoxia. Several lines of evidences suggest that the central nervous system (CNS) expresses EPO and EPOR both at the mRNA and protein levels (Maiese, 2016).

EPO was purified from the urine of patients with aplastic anemia, and the therapeutic use of EPO was approved by the US Food and Drug Administration for treatment of anemia in patients with chronic renal failure 30 years ago (Jelkmann, 2013). EPO is now widely used for the treatment of anemia associated with renal failure, cancer, cancer chemotherapy, prematurity, chronic inflammatory diseases, and human immunodeficiency virus infection (Bennett et al., 2016; Debeljak, Solar, & Sytkowski, 2014). In the last 20 years, many in vitro and in vivo studies showed that EPO has cytoprotective and tissue protective, cell-proliferative, antiapoptotic, antiinflammatory, vascular protective, angiogenic, antiedema, antioxidant, cell migration promoting, neurogenesis stimulation, and metabolism regulation effects (Maiese, 2016). It has in vitro cytoprotective effect against various insults include neurotoxic agents, irradiation, trauma, chemotherapy, hypoxia, ischemia, oxygen glucose deprivation in many cell types (Maiese, 2016). These effects have also been shown in experimental models of various acute CNS injuries, such as stroke and chronic neurodegenerative diseases including Alzheimer's disease and Parkinson's disease (Sargin, Friedrichs, El-Kordi, & Ehrenreich, 2010).

Transport of EPO via the blood–brain barrier to CNS after systemic administration has been observed both in experimental animals and humans at high doses (Brines et al., 2000; Xenocostas et al., 2005). Cerebrospinal fluid concentration of EPO highly increases following intravenous administration (Ehrenreich et al., 2007). In spite of strong preclinical evidences, many clinical studies have failed in small unrandomized and retrospective patient studies. Thus, well-designed, randomized, prospective, placebo-controlled, and larger high quality clinical trials are still needed (Kochanek & Clark, 2016; Pearl, 2014; Sargin et al., 2010). Although several recent studies in stroke, bipolar disorder, schizophrenia patients have not reported negative results, sample sizes are relatively small (Miskowiak, Ehrenreich, Christensen, Kessing, & Vinberg, 2014; Tsai et al., 2015; Wustenberg et al., 2011). EPO is already used routinely in prematurity anemia for a long time (Juul & Pet, 2015). Use of EPO for neuroprotection in very early term infants and neonates with hypoxic-ischemic encephalopathy seems a promising strategy (Juul & Pet, 2015). However, neurodevelopmental deficits and progression of any existing CNS tumor such as pediatric glioblastoma (GB) should be considered. The results of animal studies in the models of age-related chronic neurodegenerative diseases are currently not conclusive. Furthermore, EPO's serious side effects including hypertension, procoagulant, thromboembolic events, and tumor promotion can limit EPO therapy especially in elderly patients (Chong et al., 2013).

GB is the most aggressive and most prevalent primary brain tumor in adults. The World Health Organization classification system groups GB into four histological grades (Louis et al., 2016). Glioblastoma multiforme, also known as grade IV GB, is the most common and aggressive form of GB (Bush, Chang, & Berger, 2016). The standard therapy for GB is maximal surgical resection followed by radiotherapy and adjuvant temozolomide (TMZ) chemotherapy. In spite of intensive treatment, GB is associated with poor clinical outcome and currently not curable. Recent genome-wide association studies have contributed to the elucidation of the pathogenesis of GB (Bush et al., 2016). Cell surface receptors of growth factors such as epidermal growth factor receptor, signaling pathways, and transcription factors play central roles in the pathobiology of GB. Generally, almost all cancers, in this case GB, are characterized with biological and pathological processes including self-renewal, immortality, unlimited cell proliferation, hypoxia, invasiveness, and metastasis.

Many clinical and experimental studies showed that EPO stimulates tumor growth and progression in many types of cancers (Debeljak et al., 2014). Relatively rare in situ and in vitro studies reported that high EPO and EPOR expression in patients GB tissue samples (Brunotte, Bock, Bruck, Hemmerlein, & Strik, 2011; Mittelbronn et al., 2007; Mohyeldin et al., 2007; Nico et al., 2011; Said et al., 2007). In vitro studies with GB cell lines showed that these cells express EPO and EPOR (Belenkov et al., 2004; Hassouna et al., 2008; Mohyeldin et al., 2007; Peres et al., 2011; Said et al., 2007). In situ studies searched a possible concordance between EPOR expression levels with patient survival and histopathological grade of tumor. Mohyeldin et al. found that expression of EPOR correlated with the stage of the tumor (Mohyeldin et al., 2007). Unexpectedly, an inverse correlation was found between EPOR and GB grade of malignancy (Mittelbronn et al., 2007). However, EPOR expression level is directly associated with survival (Brunotte et al., 2011; Mittelbronn et al., 2007). EPO stimulates cell growth and proliferation in cultured GB cell lines (Hassouna et al., 2008; Yin et al., 2007). EPOR may mediate these effects both in vitro and in vivo experiments (Peres et al., 2015, 2011). EPOR signaling pathways are also involved in responsiveness to EPO (Belenkov et al., 2004; Cao et al., 2010; Kwon et al., 2014). EPO–EPOR signaling may also increase resistance of GB cells to chemotherapy and radiotherapy (Belenkov et al., 2004; Mohyeldin et al., 2007; Peres et al., 2015). Unexpectedly, Hassouna have reported that EPO treatment increased sensitivity to radiation and TMZ (Hassouna et al., 2008). Another interesting finding was that prevention of anemia with EPO-enhanced radiosensitivity of xenografted GB cells (Stuben et al., 2003).

Epigenetic mechanisms and posttranscriptional gene regulation by non-coding RNAs (ncRNAs) can also mediate initiation and progression of GB. MicroRNAs (miRNAs) are short, single-stranded, ncRNAs molecules that regulate gene expression at the posttranscriptional level (Hammond, 2015). miRNAs can control multiple cellular and biological processes including development, proliferation, differentiation, migration, apoptosis, and growth. Association of miRNAs with many human diseases has been extensively studied. Several functional groups of miRNAs such as oncomirs, tumor suppressor miRNAs, angiomirs, and hypoxamirs contribute to GB etiopathogenesis (Costa, Cardoso, Mano, & de Lima, 2015). miR-451 is a widely dysregulated miRNA in several human cancers including GB (Godlewski, Bronisz, Nowicki, Chiocca, & Lawler, 2010; Godlewski, Nowicki, et al., 2010; Pan, Wang, & Wang, 2013; Tian et al., 2012). Different studies shown that miR-451 inhibited cell proliferation, migration, and invasion and induced apoptosis in GB cell lines (Godlewski, Bronisz, et al., 2010; Godlewski, Nowicki, et al., 2010; Nan et al., 2010). Recently, we have determined that EPO downregulates the expression of miR-451 in SH-SY5Y neuroblastoma cell line and biological effects of EPO such as survival, proliferation, and migration depend on miR-451 suppression (Alural et al., 2014).

In the present study we examined the GB promoting effect of EPO in GB cells and the role of miR-451 in this GB promoting effect of EPO.

2. MATERIALS AND METHODS

2.1 Cell Culture and Treatment

Human U87-MG glioblastoma cells (American Type Culture Collection) were cultured in Dulbecco's Modified Eagle Medium (DMEM) (Biochrom GbmH, Berlin, Germany) and supplemented with heat-inactivated fetal bovine serum (10% v/v), L-glutamine (1% v/v), and penicillin streptomycin (1% v/v). Cells were incubated at 37°C in 5% CO2.

2.2 Presto Blue Assay

Cell viability was assessed by the reducing environment of viable cells which upon entry, converts nonfluorescent resazurin-based solution into highly fluorescents resofurin. Cells were seeded in 96-well plate at a density of 1×10^4 cells per well. Cells were treated with 0–72h with 0–10U/mL EPO for basal cell viability. For chemoresistance study, cells were pretreated with 1 U/mL EPO for 4 h, then incubated with cisplatin at 50 μM for 72 h. At the end of treatment, 10 μL of Presto Blue reagent (Thermo Fisher Scientific, USA) was added to each well containing 100 μL of medium. The plate was then incubated for 30 min at 37°C, and the absorbance of each well were measured on a microplate reader (Varioscan, Thermo, USA) at 590 nm. The relative cell viability was calculated as the percentage of untreated cells.

2.3 BrdU Assay

Cell proliferation of U87-MG glioblastoma cells in response to 72-h treatment of 1 U/mL EPO was measured by BrdU Cell Proliferation Chemiluminescent Assay Kit (Cell Signaling, USA). Cells were seeded in to 96-well plate with a density of 1×10^4 cell per well. These cells were incubated with 20 μL BrdU-labeling solution 16 h prior to the end of 72 h 1 U/mL EPO treatment. Following BrdU incorporation, the culture medium was removed in each well, and cells were fixed with fixing/denaturing solution for 30 min. Subsequently, cells were incubated with BrdU detection antibody for 1 h and HRP-conjugated secondary antibody for 30 min. After several washing steps, 100 μL of substrate solution was added to each well, and the color reaction product was quantified using microplate reader (Varioscan, Thermo, USA) at 425 nm. The relative cell proliferation was calculated as the percentage of untreated cells.

2.4 Western Blotting

The protein samples were extracted from U87-MG cells, and the concentrations were determined using BCA Protein Assay Kit (Sigma-Aldrich, USA). Equal amounts of protein were resolved by 10% SDS-PAGE, transferred to PVDF membrane, and blocked with 5% BSA. After washing three times for 10 min each in PBS (with 5% NP-40), membranes were incubated with EPOR (Abcam), CD131 (Abnova), Ephrin type-B receptor 4 (EphB4) (Abcam), or β-actin antibodies (Abcam) at 1:1000 concentration overnight at 4°C. Membranes were then washed again with PBS (with 5% NP-40), three times for 10 min each, and incubated with the horseradish peroxidase-conjugated secondary antibody for 1 h at room temperature. Immunodetection was performed using enhanced chemiluminescence (Super Signal West Pico, Thermo Fisher Scientific, USA) according to the manufacturer's instructions. The density of the protein bands was analyzed by Image J software (Schneider, Rasband, & Eliceiri, 2012) and normalized to β-actin loading control.

2.5 Immunofluorescent Staining

Cells were fixed with 4% paraformaldehyde in PBS and permeabilized with 0.1% Triton X-100 in PBS for 10 min at room temperature and blocked for 30 min in 10% FBS in PBS. Cells were stained with primary antibodies for 2 h at room temperature and the respective Alexafluor-488 or -555 conjugated secondary antibodies (Invitrogen, Germany) for 50 min. Finally, cells were mounted with antifade reagent including 4′,6-diamidino-2-phenylindole (Millipore) for counterstaining and observed with BX61 fluorescent microscope (Olympus, Japan).

2.6 Migration Assay

The U87-MG glioblastoma cells were seeded at a density of 1×10^5 cells/well in a six-well plate and incubated in DMEM containing 10% FBS for 24 h to confluence. A confluent monolayer of each well was scratched with a 1000-μL pipette tip, and cells were washed once with PBS and replaced with fresh growth medium containing 1% FBS, with or without 1 U/mL EPO. Immediately after the scratch (0 h), at 24 and 48 h, wounded areas were marked in each well and images obtained using phase-contrast inverted microscope (CKX41, Olympus) with a 4 × magnification. The number of cells that migrated over the margins of the wounds was counted by using Image J software (Schneider et al., 2012).

2.7 Invasion Assay

Thincert cell culture inserts with an 8-μm pore diameter (Greiner Bio-One, Austria) and Matrigel Basement Membrane Matrix (Corning, NY, USA) were used to perform invasion experiments. Prior to experiments, all inserts placed in wells of 24-well plate with sterile forceps. Inserts were coated with matrigel membrane matrix and incubated for 2 h at 37°C. Following incubation, each insert was fixed with 0.1% BSA. U87-MG glioblastoma cells (2×10^4 cells suspended in 1% FBS supplemented media with or without 1 U/mL EPO) were plated in each insert. Media supplemented with 10% FBS was used as chemoattractant in each well, below the insert. The cells were allowed to invade for 24 h at 37°C in a humidified incubator. At the end of the incubation period, cells were stained with Diff-Quick Staining Set (Siemens, Germany), and cells in the upper transfilter were removed with cotton swabs. The images were obtained using phase-contrast inverted microscope (CKX41, Olympus) with a 20× and 40× magnification. Invasion was quantified by cell counting with Image J software (Schneider et al., 2012) in at least three different fields per membrane.

2.8 Real-Time PCR for miR-451

U87-MG cells were seeded in six-well plate at a density of 1×10^5 cells per well. After treatment with 1 U/mL EPO for 18 h, total RNA was isolated from cells using Qiagen miRNeasy Mini Kit. Quality and quantity of obtained RNAs were measured with the NanoDrop Spectrophotometer (Thermo Scientific, USA). 500 ng RNA was reverse-transcribed using miScript II RT Kit (Qiagen, Valencia, CA, USA). Subsequently, miScript SYBR Green PCR Kit was used to perform quantitative real-time PCR (qRT-PCR) on a Lightcycler 1.5 Real-Time PCR System (Roche Diagnostics, Germany). Both primers (miR-451 and U6) for mature miRNAs were purchased from Qiagen (Valencia, CA, USA). The conditions of qRT-PCR reactions are as follows: 95°C for 10 min, 40 cycles at 95°C for 15 s and at 60°C for 30 s. PCR products were evaluated for their specificity by melting curve analysis. The $2^{-\Delta\Delta Ct}$ method was used to calculate the relative expression levels of miR-451 normalized with U6.

2.9 Transfection With miR-451 Mimic

MiR-451 mimic and negative controls were purchased from Qiagen. U87-MG glioblastoma cells were transfected 24 h after seeding in cell culture plates with a density of 1.5×10^3 cells per well in 96-well plate. Transfection of cells with miR-451 mimic and negative control oligomers was performed using the HiPerFect transfection reagent (Qiagen) according to the manufacturer's protocol. The final concentration of the mimics was 50 nM. After 24 h, media from each well was replaced with DMEM supplemented with 10% FBS or DMEM supplemented with 10% FBS and 1 U/mL EPO and left for 72 h in the humidified incubator.

Transfection efficiency was determined by fluorescence microscope with the use of Cy3-labeled control mimic oligonucleotide.

2.10 Statistical Analysis

Statistical analyses were performed with SPSS 18.0. Results are expressed as mean ± standard error of the mean. Comparison of two groups was analyzed by using Mann–Whitney U test.

3. RESULTS

3.1 EPO Increased Viability of Glioblastoma Cells

To evaluate EPO effect on cell viability, U87-MG cells were stimulated with recombinant human erythropoietin (rhEPO) at different concentrations (0.1–10 U/mL). EPO increased cell viability in a time- and dose-dependent manner as shown by Presto Blue assay (Fig. 1A–C). There was no significant difference between 1 and 10 mL EPO (23% vs 26%) on cell viability at 72 h. Thus, we performed all experiments using only 1 U/mL concentration of EPO.

3.2 miR-451 Overexpression Reversed Cell Viability Promoting Effect of EPO

Using qPCR, we demonstrated that 1 U/mL EPO treatment of U87-MG glioma cells for 18 h significantly decreased basal expression of miR-451 (Fig. 1D). To evaluate a possible functional role of miR-451 on EPO's viability promoting effect U87-MG cells were transfected with miR-451 mimic. High transfection efficiency (over 95%) was confirmed by fluorescent detection of Cy3-tagged negative control miRNA mimic (Fig. 1E). qPCR showed miR-451 mimic robustly upregulated miR-451 level (Fig. 1D). However, contrary to basal miR-451 levels, EPO did not affect ectopic miR-451 expression. We found that the EPO-mediated increase of cell viability was significantly reduced in the mimic transfected cells as compared with cells that were transfected with control mimics (Fig. 1F).

3.3 EPO Promoted Proliferation of Glioblastoma Cells

EPO effect on cell proliferation was determined by measuring the nuclear incorporation of BrdU. 1 U/mL EPO treatment for 72 h led 38% increase in BrdU incorporation suggesting that EPO increased proliferation of U87 cells (Fig. 2A). We further investigated the role of miR-451 on the proliferative effect of EPO in GB cells. We found that the miR-451 mimics transfection significantly reversed the proliferation promoting effect of EPO (Fig. 2A).

3.4 EPO Treatment Increased Resistance to Cisplatin in Glioblastoma Cell Line

We also assessed that EPO treatment has any effect on the response of chemotherapeutic drug cisplatin. Cisplatin application at 50 μM dose

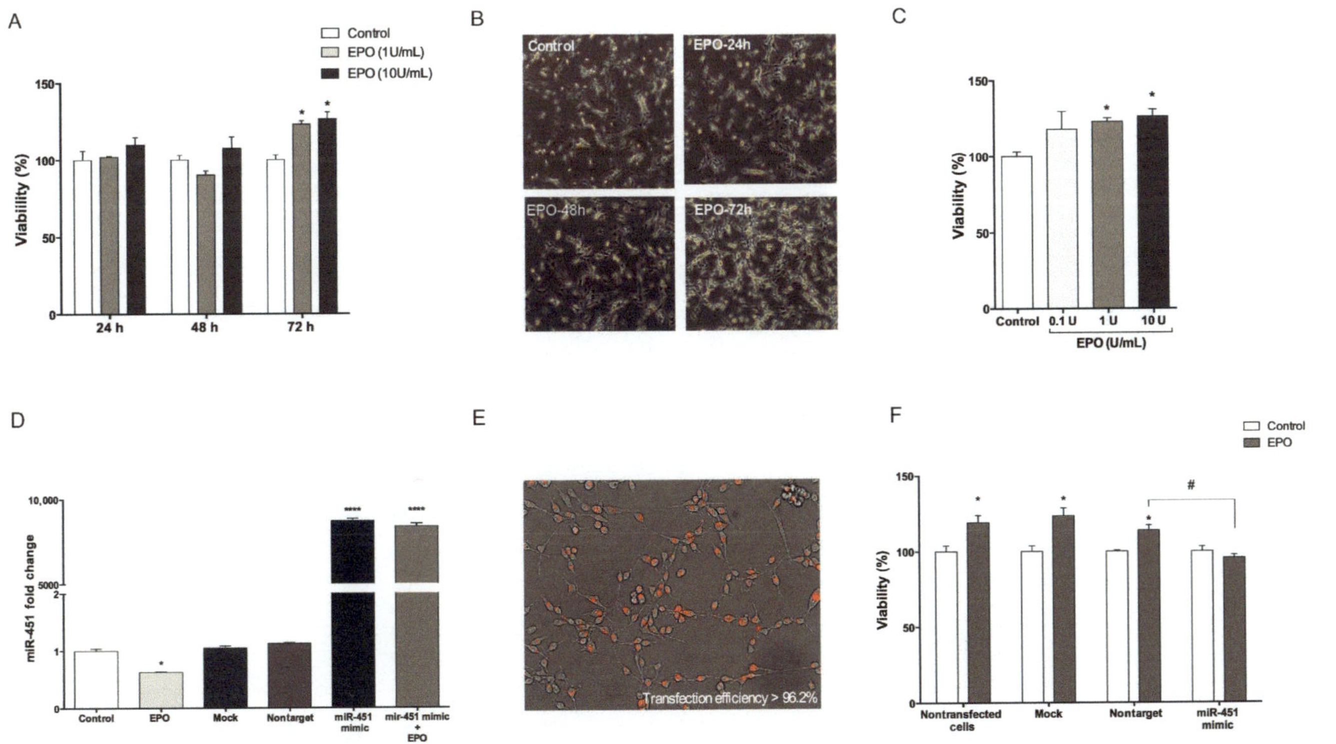

Fig. 1 Erythropoietin increases viability of U87-MG glioma cells. Glioma cells were treated with different concentrations of EPO for 0–72 h, and cell viability was assessed by Presto Blue assay. EPO treatment significantly increased cell viability (A and B) in a time- and (C) dose-dependent fashion. (D) EPO downregulated basal miR-451 expression and miR-451 mimic transfection efficiently induced the miR-451 expression levels in both nontreated (control) and EPO-treated glioma cells. (E) Transfection efficiency was calculated in cells transfected with Cy3-labeled control mimic. (F) Transfection of miR-451 mimic reversed EPO-induced cell viability (*$P < 0.05$ compared to control cells; $^{\#}P < 0.05$ com-

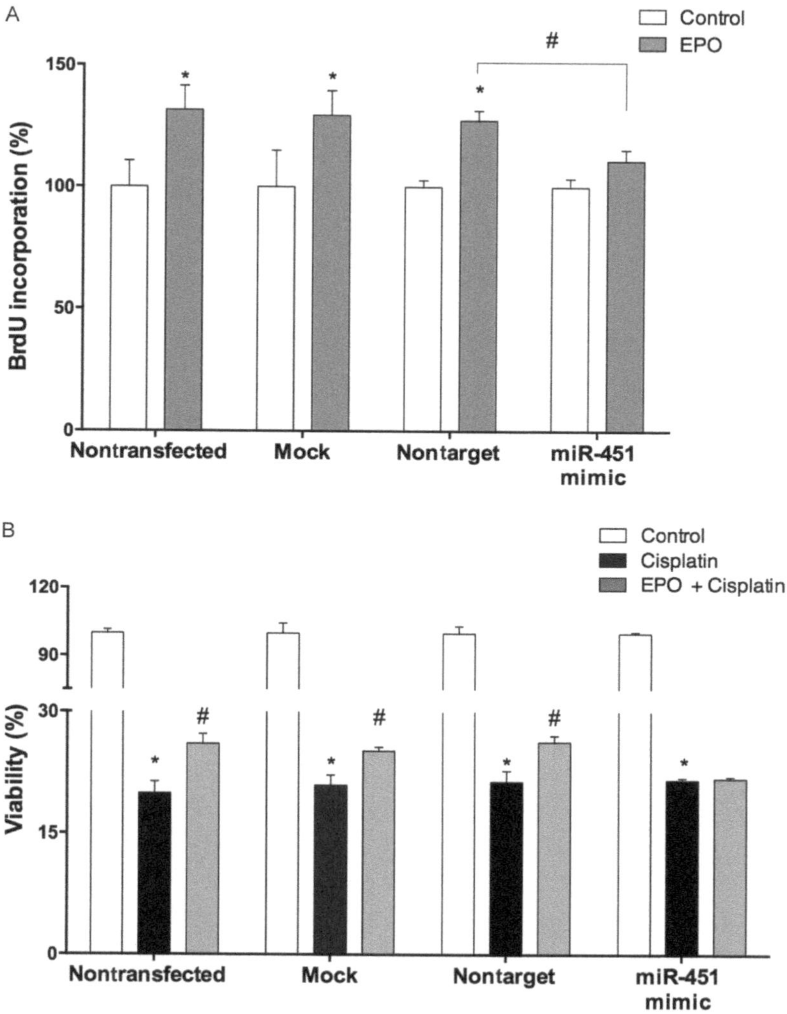

Fig. 2 miR-451 overexpression reversed EPO-induced proliferation and chemoresistance of U87-MG cells. (A) For proliferation analysis, cells were inoculated with 1 U/mL EPO for 72 h, and proliferation was analyzed by BrdU incorporation assay. Transfection of miR-451 mimic reversed EPO-induced cell proliferation. (B) For chemoresistance analysis, cells were pretreated with EPO for 4 h, then cisplatin was added (final concentration: 50 μM). EPO pretreatment increased the cell viability to cisplatin response and miR-451 mimic transfection reversed this effect significantly (*$P < 0.05$ compared to control cells; $^{\#}P < 0.05$ compared to cisplatin-treated cells).

reduces viability to 19%, whereas EPO pretreatment increased the viability to 26% (Fig. 2B). This significant increase in the viability to cisplatin response in EPO pretreated U87-MG cells is a result of EPO-induced chemoresistance to cisplatin. To investigate the effect of miR-451 on EPO-induced chemoresistance to cisplatin, we used miR-451 mimic to simulate overexpression of miR-451. We found that EPO-induced chemoresistance to cisplatin was reversed by miR-451 overexpression in U87-MG cells.

3.5 EPO-Induced Migration of Glioblastoma Cells

Next, we assessed if rhuEPO at the same concentration may have any influence on migration using a scratch assay. EPO treatment increased percentage of migrated cell numbers at 1 U/mL concentration for 48 h (about 80%) (Fig. 3A and B). To investigate the role of miR-451 upregulation on EPO-mediated migration in U87-MG cells, we used mimics to simulate overexpression of miR-451. We found that miR-451 mimic decreased migration of GB cells in both control and EPO-treated groups (Fig. 3A and B).

3.6 Glioblastoma Cells Gain Invasiveness Under Stimulation of EPO Treatment

We evaluated the effects of rhuEPO on cancer cells invasion. Therefore, we examined the effects of rhEPO on the invasion of U87-MG cells across matrigel-coated inserts. 1 U/mL rhEPO treatment for 72 h significantly promoted invasion of U87-MG cells through matrigel (Fig. 3C and D). EPO treatment for 72 h led 137% increase in invaded cells suggesting that EPO increased invasion of U87 cells (Fig. 3C and D). Cell invasion in matrigel was evaluated after transfection with mimics to determine the impact of miR-451- on EPO-mediated invasiveness. miR-451 mimic decreased invasion of GB cells in both control and EPO-treated groups (Fig. 3C).

3.7 miR-451 Overexpression Downregulated the Expressions of Its Own Target Genes

EPO upregulated gene expressions of *BCL-2, MMP-2*, MMP-9, and *VEGF* under basal condition (Fig. 4). Overexpression of miR-451 significantly reduced EPO-upregulated mRNA levels of *BCL-2, MMP-2, MMP-9* in U87-MG cells.

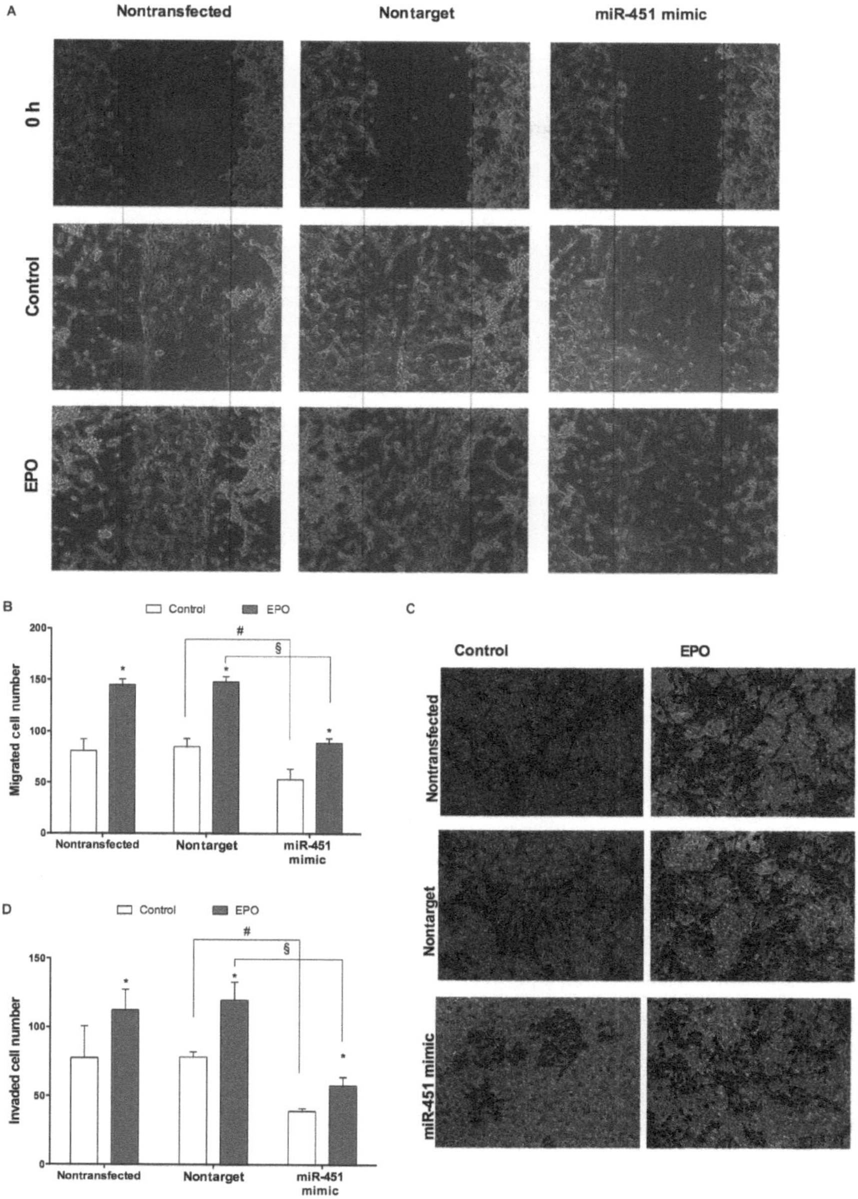

Fig. 3 miR-451 overexpression reversed EPO-induced migration and invasion of U87-MG cells. Treatment with 1 U/mL EPO significantly increased (A and B) migration and (C and D) invasion of U87-MG cells. Transfection with miR-451 mimic decreased migration and invasion of GB cells in both control and EPO-treated conditions (*$P < 0.05$ compared to control; $^{\#}P < 0.05$ compared to untreated nontarget transfected cells and $^{\S}P < 0.05$ compared to EPO-treated nontarget transfected cells).

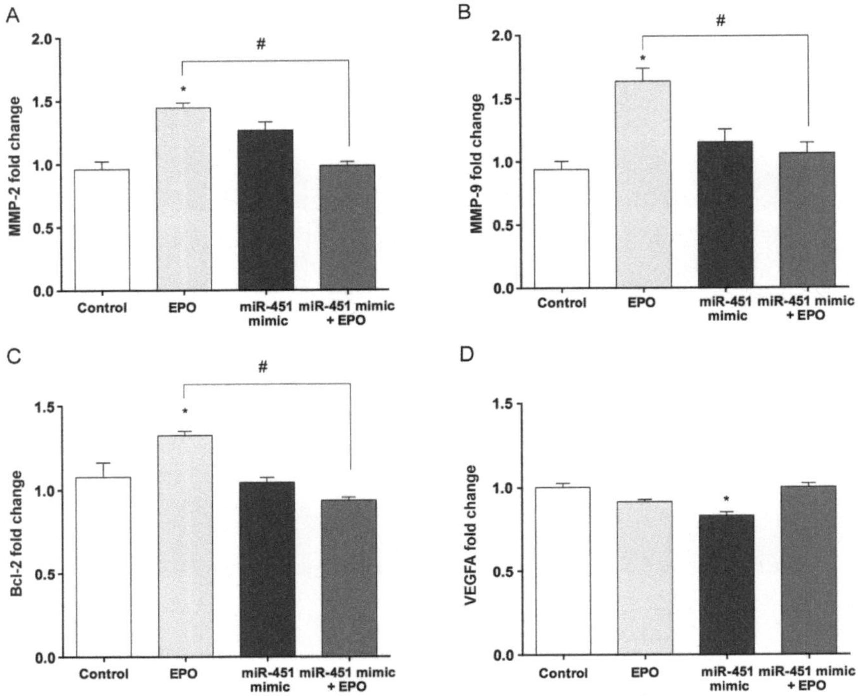

Fig. 4 miR-451 overexpression downregulated target genes induced by EPO. After transfection with miR-451 mimic, cells were treated with 1 U/mL EPO. Expression levels of (A) MMP-2, (B), MMP-9, (C) Bcl-2, and (D) VEGFA were quantified by qRT-PCR (*, #$P < 0.05$).

3.8 U87 Glioblastoma Cells Expressed Different EPORs

Western Blotting and immunofluorescence staining were used to determine the expression of EPORs in U87-MG at the basal level. No primary antibody controls in immunofluorescent staining led to elimination of nonspecific binding of primary antibody to antigen and possible effects of fixation and detergent treatment. We found that U87-MG Glioblastoma cells express three EPO receptors: EPOR, CD131, and EphB4 at the protein levels (Fig. 5).

4. DISCUSSION

The aim of present study was to examine the influences of EPO on cell survival, proliferation, invasiveness, and migration in U87 glioblastoma cell

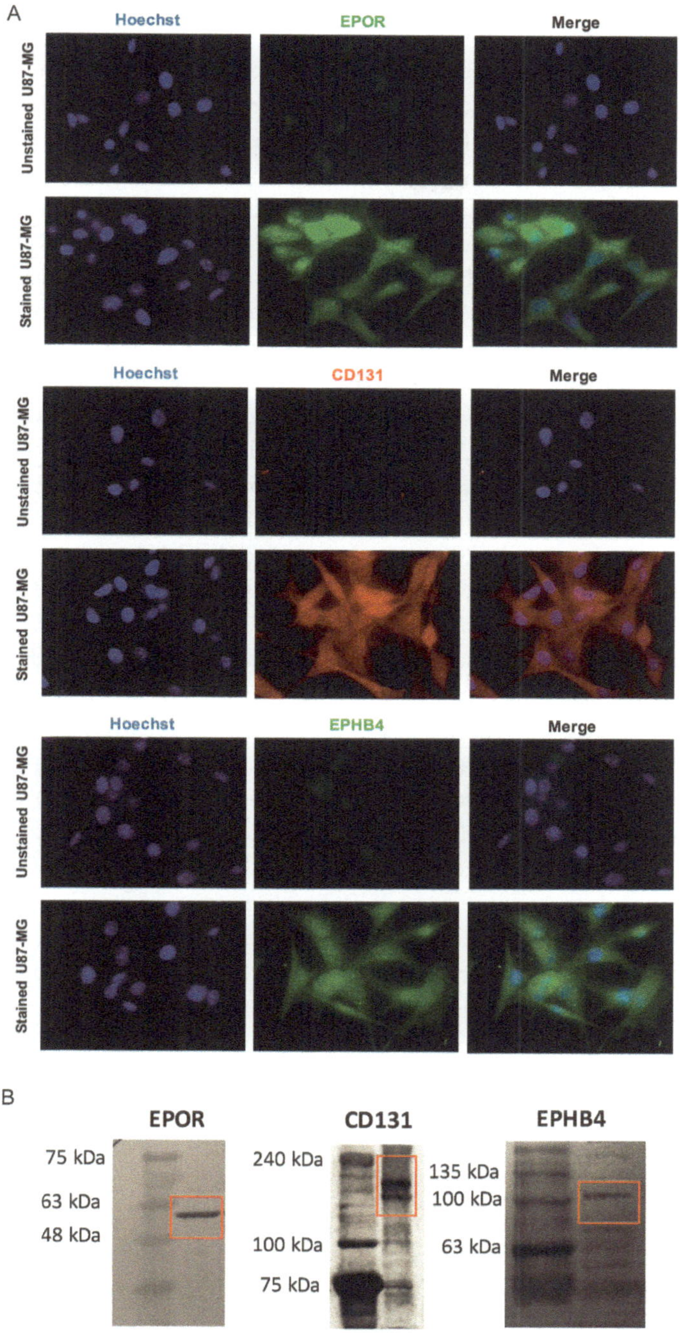

Fig. 5 EPOR, CD131, and Ephrin B4 receptors are present in U87-MG cell line. (A) Cells were plated and stained with related primary antibodies. Cells were visualized under a fluorescence microscope. (B) Protein levels of EPOR, CD131, and Ephrin B4 were analyzed with Western blotting.

line. In addition, we searched whether EPO restores cell survival against cisplatin treatment. Finally, functional experiments were performed to examine the possible reversal effect of miR-451 ectopic expression upon EPO exposure.

We found that 1 U/mL EPO treatment increased basal U87-MG cell survival in a dose- and time-dependent manner. However, EPO significantly increased cell viability only at 72 h. There is no difference between 1 and 10 U/mL doses. We chose 1 U/mL dose for further experiments. Glioblastoma cell proliferation was also increased by EPO. Additionally, our results also demonstrated that EPO rescued GB cells against cisplatin cytotoxicity. The mechanisms of EPO's positive effect on U87-MG cell viability are not fully understood. EPO also protects GB cells from staurosporine, cisplatin, and radiation cytotoxicity, and following EPO treatment, GB cells become more resistance to ionizing radiation and to chemotherapy (Belenkov et al., 2004; Kwon et al., 2014; Mohyeldin et al., 2007). Preincubation of GB cells with EPO resulted in protection against irradiation and TMZ-induced cytotoxicity (Hassouna et al., 2008). rhEPO-induced resistance to ionizing radiation and cisplatin is Janus kinase 2 (JAK2) dependent. Phosphatidylinositol-4,5-bisphosphate 3-kinase (PI3K)/Akt and mitogen-activated protein kinases (MAPK) pathways are responsible for survival effect of EPO against staurosporine cytotoxicity in C6 glioma cells (Kwon et al., 2014). These intracellular signaling pathways also contribute to GB pathogenesis and EPO's signaling (Alifieris & Trafalis, 2015).

We examined the effects of rhEPO on the invasion of U87-MG cells across matrigel-coated inserts. rhuEPO treatment at 1 U/mL concentration for 72 h has significantly promoted invasion of U87-MG cells through matrigel. EPO promotes matrigel invasion of U251 cells dose dependently (Mohyeldin et al., 2007). However, Hassauna et al. did not find any migratory effect with EPO treatment at low concentration (0.3 U/mL) in four GB cell lines including U87 and U251. Functional studies with matrix metalloproteinases-2 (MMP-2) inhibitor confirmed that EPO performs invasion promoting effect in GB cells via MMP-2 upregulation (Mohyeldin et al., 2007). We previously reported EPO-induced expression of migratory and invasiveness genes such as MMP-2, matrix metalloproteinases-9 (MMP-9), and C-X-C chemokine receptor type 4 (CXCR4) in SH-SY5Y neuroblastoma cell line (Alural et al., 2014). GB cells can degrade extracellular matrix via MMPs to gain migration and invasion capacity. Many studies have reported the overexpression of several MMPs, especially MMP-2 and -9 in GB cells compared to their normal cell counterparts (Paw, Carpenter, Watabe, Debinski, & Lo, 2015). Degradation of the extracellular matrix is also a key event in the tumor angiogenesis allowing the proteolytic destruction of basement membranes by activated endothelial cells and their migration into the tumor tissue for formation of new blood vessels (Onishi, Ichikawa, Kurozumi, & Date, 2011). EPO contributes to GB angiogenesis via a direct effect on endothelial cells, and indirectly by modulating the release of other angiogenic factors such as vascular endothelial growth factor (VEGF) and CXCR4 (Cao et al., 2010; Nico et al., 2011).

MiRNAs and their capacity of simultaneously regulating multiple target genes may play a key role in explaining the complex mechanisms underlying GB formation. They can simultaneously modulate distinct processes such as tumor growth, invasion, angiogenesis, and drug resistance in GB. miRNA-451 is one of the most deregulated miRNAs in GB and has a critical role in GB tumorogenesis and progression. miR-451 maturation process occurs via Dicer-independent, Ago2-mediated noncanonical miRNA biogenesis (Pan et al., 2013). Transcriptional regulation of miR-451 is regulated by GATA-1 and c-myc transcription factors.

In the present study, functional studies performed by ectopic miR-451 expression showed that this tumor suppressor miRNA substantially or completely counteracts promoting effects of EPO on U87-MG cell viability, proliferation, migration, invasion, and chemoresistance to cisplatin. The exact mechanisms of apparent miR-451-mediated reversal of EPO's on tumor cell properties remain to be clarified. However, it seems that miR-451 targets tumor promoting and EPO upregulated survival genes [B-cell lymphoma 2 (Bcl-2), survivin, and Akt)], chemoattractant genes such as CXCR4, tissue remodeling genes that play an important role in tumor invasion (MMP-2 and -9), cell cycle genes [Cyclin D1, cyclin-dependent kinase inhibitor 2D (CDKN2D), and cyclin-dependent kinase inhibitor 2B (CDKN2B)] in various human cancer cell lines including U87-MG (Ehtesham et al., 2013; Godlewski, Bronisz, et al., 2010; Kwon et al., 2014; Nan et al., 2010; Tian et al., 2012; Yin et al., 2015; Zang et al., 2015). c-Myc, transcriptional activator of miR-451, plays a regulator role in cell cycle and also is a target of this miRNA (Pan et al., 2013). We showed that miR-451 inhibits cell survival, proliferative, and migratory effects of EPO in neuroblastoma cell line (Alural et al., 2014). Our ongoing study aims to clarify between EPO/EPOR signaling pathways and miR-451 target genes.

Tumor suppressor miR-451 and growth factor EPO show opposite effects on chemosensitivity and radiosensitivity in different types of cancer cell lines. EPO induces radio resistance to ionizing radiation and chemoresistance to cisplatin (Belenkov et al., 2004). On the other hand, EPOR knockdown using RNA interference showed that silencing EPOR increases TMZ efficiency and radiosensitivity in GB cell lines (Peres et al., 2015). Transfection of miR-451 mimic sensitizes A549 lung cancer cell line to cisplatin (Bian, Pan, Yang, Wang, & De, 2011). miR-451 also targets MDR-1/p-glycoprotein and increases chemosensitivity to doxorubicin and vinblastine to ovarian and cervical cell lines (Zhu et al., 2008). High levels of miR-451 expression enhanced radio sensitivity of nasopharyngeal carcinoma by directly targeting ras-related protein 14 (RAB 14) (Zang et al., 2015). In our study, miR-451 ectopic expression reversed EPO-induced cisplatin resistance of the U87-MG cell line. The exact mechanisms of chemoresistance and radioresistance of GB cells remain to be clarified. However, our findings may link diverse effects of EPO and miR-451. Multiple signaling pathways, multidrug resistance genes, survival genes, DNA repair mechanisms, and miRNAs may have impact on these processes underlying sensitivity to GB therapy (Koshkin, Chistiakov, & Chekhonin, 2013).

Several studies show that EPOR overexpression in patient GB samples and EPOR silencing experiments may suggest an important role for functional EPOR in GB (Peres et al., 2015, 2011). When EPOR silenced cells were inoculated into nude mice, tumor growth decreased and animal survival increased in preclinical experiments (Peres et al., 2015, 2011). Recently, a modified variant of EPO, carbamylated EPO (cEPO), has been produced and demonstrated that CD131 may mediate cEPO's tissue protective effects (Chen, Yang, & Zhang, 2015). However, any glioma promoting effect of cEPO and CD131 signaling has not been reported to date. In addition to classical EPOR and CD131, EphB4 has been reported as another EPOR and tumor growth and promoting factor in human ovarian and breast cancers (Pradeep et al., 2015). This finding should be verified in GB cell lines and in vivo experimental models. Ephrin and ephrin receptors are involved in glioma and tumor angiogenesis (Ferluga & Debinski, 2014). A possible relationship between these novel noncanonical EPORs and miR-451 should also be considered in GB pathogenesis.

Our present study has some limitations. Comparison of different cell lines would reveal cell-type specific responses (Hong, Chedid, & Kalkanis, 2012). For instance, Hassauna et al. found that low concentration of EPO (0.3 U/mL) results in proliferation in only one (G44 cell line) of four GB cell lines. Second, in vivo xenografted GB model are valuable to dissect EPO–miR-451 link in living organism. Glioma microenvironment consists of various tumor surrounding cells including tumor-associated macrophage, activated microglia, peripheral immune cells migrated to CNS, vascular cells, glioma, and mesenchymal stem cells (Godlewski, Krichevsky, Johnson, Chiocca, & Bronisz, 2015). Coculture studies using these different types of cells provide a means to study cellular interactions. Intercellular transfer of extracellular vesicles (EVs) loaded with mRNAs, proteins, and miRNAs is also implicated in glioma pathogenesis (Godlewski et al., 2015). miR-451 is enriched in these cargoes and may play an important role in very complex GB microenviroment (Guduric-Fuchs et al., 2012). Indeed, miR-451 contained EVs are transferred from GB cells to microglia and uptaken miR-451 targets c-myc in recipient cells (van der Vos et al., 2016). Such studies highlight the importance of EVs as intercellular mediators. It is very likely many other GB- and cancer-related miRNAs are mediators of EPO's actions. Thus, high-throughput assays and bioinformatics studies should be done before further functional studies.

A growing body of evidence indicates an association between exogenous EPO treatment and tumor progression. However, the mechanisms of tumor growth effect of EPO remain still elusive. In addition to classical EPOR, noncanonical EPORs are also likely mediated to EPO's cell and tissue protective effects and further functional experiments targeting EPORs' are warranted. EPO treatment in cancer patients with anemia and chemotherapy-induced anemia is still a good alternative to blood transfusion. Moreover, the extension of EPO treatment to several neurological diseases and injury is not unlikely in the future and local undesirable effects of EPO on GB may also come up. Long-term EPO administration is a serious problem with regard to tumor progression. Our results have revealed a link between cellular characteristics of GB cells and EPO-mediated inhibition of tumor suppressor miR-451. MiRNA mimic therapy targeting tumor-promoting genes may be a good candidate as an adjuvant therapy in GB.

ACKNOWLEDGMENT

The authors wish to thank Dr. William Gault for his careful and critical reading of the manuscript. This study was supported by Dokuz Eylul University (Grant Number: 2016. KB.SAG.008).

Conflict of interest disclosure: The authors declare no competing financial interests.

REFERENCES

Alifieris, C., & Trafalis, D. T. (2015). Glioblastoma multiforme: Pathogenesis and treatment. *Pharmacology & Therapeutics*, *152*, 63–82. http://dx.doi.org/10.1016/j.pharmthera.2015.05.005.

Alural, B., Duran, G. A., Tufekci, K. U., Allmer, J., Onkal, Z., Tunali, D., et al. (2014). EPO mediates neurotrophic, neuroprotective, anti-oxidant, and anti-apoptotic effects via downregulation of miR-451 and miR-885-5p in SH-SY5Y neuron-like cells. *Frontiers in Immunology*, *5*, 475. http://dx.doi.org/10.3389/fimmu.2014.00475.

Belenkov, A. I., Shenouda, G., Rizhevskaya, E., Cournoyer, D., Belzile, J. P., Souhami, L., et al. (2004). Erythropoietin induces cancer cell resistance to ionizing radiation and to cisplatin. *Molecular Cancer Therapeutics*, *3*(12), 1525–1532.

Bennett, C. L., Lai, S. Y., Sartor, O., Georgantopoulos, P., Hrushesky, W. J., Henke, M., et al. (2016). Consensus on the existence of functional erythropoietin receptors on cancer cells. *JAMA Oncology*, *2*(1), 134–136. http://dx.doi.org/10.1001/jamaoncol.2015.3940.

Bian, H. B., Pan, X., Yang, J. S., Wang, Z. X., & De, W. (2011). Upregulation of microRNA-451 increases cisplatin sensitivity of non-small cell lung cancer cell line (A549). *Journal of Experimental & Clinical Cancer Research*, *30*, 20. http://dx.doi.org/10.1186/1756-9966-30-20.

Brines, M. L., Ghezzi, P., Keenan, S., Agnello, D., de Lanerolle, N. C., Cerami, C., et al. (2000). Erythropoietin crosses the blood-brain barrier to protect against experimental brain injury. *Proceedings of the National Academy of Sciences of the United States of America*, *97*(19), 10526–10531.

Brunotte, J., Bock, H. C., Bruck, W., Hemmerlein, B., & Strik, H. (2011). High expression levels of erythropoietin and its receptor are not correlated with shorter survival in human glioblastoma. *Experimental and Therapeutic Medicine*, *2*(2), 295–299. http://dx.doi.org/10.3892/etm.2011.198.

Bush, N. A., Chang, S. M., & Berger, M. S. (2016). Current and future strategies for treatment of glioma. *Neurosurgical Review*, *40*, 1–14. http://dx.doi.org/10.1007/s10143-016-0709-8.

Cao, Y., Lathia, J. D., Eyler, C. E., Wu, Q., Li, Z., Wang, H., et al. (2010). Erythropoietin receptor signaling through STAT3 is required for glioma stem cell maintenance. *Genes & Cancer*, *1*(1), 50–61. http://dx.doi.org/10.1177/1947601909356352.

Chen, J., Yang, Z., & Zhang, X. (2015). Carbamylated erythropoietin: A prospective drug candidate for neuroprotection. *Biochemistry Insights*, *8*(Suppl. 1), 25–29. http://dx.doi.org/10.4137/BCI.S30753.

Chong, Z. Z., Shang, Y. C., Mu, Y., Cui, S., Yao, Q., & Maiese, K. (2013). Targeting erythropoietin for chronic neurodegenerative diseases. *Expert Opinion on Therapeutic Targets*, *17*(6), 707–720. http://dx.doi.org/10.1517/14728222.2013.780599.

Costa, P. M., Cardoso, A. L., Mano, M., & de Lima, M. C. (2015). MicroRNAs in glioblastoma: Role in pathogenesis and opportunities for targeted therapies. *CNS & Neurological Disorders Drug Targets*, *14*(2), 222–238.

Debeljak, N., Solar, P., & Sytkowski, A. J. (2014). Erythropoietin and cancer: The unintended consequences of anemia correction. *Frontiers in Immunology*, *5*, 563. http://dx.doi.org/10.3389/fimmu.2014.00563.

Ehrenreich, H., Hinze-Selch, D., Stawicki, S., Aust, C., Knolle-Veentjer, S., Wilms, S., et al. (2007). Improvement of cognitive functions in chronic schizophrenic patients by recombinant human erythropoietin. *Molecular Psychiatry*, *12*(2), 206–220. http://dx.doi.org/10.1038/sj.mp.4001907.

Ehtesham, M., Min, E., Issar, N. M., Kasl, R. A., Khan, I. S., & Thompson, R. C. (2013). The role of the CXCR4 cell surface chemokine receptor in glioma biology. *Journal of Neuro-Oncology*, *113*(2), 153–162. http://dx.doi.org/10.1007/s11060-013-1108-4.

Ferluga, S., & Debinski, W. (2014). Ephs and Ephrins in malignant gliomas. *Growth Factors*, *32*(6), 190–201. http://dx.doi.org/10.3109/08977194.2014.985787.

Godlewski, J., Bronisz, A., Nowicki, M. O., Chiocca, E. A., & Lawler, S. (2010a). microRNA-451: A conditional switch controlling glioma cell proliferation and migration. *Cell Cycle*, *9*(14), 2742–2748.

Godlewski, J., Krichevsky, A. M., Johnson, M. D., Chiocca, E. A., & Bronisz, A. (2015). Belonging to a network—microRNAs, extracellular vesicles, and the glioblastoma microenvironment. *Neuro-Oncology*, *17*(5), 652–662. http://dx.doi.org/10.1093/neuonc/nou292.

Godlewski, J., Nowicki, M. O., Bronisz, A., Nuovo, G., Palatini, J., De Lay, M., et al. (2010b). MicroRNA-451 regulates LKB1/AMPK signaling and allows adaptation to metabolic stress in glioma cells. *Molecular Cell*, *37*(5), 620–632. http://dx.doi.org/10.1016/j.molcel.2010.02.018.

Guduric-Fuchs, J., O'Connor, A., Camp, B., O'Neill, C. L., Medina, R. J., & Simpson, D. A. (2012). Selective extracellular vesicle-mediated export of an overlapping set of microRNAs from multiple cell types. *BMC Genomics*, *13*, 357. http://dx.doi.org/10.1186/1471-2164-13-357.

Hammond, S. M. (2015). An overview of microRNAs. *Advanced Drug Delivery Reviews*, *87*, 3–14. http://dx.doi.org/10.1016/j.addr.2015.05.001.

Hassouna, I., Sperling, S., Kim, E., Schulz-Schaeffer, W., Rave-Frank, M., Hasselblatt, M., et al. (2008). Erythropoietin augments survival of glioma cells after radiation and temozolomide. *International Journal of Radiation Oncology Biology Physics*, *72*(3), 927–934. http://dx.doi.org/10.1016/j.ijrobp.2008.06.1923.

Hong, X., Chedid, K., & Kalkanis, S. N. (2012). Glioblastoma cell line-derived spheres in serum-containing medium versus serum-free medium: A comparison of cancer stem cell properties. *International Journal of Oncology*, *41*(5), 1693–1700. http://dx.doi.org/10.3892/ijo.2012.1592.

Jelkmann, W. (2013). Physiology and pharmacology of erythropoietin. *Transfusion Medicine and Hemotherapy*, *40*(5), 302–309. http://dx.doi.org/10.1159/000356193.

Juul, S. E., & Pet, G. C. (2015). Erythropoietin and neonatal neuroprotection. *Clinics in Perinatology*, *42*(3), 469–481. http://dx.doi.org/10.1016/j.clp.2015.04.004.

Kochanek, P. M., & Clark, R. S. (2016). Traumatic brain injury research highlights in 2015. *The Lancet Neurology*, *15*(1), 13–15. http://dx.doi.org/10.1016/S1474-4422(15)00339-7.

Koshkin, P. A., Chistiakov, D. A., & Chekhonin, V. P. (2013). Role of microRNAs in mechanisms of glioblastoma resistance to radio- and chemotherapy. *Biochemistry (Moscow)*, *78*(4), 325–334. http://dx.doi.org/10.1134/S0006297913040019.

Kwon, M. S., Kim, M. H., Kim, S. H., Park, K. D., Yoo, S. H., Oh, I. U., et al. (2014). Erythropoietin exerts cell protective effect by activating PI3K/Akt and MAPK pathways in C6 cells. *Neurological Research*, *36*(3), 215–223. http://dx.doi.org/10.1179/1743132813Y.0000000284.

Louis, D. N., Perry, A., Reifenberger, G., von Deimling, A., Figarella-Branger, D., Cavenee, W. K., et al. (2016). The 2016 World Health Organization classification of tumors of the central nervous system: A summary. *Acta Neuropathologica*, *131*(6), 803–820. http://dx.doi.org/10.1007/s00401-016-1545-1.

Maiese, K. (2016). Regeneration in the nervous system with erythropoietin. *Frontiers in Bioscience*, *21*, 561–596.

Miskowiak, K. W., Ehrenreich, H., Christensen, E. M., Kessing, L. V., & Vinberg, M. (2014). Recombinant human erythropoietin to target cognitive dysfunction in bipolar disorder: A double-blind, randomized, placebo-controlled phase 2 trial. *The Journal of Clinical Psychiatry*, *75*(12), 1347–1355. http://dx.doi.org/10.4088/JCP.13m08839.

Mittelbronn, M., Capper, D., Bunz, B., Dietz, K., Goeppert, B., Ajaaj, R., et al. (2007). De novo erythropoietin receptor (EPO-R) expression in human neoplastic glial cells decreases with grade of malignancy but is favourably associated with patient survival. *Neuropathology and Applied Neurobiology*, *33*(3), 299–307. http://dx.doi.org/10.1111/j.1365-2990.2006.00820.x.

Mohyeldin, A., Dalgard, C. L., Lu, H., McFate, T., Tait, A. S., Patel, V. C., et al. (2007). Survival and invasiveness of astrocytomas promoted by erythropoietin. *Journal of Neurosurgery*, *106*(2), 338–350. http://dx.doi.org/10.3171/jns.2007.106.2.338.

Nan, Y., Han, L., Zhang, A., Wang, G., Jia, Z., Yang, Y., et al. (2010). MiRNA-451 plays a role as tumor suppressor in human glioma cells. *Brain Research*, *1359*, 14–21. http://dx.doi.org/10.1016/j.brainres.2010.08.074.

Nico, B., Annese, T., Guidolin, D., Finato, N., Crivellato, E., & Ribatti, D. (2011). Epo is involved in angiogenesis in human glioma. *Journal of Neuro-Oncology*, *102*(1), 51–58. http://dx.doi.org/10.1007/s11060-010-0294-6.

Onishi, M., Ichikawa, T., Kurozumi, K., & Date, I. (2011). Angiogenesis and invasion in glioma. *Brain Tumor Pathology*, *28*(1), 13–24. http://dx.doi.org/10.1007/s10014-010-0007-z.

Pan, X., Wang, R., & Wang, Z. X. (2013). The potential role of miR-451 in cancer diagnosis, prognosis, and therapy. *Molecular Cancer Therapeutics*, *12*(7), 1153–1162. http://dx.doi.org/10.1158/1535-7163.MCT-12-0802.

Paw, I., Carpenter, R. C., Watabe, K., Debinski, W., & Lo, H. W. (2015). Mechanisms regulating glioma invasion. *Cancer Letters*, *362*(1), 1–7. http://dx.doi.org/10.1016/j.canlet.2015.03.015.

Pearl, R. G. (2014). Erythropoietin and organ protection: Lessons from negative clinical trials. *Critical Care*, *18*(5), 526. http://dx.doi.org/10.1186/s13054-014-0526-9.

Peres, E. A., Gerault, A. N., Valable, S., Roussel, S., Toutain, J., Divoux, D., et al. (2015). Silencing erythropoietin receptor on glioma cells reinforces efficacy of temozolomide and X-rays through senescence and mitotic catastrophe. *Oncotarget*, *6*(4), 2101–2119. http://dx.doi.org/10.18632/oncotarget.2937.

Peres, E. A., Valable, S., Guillamo, J. S., Marteau, L., Bernaudin, J. F., Roussel, S., et al. (2011). Targeting the erythropoietin receptor on glioma cells reduces tumour growth. *Experimental Cell Research*, *317*(16), 2321–2332. http://dx.doi.org/10.1016/j.yexcr.2011.06.011.

Pradeep, S., Huang, J., Mora, E. M., Nick, A. M., Cho, M. S., Wu, S. Y., et al. (2015). Erythropoietin stimulates tumor growth via EphB4. *Cancer Cell*, *28*(5), 610–622. http://dx.doi.org/10.1016/j.ccell.2015.09.008.

Said, H. M., Hagemann, C., Staab, A., Stojic, J., Kuhnel, S., Vince, G. H., et al. (2007). Expression patterns of the hypoxia-related genes osteopontin, CA9, erythropoietin, VEGF and HIF-1alpha in human glioma in vitro and in vivo. *Radiotherapy and Oncology*, *83*(3), 398–405. http://dx.doi.org/10.1016/j.radonc.2007.05.003.

Sargin, D., Friedrichs, H., El-Kordi, A., & Ehrenreich, H. (2010). Erythropoietin as neuroprotective and neuroregenerative treatment strategy: Comprehensive overview of 12 years of preclinical and clinical research. *Best Practice & Research Clinical Anaesthesiology*, *24*(4), 573–594. http://dx.doi.org/10.1016/j.bpa.2010.10.005.

Schneider, C. A., Rasband, W. S., & Eliceiri, K. W. (2012). NIH image to ImageJ: 25 years of image analysis. *Nature Methods*, *9*(7), 671–675.

Stuben, G., Thews, O., Pottgen, C., Knuhmann, K., Sack, H., Stuschke, M., et al. (2003). Impact of anemia prevention by recombinant human erythropoietin on the sensitivity of xenografted glioblastomas to fractionated irradiation. *Strahlentherapie und Onkologie*, *179*(9), 620–625. http://dx.doi.org/10.1007/s00066-003-1110-4.

Suzuki, N., & Yamamoto, M. (2016). Roles of renal erythropoietin-producing (REP) cells in the maintenance of systemic oxygen homeostasis. *Pflügers Archiv: European Journal of Physiology*, *468*(1), 3–12. http://dx.doi.org/10.1007/s00424-015-1740-2.

Tian, Y., Nan, Y., Han, L., Zhang, A., Wang, G., Jia, Z., et al. (2012). MicroRNA miR-451 downregulates the PI3K/AKT pathway through CAB39 in human glioma. *International Journal of Oncology*, *40*(4), 1105–1112. http://dx.doi.org/10.3892/ijo.2011.1306.

Tsai, T. H., Lu, C. H., Wallace, C. G., Chang, W. N., Chen, S. F., Huang, C. R., et al. (2015). Erythropoietin improves long-term neurological outcome in acute ischemic stroke patients: A randomized, prospective, placebo-controlled clinical trial. *Critical Care*, *19*, 49. http://dx.doi.org/10.1186/s13054-015-0761-8.

van der Vos, K. E., Abels, E. R., Zhang, X., Lai, C., Carrizosa, E., Oakley, D., et al. (2016). Directly visualized glioblastoma-derived extracellular vesicles transfer RNA to microglia/macrophages in the brain. *Neuro-Oncology*, *18*(1), 58–69. http://dx.doi.org/10.1093/neuonc/nov244.

Wustenberg, T., Begemann, M., Bartels, C., Gefeller, O., Stawicki, S., Hinze-Selch, D., et al. (2011). Recombinant human erythropoietin delays loss of gray matter in chronic schizophrenia. *Molecular Psychiatry*, *16*(1), 26–36. http://dx.doi.org/10.1038/mp.2010.51. 21.

Xenocostas, A., Cheung, W. K., Farrell, F., Zakszewski, C., Kelley, M., Lutynski, A., et al. (2005). The pharmacokinetics of erythropoietin in the cerebrospinal fluid after intravenous administration of recombinant human erythropoietin. *European Journal of Clinical Pharmacology*, *61*(3), 189–195. http://dx.doi.org/10.1007/s00228-005-0896-7.

Yin, D., Kawabata, H., Tcherniamtchouk, O., Huynh, T., Black, K. L., & Koeffler, H. P. (2007). Glioblastoma multiforme cells: Expression of erythropoietin receptor and response to erythropoietin. *International Journal of Oncology*, *31*(5), 1193–1198.

Yin, P., Peng, R., Peng, H., Yao, L., Sun, Y., Wen, L., et al. (2015). MiR-451 suppresses cell proliferation and metastasis in A549 lung cancer cells. *Molecular Biotechnology*, *57*(1), 1–11. http://dx.doi.org/10.1007/s12033-014-9796-3.

Zang, W. Q., Yang, X., Wang, T., Wang, Y. Y., Du, Y. W., Chen, X. N., et al. (2015). MiR-451 inhibits proliferation of esophageal carcinoma cell line EC9706 by targeting CDKN2D and MAP3K1. *World Journal of Gastroenterology*, *21*(19), 5867–5876. http://dx.doi.org/10.3748/wjg.v21.i19.5867.

Zhu, H., Wu, H., Liu, X., Evans, B. R., Medina, D. J., Liu, C. G., et al. (2008). Role of MicroRNA miR-27a and miR-451 in the regulation of MDR1/P-glycoprotein expression in human cancer cells. *Biochemical Pharmacology*, *76*(5), 582–588. http://dx.doi.org/10.1016/j.bcp.2008.06.007.

CHAPTER FOURTEEN

Erythropoietin, Stem Cell Factor, and Cancer Cell Migration

Maria J. Vazquez-Mellado*, Victor Monjaras-Embriz†, Leticia Rocha-Zavaleta*,†,1

*Instituto de Investigaciones Biomédicas, Universidad Nacional Autónoma de México, Ciudad de Mexico, Mexico
†Facultad de Ciencias de la Salud, Universidad Panamericana, Ciudad de Mexico, Mexico
[1]Corresponding author: e-mail address: lrochaz@biomedicas.unam.mx

Contents

Abstract

Cell migration of normal cells is tightly regulated. However, tumor cells are exposed to a modified microenvironment that promotes cell migration. Invasive migration of tumor cells is stimulated by receptor tyrosine kinases (RTKs) and is regulated by growth factors. Erythropoietin (Epo) is a glycoprotein hormone that regulates erythropoiesis and is also known to be a potent chemotactic agent that induces cell migration by binding to its receptor (EpoR). Expression of EpoR has been documented in tumor cells, and the potential of Epo to induce cell migration has been explored. Stem cell factor (SCF) is

Vitamins and Hormones, Volume 105
ISSN 0083-6729
http://dx.doi.org/10.1016/bs.vh.2017.02.008

© 2017 Elsevier Inc.
All rights reserved.

a cytokine that synergizes the effects of Epo during erythropoiesis. SCF is the ligand of c-Kit, a member of the RTKs family. Molecular activity of RTKs is a primary stimulus of cell motility. Thus, expression of the SCF/c-Kit axis is associated with cell migration. In this chapter, we summarize data describing the potential effect of Epo/EpoR and SCF/c-Kit as promoters of cancer cell migration. We also integrate recent findings on molecular mechanisms of Epo/EpoR- and SCF/c-Kit-mediated migration described in various cancer models.

1. INTRODUCTION

Cell migration is a fundamental process during development and adult life. Primordial germ cells migrate to populate gonads; melanocytes and their precursor cells migrate to localize to specific body sites, establishing a stem cell population for self-renewal (Mort, Jackson, & Patton, 2015); and hematopoietic cells migrate from the bone marrow into the circulation and are then capable of rehoming back into the bone marrow (Abkowitz, Robinson, Kale, Long, & Chen, 2003). Cell migration is promoted by external signals that lead the cell to produce polarized cytoplasmic protrusions, which help the cell to detach from the primary tissue and move toward a gradient of chemokines or growth factors. This is known as directional migration. In a living organism, migrating cells encounter a variety of environments; in some cases, they are forced to move through complex extracellular matrices or even to traverse cellular barriers, as is the case during transendothelial migration of leukocytes (Gerard, van der Kammen, Janssen, Ellenbroek, & Collard, 2009). This type of migration is mediated by the formation of podosomes, which are actin-rich protrusions of the plasma membrane, and which contain adhesion molecules, regulators of actin polymerization, and proteolytic enzymes (Linder & Aepfelbacher, 2003). Migration is not restricted to normal cells. Indeed, invasive migration is a hallmark of tumor cells that have acquired a metastatic phenotype. The basic dynamic structures in cancer cells have been termed invadopodia. Invadopodia and podosomes share a common molecular machinery that ultimately regulates their assembly and function. Both invadopodia and podosomes are induced by receptor tyrosine kinase (RTK) signaling. This implies that cancer cell and normal cell migration are strongly dependent on the presence and activity of growth factors and RTKs.

It is clear that migration of normal cells is tightly regulated. However, tumor cells are exposed to a highly modified microenvironment that may promote and enhance the migratory potential of particular groups of cells. The tumor microenvironment is known to be rich in growth factors. Additionally, cancer patients may sometimes be exposed to growth factors. This is the case of patients with cancer-associated anemia, who are treated with recombinant erythropoietin (Epo). Epidemiological data suggest that administration of erythropoietin is associated with disease progression and decreased overall survival (Temkin, Hellmann, Serur, Lee, & Abulafia, 2006). In addition, the expression of RTKs, such as c-Kit, has been reported in many types of tumors (Went et al., 2004). In this chapter, we review data related to the potential effect of Epo and c-Kit as promoters of cancer cell migration. We also consider recent findings regarding molecular mechanisms of Epo- and c-Kit-mediated migration in various cancer models.

2. ERYTHROPOIETIN AND STEM CELL FACTOR PROMOTE CELL MIGRATION

Directional migration of normal cells is regulated by a complex milieu of chemokines, cytokines, and growth factors. The bone marrow is perhaps one of the most active sites involved in cell migration. A constant flow of hematopoietic and nonhematopoietic cells from the bone marrow into the systemic circulation occurs as a response to hypoxia, organ injury, and pathogen insult. Reduced levels of oxygen trigger the transcriptional activation of Epo in order to stimulate erythropoiesis. Proliferation, differentiation, and migration of erythroid progenitors depend on the concomitant activation of signaling events initiated by the binding of Epo to its receptor (EpoR) and of stem cell factor (SCF) to its receptor (c-Kit) (Munugalavadla & Kapur, 2005).

Epo is a glycoprotein hormone produced primarily in the kidneys. The human erythropoietin gene encodes a 193-amino acid polypeptide chain, which is processed to yield a mature, circulating 165-amino acid protein (Sasaki, Masuda, & Nagao, 2000). Mature Epo is heavily glycosylated; it has three N-linked and one O-linked acidic oligosaccharide side chains (Jelkmann, 2003). Glycosylated chains maintain the biological activity of Epo (Maiese, Chong, Hou, & Shang, 2008) and provide stability (Toyoda, Itai, Arakawa, Aoki, & Yamaguchi, 2000). Production of Epo is stimulated by changes in oxygen tension, and Epo gene transcription is regulated by hypoxia-inducible factor-1 (HIF-1). HIF-1 also activates transcription of the EpoR.

EpoR belongs to the class I cytokine receptor family. It consists of an extracellular domain, a transmembrane domain, and an intracellular domain. As shown in Fig. 1, the binding of one erythropoietin molecule induces the formation of a homodimeric, functional receptor that initiates signaling by transactivation of the Janus Kinase-2 (JAK2) molecules bound to its intracellular domain (Witthuhn et al., 1993). Activated JAK2 phosphorylates various EpoR tyrosine residues that serve as docking sites for signal molecules, such as Src homology-2 (SH2) domain-containing proteins, and signal transducer and activation of transcription (STAT) 1, 3 (Kirito et al., 2002), and 5 (Klingmuller, Bergelson, Hsiao, & Lodish, 1996). As a result, in addition to activating the JAK/STAT pathway, Epo stimulates other relevant signaling pathways, including the phosphatidylinositol-3 kinase (PI3K)/AKT pathway and the mitogen-activated protein kinase (MAPK)/extracellular signal-related kinase (ERK) 1/2 pathway, promoting cell survival (Bao et al., 1999) and proliferation (Damen et al., 1995).

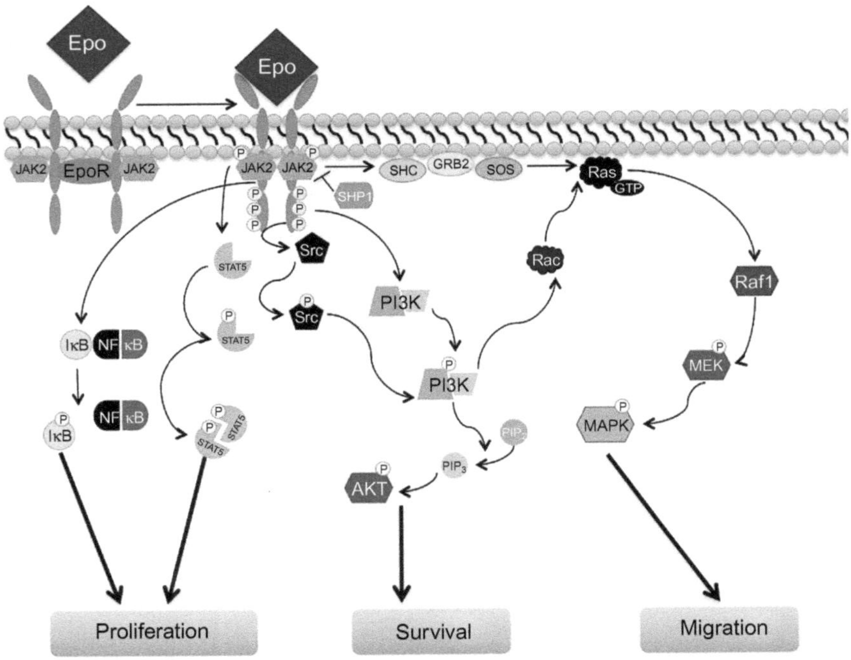

Fig. 1 Summary of the signal transduction pathways activated by erythropoietin (Epo). Three different signal transduction pathways are predominantly activated after binding of Epo to its receptor (EpoR): JAK2/STAT5, which is associated with cell proliferation; PI3K/AKT, associated with cell survival; and Ras/Raf/MAPK, associated with cell migration.

In addition to its role as an erythropoietic factor, Epo has been recognized as a potent chemotactic agent. It has been reported that Epo induces directional migration of bone marrow-derived mesenchymal stem cells to the microenvironment of organs undergoing acute injury, such as the kidney (Liu, Tian, Cheng, & Zhang, 2013) and the heart (Lin, Chen, Chiang, & Ma, 2008). Reported data suggest that Epo promotes bone marrow-derived mesenchymal cell mobilization by increasing levels of the chemokine stromal cell-derived factor-1 (SDF-1). SDF-1 is the natural ligand of CXCR4, which is expressed on the surface of bone marrow-derived mesenchymal stem cells. Thus, increased concentrations of SDF-1 during kidney or heart injury are capable of directing CXCR4-expressing cells to the damaged site. Due to its chemotactic properties, Epo has also been implicated in the migration of endothelial precursor cells from the bone marrow to the pulmonary vasculature in a model of hypoxia-induced pulmonary hypertension (Satoh et al., 2006). Furthermore, Epo-induced mobilization of bone marrow-derived endothelial progenitor cells promotes neovascularization, and there is evidence showing that Epo regulates endothelial progenitor cell mobilization in patients with ischemic heart disease (Heeschen et al., 2003).

In the bone marrow, Epo acts synergistically with SCF to regulate erythropoiesis. SCF is the product of the *SI* locus on chromosome 12 (Ropers & Craig, 1989). In humans, SCF can be found as a 248-amino acid membrane-bound homodimer (mSCF). mSCF contains a proteolytic cleavage site that yields a monomeric, soluble 165-amino acid SCF after posttranscriptional modification. Although both forms are ligands of c-Kit, dimeric mSCF is remarkably more active (Hsu et al., 1997) and induces a more persistent activation of the receptor, whereas the soluble counterpart produces a transient activation of c-Kit and promotes its degradation (Langley et al., 1993). SCF is the natural ligand of c-Kit. c-Kit is a member of the class III RTKs (PDGFR family). It is produced as a glycosylated protein with an extracellular domain containing five immunoglobulin-like repeats, a transmembrane region, an autoinhibitory domain, and two intracellular tyrosine kinase domains (Liu, Chen, Focia, & He, 2007). The binding of SCF to c-Kit produces a conformational change, the formation of a homodimer, and the consequent activation of its enzymatic activity. Autophosphorylation of the receptor occurs, generating docking sites for signal transduction proteins containing SH2 domains. Downstream signaling pathways, such as the Ras/ERK, PI3K/AKT, and JAK/STAT pathways along with Src kinases, are activated as a response to SCF (Fig. 2) (Ronnstrand, 2004).

Molecular activity of SCF/c-Kit is closely associated with directional cell migration. In the human embryo, c-Kit and SCF expression is detected along the migratory pathways of various cell types, including melanocytes, hematopoietic cells, and primordial germ cells (Matsui, Zsebo, & Hogan, 1990). In adults, activation of the c-Kit receptor by SCF stimulates directional mast cell motility (Meininger et al., 1992), migration of melanocytes (Alexeev & Yoon, 2006), as well as migration and capillary tube formation of human umbilical vein endothelial cells (Matsui, Wakabayashi, Asada, Yoshimatsu, & Okada, 2004). More recently, Guo and colleagues showed that c-Kit-expressing cardiac stem cells migrate in response to SCF. SCF-induced chemotaxis was mediated by the activation of PI3K/AKT signaling, which in turn led to the upregulation of matrix metalloproteinase (MMP)-2 and MMP-9 expression and activity (Guo et al., 2014).

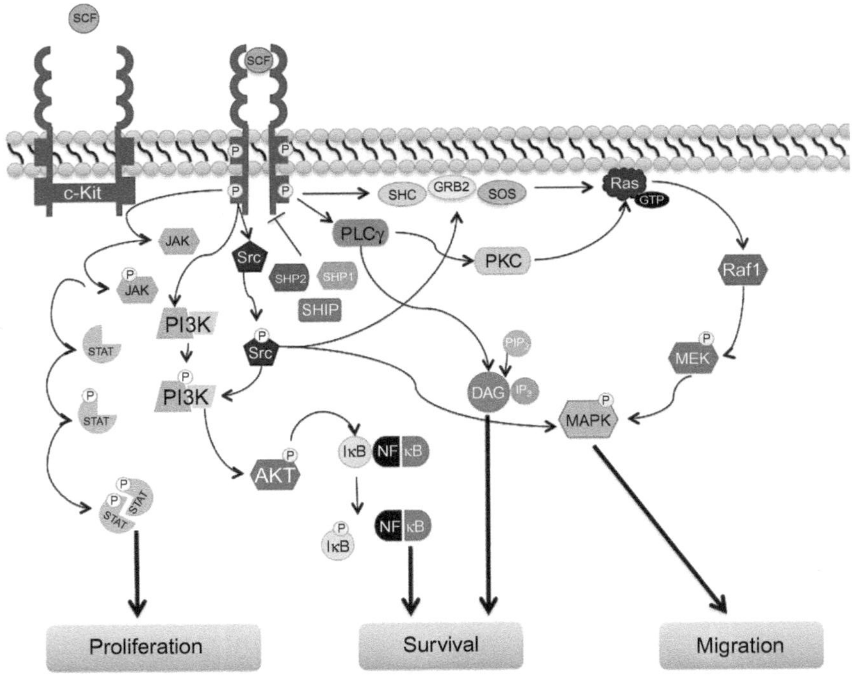

Fig. 2 Summary of the signal transduction pathways activated by the stem cell factor (SCF)/c-Kit system. Downstream signal transduction pathways activated by SCF include JAK/STAT pathway, associated with cell proliferation; PI3K/AKT, associated with cell survival; PLCγ/PKC, associated with cell survival and migration; and Ras/Raf/MAPK, associated with cell migration.

3. ERYTHROPOIETIN AND CANCER CELL MIGRATION

The expression and physiological significance of the Epo/EpoR axis in cancer cells have been controversial, since techniques and reagents used in pioneer studies for the detection of active EpoR in human cell lines and tissues were demonstrated to be poorly specific (Brown et al., 2007; Elliott et al., 2006). More recently, a number of specific anti-EpoR antibodies have been prepared and used in a number of trials, producing new evidence of the expression of EpoR in tumor cells. In addition, innovative technical approaches for the specific detection of EpoR in vivo are being developed. For instance, recombinant human Epo was coupled to the near-infrared dye Cy5.5; this probe has been tested in order to target the EpoR in vivo, using noninvasive fluorescence-mediated tomography (Doleschel et al., 2012). Similarly, a radiotracer for monitoring EpoR status in tumors was developed by coupling (68)Ga-DOTA to carbohydrate side chains of recombinant Epo. This reagent was shown to be useful for assessing EpoR status in vivo by positron emission tomography (Fuge et al., 2015). In this section, we review various studies describing the expression of the Epo/EpoR system and its potential role as a migration factor in various types of cancer. In order to avoid questionable data, we have only included figures obtained from experiments using specific reagents and techniques.

3.1 Breast Cancer

Quantitative real-time reverse transcriptase-PCR analysis has demonstrated that several breast cancer-derived cell lines express EpoR. Receptor copy numbers ranging from 104 to 1608 per 10 ng RNA were detected in the cell lines (Shi et al., 2010). In a previous study, Um and coworkers demonstrated that just 50 high-affinity Epo binding sites are sufficient for the activation of Epo-mediated signal transduction in cancer cells (Um, Gross, & Lodish, 2007). It has also been reported that breast cancer-derived cell lines SKBR3, MDA468, MDA453, and MCF7 synthesize and secrete Epo into the extracellular medium, suggesting the existence of an autocrine/paracrine Epo/EpoR system in these cells. In particular, a study of the SKBR3 cell line showed that stimulation with recombinant Epo leads to the phosphorylation of AKT and ERK, indicating that the EpoR is functional. SKBR3 cells are able to migrate in both transwell and scratch assays. However, inhibition of Epo expression by transfecting the cells with Epo siRNA significantly decreases the number of migrating cells. Similarly, interruption of the autocrine/paracrine loop, by knocking down EpoR expression through the transduction of the cells with lentivirus-based short hairpin RNA (shRNA) constructs, inhibits cell migration. These data strongly suggest that autocrine/paracrine Epo/EpoR activity is associated with breast cancer cell migration potential (Liang, Qiu, Lu, & Fan, 2014).

In an earlier work (Fu, Jiang, & Arcasoy, 2009), MCF-7 cells were transfected with a plasmid encoding a constitutively active mutant EpoR (EpoR-R129C) (Yoshimura, Longmore, & Lodish, 1990). Overexpression of the mutant EpoR-R129C was associated with an increase in cell migration. In this case, migration was dependent on the activation of JNK kinase (Fu et al., 2009). Likewise, inducing the overexpression of EpoR in an EpoR-negative rat mammary epithelial cell line significantly increased migration as a response to Epo compared to parental control cells (Shi et al., 2010).

The mechanism controlling EpoR-mediated migration in breast cancer cells has been explored. Epo promotes migration of MCF-7 cells under normoxic conditions, but the effect is even stronger under hypoxic conditions. Migration was abolished by either the administration of MEK inhibitors or inducing the expression of a dominant negative MEK-1 in the cells. These observations indicate that migration is regulated by an MAPK-dependent pathway (Lester, Jo, Campana, & Gonias, 2005). In agreement with these findings, a subsequent report showed that migration induced by a constitutively active EpoR in MCF-7 cells was found to depend on the activation of the MAP kinase pathway (Fu et al., 2009). Further understanding of the mechanism controlling Epo-driven migration was provided by Inbar (Inbar, Cohen-Armon, & Neumann, 2012), who published evidence showing that polyADP-ribose polymerase-1 (PARP-1) regulates Epo-induced migration in the breast cancer cell line MDA-MD-435. PARP-1 catalyzes protein modification by polyADP-ribosylation, and it is activated by phosphorylated extracellular signal-regulated kinase 2 (ERK2), a critical contributor to the intracellular signaling of the MAPK cascade. Recently, PARP-1 has been considered an important therapeutic target in ovarian cancer. Thus, this observation opens a new avenue for the use of PARP-1 inhibitors in combination with Epo in order to prevent the undesired effects of Epo in the treatment of cancer-associated anemia (Inbar et al., 2012).

3.2 Cervical Cancer

Female reproductive organs are nonclassical sites of Epo production (Yasuda et al., 2001). Epo participates in the regulation of cyclic changes of human endometrium (Yokomizo et al., 2002), and in the control of estrogen-dependent uterine angiogenesis (Yasuda et al., 1998). The expression of EpoR has been established in various cervical cancer-derived cell lines (HeLa, C33, SiHa, InBl, and CaLo) and validated by a combination of techniques, i.e., RT-PCR, flow cytometry, and immunocytochemistry, using specific antibodies (Aguilar, Aguilar, Lopez-Marure, Jimenez-Sanchez, & Rocha-Zavaleta, 2014; Lopez et al., 2011). In addition, it has been shown that SiHa, HeLa, and C33 produce and secrete Epo, resulting in the activation of an endogenous Epo/EpoR axis (Lopez et al., 2011).

Two independent reports have provided evidence showing that Epo induces migration of cervical cancer cells (Aguilar et al., 2014; Hamadmad & Hohl, 2008). Using the transwell migration system, based on the Boyden chamber migration assay, it was observed that Epo promotes cell migration in a dose-dependent manner. Although different cell lines were analyzed in the aforementioned reports, the authors found that Epo-mediated migration was sustained by the activation or ERK1/2. Moreover, the participation of the RhoA GTPase in Epo-driven migration was also identified (Hamadmad & Hohl, 2008). This is an important observation since RhoA is a key player in invadosome formation (Spuul et al., 2014).

3.3 Ovarian Cancer

Molecular analysis, using semiquantitative RT-PCR, restriction digestion of the PCR products, and DNA sequencing, confirmed the expression of EpoR in CaOV, SKOV, OVCAR-3, and A2780 human ovarian cancer cell lines (Jeong, Feldman, Solar, Szenajch, & Sytkowski, 2008). Nevertheless, fluorescence microscopy of A2780 cells revealed that most of the EpoR was found in the cytoplasm. Notably, silencing EpoR expression by lentiviral-mediated shRNA resulted in reduced A2780 proliferation and Epo-induced phosphorylation of ERK1/2 (Solar et al., 2012), suggesting that the intracellular receptor was still functional.

The case of the A2780 ovarian cells is perhaps one of the most intriguing because it was previously believed that to be active, EpoR must be located in the cell membrane. In this cell line, EpoR is mainly located in the cytoplasm. This may be the reason why exogenous Epo does not stimulate EpoR-mediated signaling, proliferation, invasiveness, or resistance to cytotoxic drugs in A2780 cells. However, specific inhibition of EpoR expression not only results in a reduction of proliferation but also restricts the intrinsic motility potential of the cells (Paragh et al., 2009). These observations suggest that EpoR may be constitutively active in the absence of Epo, and provide evidence for the potential role of an Epo-independent, EpoR-mediated pathway in ovarian cancer.

3.4 Prostate Cancer

The analysis of commercial tissue microarrays permits easy detection of protein expression, and a direct comparison of expression levels between tumors and their normal tissue counterparts. By analyzing a commercial tissue microarray constructed from clinical samples obtained from a cohort of 400 malignant tumors and 100 benign tissues from 20 different organs, it was determined that lymph node, thyroid, uterus, lung, and prostate malignant tissues had a significantly higher expression of EpoR compared with their corresponding benign tissue (Miyake et al., 2013). In agreement with this report, a comparative analysis of EpoR expression in 30 cases of prostate cancer, 16 high-grade prostate intraepithelial neoplasia, 50 benign prostatic hyperplasia, and 30 normal prostate samples showed that overexpression of EpoR was only detected in prostate cancer and prostate intraepithelial neoplasia (Xu, Zhou, He, & Sun, 2010).

Although the expression of EpoR in prostate cancer has been described, studies on the biological significance of Epo/EpoR are scarce. Examination of the PC-3 and LNCaP prostate cancer-derived cell lines confirmed that exogenous Epo stimulates cell proliferation (Feldman et al., 2006). Moreover, coexpression of Epo and EpoR in PC-3 and LNCaP was later identified, and it was reported to modulate cell growth and survival (Jeong, Hoxhaj, Socha, Sytkowski, & Feldman, 2009). Nevertheless, in an in vivo study, Epo was unable to stimulate the migration of tumor cells to the bone (Shiozawa et al., 2013).

3.5 Lung Cancer

In recent years, the expression of EpoR has been demonstrated in lung cancer cell lines and tumor samples. By studying the expression of EpoR in bronchoscopy brushes of patients with stage III–IV non-small-cell lung adenocarcinoma, using quantitative real-time polymerase chain reaction, Rozsas et al. (2013) found that 35% of the tumors had a high expression of the receptor, while the other 65% of the tumors exhibited low levels of EpoR expression. Surprisingly, high intratumoral EpoR expression was associated with significantly longer survival. This observation may be explained, at least in part, by the capacity of Epo to induce proliferation and migration of endothelial cells, as demonstrated by Doleschel et al., who used Cy5.5-coupled Epo as a theranostic near-infrared fluorescent probe to analyze the effects of Epo when coadministered with carboplatin in a non-small-cell lung cancer mouse model (Doleschel et al., 2015). Intratumoral EpoR was detected by combined fluorescence molecular tomography/microcomputed tomography hybrid imaging. Hybrid images showed that Epo acts on the tumor endothelium, inducing vascular remodeling with a subsequent improvement of carboplatin delivery, resulting in enhanced tumor cell apoptosis. Taken together, these observations may suggest that in the particular case of lung cancer, proper doses of Epo may improve chemotherapy.

3.6 Melanoma

In 2014, Mirkina and colleagues published a crucial analysis of the composition of relevant cell receptors in melanoma cells. They evaluated the expression of ErbB3/Her3, CD44/Pgp-1, ICAM-1/CD54, IGF-1-R/CD221, CD20, ErbB2/Her2, c-KIT/CD117, AC133/CD133, MDR-1/CD243, ErbB4, PD-1, NGF-R/CD271, and EpoR in freshly isolated skin melanoma cells, cultured patient-derived melanoma cells, four human melanoma cell lines (A375, Mel-Juso, SKMel28, 607B), and patient-derived melanoma cells grown in NOD/SCID/IL-2cnull (NSG) mice, by multicolor flow cytometry (Mirkina et al., 2014). The authors found that EpoR was expressed in freshly isolated melanoma cells for all patients examined. The levels of EpoR expression in individual patients ranged from 4% to 40%.

The promigratory effects of Epo were initially tested in the MV3 melanoma cell line using the transwell migration system. Epo promoted migration of MV3 cells. Migration was demonstrated to be mediated by the activation of the JAK/STAT and ERK1/2 pathways (Mirmohammadsadegh et al., 2010). A further study included the evaluation of three different human melanoma cell lines, obtained from patients with tumors in different clinical stages, WM35 (radial growth phase), WM793 (vertical growth phase), and 1205Lu (metastatic melanoma). As expected, cells from the more advanced lesions (WM793 and 1205Lu) were able to migrate in a Matrigel migration assay. When the expression of EpoR was silenced, by transfecting the cells with specific shRNA oligonucleotides, migration was inhibited, in particular in the metastatic 1250Lu cells (Kumar et al., 2012), indicating that EpoR plays a role in migration and invasion of melanoma cells.

4. STEM CELL FACTOR/C-KIT AND CANCER CELL MIGRATION

Human c-Kit has been recognized as a proto-oncogen for more than two decades (Yarden et al., 1987). Mutations and overexpression of c-Kit have been associated with a variety of tumors. Gain-of-function mutations have been detected in more than 90% of gastrointestinal stromal tumors (GIST), in 70% of mast cell tumors, and in a smaller proportion of nasal T-cell lymphomas, seminoma/dysgerminoma (Heinrich, Blanke, Druker, & Corless, 2002), and acute myeloid leukemia (Paschka et al., 2006). Mutations often lead to sustained c-Kit receptor activation. In addition, the analysis of more than 3000 tumors from more than 120 different tumor categories revealed that c-Kit can be detected in GIST (100%), seminomas (84%), adenoid-cystic carcinomas (65%), melanomas (35%), large-cell lung carcinomas (17%), and 47 additional tumor types (Went et al., 2004), suggesting that paracrine or autocrine activation of c-kit receptor may play a role in many other human malignancies.

Although the expression of c-Kit has been well documented in human cancer, the number of reports evaluating the effect of c-Kit/SCF on the migration of tumor cells is rather limited. Here, we will discuss the experimental evidence reported up to the present time.

4.1 Lung Cancer

The expression and biological significance of c-Kit in lung cancer have been extensively studied. In 1993, a pioneer report showed the expression of c-Kit in four small-cell lung cancer cell lines. By using the Boyden chambers assay, the authors demonstrated that tumor cell motility was stimulated by a positive gradient of SCF (Sekido et al., 1993). Later, the expression of c-Kit was also proven in cell lines obtained from non-small-cell lung cancer. Assessment of these cells in a Boyden chamber-based migration assay showed that SCF at concentrations ranging from 1 to 50 ng/mL stimulated migration of all the cells tested (Bredin, Liu, Hauzenberger, & Klominek, 1999).

Furthermore, a number of reports have demonstrated that coexpression of c-Kit and SCF is a common feature in small-cell lung cancer, and have suggested the presence of an autocrine or paracrine SCF/c-Kit system (Hibi et al., 1991; Krystal, Hines, & Organ, 1996; Rygaard, Nakamura, &

Spang-Thomsen, 1993). Accordingly, inhibition of SCF expression in cells derived from Lewis lung carcinoma (LLC) by using a specific SCF-shRNA in a lentivirus-mediated silencing system abolished the capacity of LLC to migrate in a transwell assay (Wang et al., 2015). However, the mechanism regulating SCF/c-Kit-mediated migration seems to be more complex and to include more molecules. It was found that small-cell lung cancer cells coexpress c-Kit and CXCR4, the chemokine receptor for SDF-1α, which is an important regulator of migration and metastasis. SCF and SDF-1α cooperatively promoted cell migration by inducing the downstream signaling targets AKT and p70 S6 kinase. Activation of c-Kit and CXCR4 by their ligands produced morphological cell changes, the formation of filopodia, and cytoplasmic protrusions associated with cell motility (Kijima et al., 2002). Taken together, these observations suggest that SCF/c-Kit may interact with other molecular systems to create a network of signals that regulate tumor cell motility.

4.2 Cervical Cancer

In a pioneering attempt to explore the expression of c-Kit and SCF in the human female genital tract, mRNA detection by Northern blot and immunohistochemical staining was used to analyze cell lines obtained from gynecological malignancies and samples from normal tissue. Expression of c-Kit was only found in one of the seven uterine cervix tumor cell lines, and in one of the eight endometrium cancer cell lines. In contrast, SCF was observed in most cell lines (Inoue, Kyo, Fujita, Enomoto, & Kondoh, 1994). The prevalence of c-Kit expression in normal tissue was equally uncommon. In agreement with the previous report, an infrequent overexpression of c-Kit protein was found in primary small-cell carcinomas of the uterine cervix (Wang & Lu, 2004) and in primary cervical squamous cell carcinoma (Han et al., 2011). An independent report demonstrated the expression of c-Kit in 13% of uterine adenosquamous carcinoma, with 7% of samples coexpressing SCF. No activating mutations or gene amplifications were detected in the c-Kit gene. However, in those samples coexpressing c-Kit and SCF, the receptor was found to be constitutively activated (Martinho et al., 2008).

At present, little is known about the effect of SCF in cervical cancer cells. However, activation of c-Kit by exogenous SCF has been shown to enhance anchorage-independent growth and migration of cervical cancer cells. By testing two c-Kit-expressing cell lines, it was shown that SCF promotes migration via the activation of the JAK2/STAT5 and the MAPK cascades (Aguilar et al., 2014).

4.3 Prostate Cancer

Human prostate cancer develops from hyperproliferating prostatic gland epithelial cells. Immunohistochemical analysis detected an impaired pattern of expression of c-Kit and SCF in prostate cancer cells compared with that of normal prostate tissue (Simak et al., 2000). Further studies have confirmed the existence of an alternative c-Kit transcript that produces a truncated form of the receptor in 28% of prostate tumors at early stages (Gleason grades 4–6) and 66% of more advanced tumors (Gleason grades 7–9), and also in a cell line (LNCaP) (Paronetto et al., 2004).

Prostate cancer frequently spreads to the bone. By using the prostate cancer cell line PC3, which expresses extremely low levels of c-Kit, in an experimental model of bone metastasis, it was observed that intraosseous bone tumor cells strongly express c-Kit, while those of subcutaneous tumors remain almost c-Kit negative. Evaluation of c-Kit expression in human samples revealed positive immunostaining in 5% of benign prostatic hyperplasias, 14% of prostate tumors, and 40% of prostate cancer bone metastases, suggesting that the bone microenvironment upregulates c-Kit expression, contributing to their intraosseous proliferation (Wiesner et al., 2008). It has been suggested that the complex prostatic microenvironment may also regulate the expression of c-Kit in prostate cancer cells. In a recent report, it was demonstrated that mixing DU145 prostate cancer-derived cells with stromal cells obtained from the prostatic peripheral zone induced a significant increase of c-Kit expression in cancer cells, resulting in an increased ability to migrate. In sharp contrast, mixing DU145 cells with stromal cells from the prostatic transitional zone induced downregulation of c-Kit expression. Considered as a whole, these data indicate that c-Kit promotes migration of prostate cancer cells, and that its expression may be regulated by cellular components of the prostatic and bone marrow microenvironments (Peng et al., 2015).

4.4 Ewing's Sarcoma

Ewing's sarcoma is an extremely aggressive tumor of the bone and soft tissues. The expression of c-Kit and SCF was analyzed in a panel of six Ewing's sarcoma-derived cell lines. All of them exhibited surface c-Kit, while five of six were also positive for the expression of transmembrane SCF. Direct incubation of the cells with exogenous SCF induced the downmodulation of c-Kit expression. However, when the exogenous SCF was placed in the lower chamber of a transwell migration assay, it was demonstrated to be a strong chemoattractant, suggesting the participation of the SCF/c-Kit axis in the migration and metastatic ability of Ewing's sarcoma cells (Landuzzi et al., 2000).

4.5 Colorectal Cancer

Evaluation of c-Kit expression in colorectal cancer samples has produced divergent results. Some groups have reported that the expression of c-Kit in colorectal carcinoma biopsies is very low (Friederichs et al., 2010; Reed et al., 2002; Yorke, Chirala, & Younes, 2003). Other studies have found c-Kit expression in up to 59% of stage II colorectal cancer patients (El-Serafi et al., 2010), 30% of colorectal neoplastic tissues (Sammarco et al., 2004), and 14% of distant metastases (Preto et al., 2007). In contrast, colorectal cancer cell lines exhibit elevated levels of c-Kit expression, and stimulation with SCF increases its migratory potential by activating the PI3K/AKT pathway (Yasuda et al., 2007). Further insights into the mechanism that regulates SCF/c-Kit-driven cell migration were provided by Siemens, Jackstadt, Kaller, and Hermeking (2013), who found that p53 is able to repress c-Kit by the induction of the miR-34 microRNA. miR-34 suppresses c-Kit via a conserved seed-matching sequence in the c-Kit 3′-UTR. The authors demonstrated that ectopic expression of miR-34 in the colorectal cancer cell line Colo320 inhibited SCF-induced migration and reduced the invasive potential of the cells.

4.6 Pancreatic Cancer

Pancreatic ductal adenocarcinoma is a malignant tumor with poor prognosis. Immunohistochemistry analysis of the expression of SCF in 95 tissue samples from pancreatic adenocarcinoma patients showed that SCF is expressed in this type of tumor, and that patients with tumors showing an overexpression of SCF have significantly lower survival rates compared to patients showing a low expression of SCF (Gao et al., 2015). By using two pancreatic adenocarcinoma cell lines, PANC-1 and BxPC-3, it was observed that SCF expression is regulated by the HIF-1α, and that exogenous SCF induces migration of both cell lines under hypoxia (Gao et al., 2015). Silencing the expression of endogenous SCF by using specific miRNA or incubation of the cells with a neutralizing anti-SCF antibody reduced cell migration. Accordingly, inhibition of HIF-1α expression by RNA interference abolished SCF-driven migration. These data demonstrate the clinicopathological importance of SCF in pancreatic adenocarcinoma and also indicate that SCF is a downstream gene of HIF-1α and may promote the development of this type of cancer under hypoxic conditions.

5. CLINICAL RELEVANCE

Anemia is a frequent condition in cancer patients. Cancer-associated anemia is multifactorial. It may be produced by chronic bleeding, by bone marrow invasion of tumor cells, and as a consequence of chemotherapy. About 40% of cancer patients will develop anemia at some time (Knight, Wade, & Balducci, 2004), but the prevalence of anemia varies depending on the type of tumor and the therapy administered (Ferrario et al., 2004). Cancer-associated anemia is considered a negative prognostic factor. Therefore, in most cases, it is treated using recombinant human Epo. Here, we have presented experimental evidence, suggesting that exogenous Epo may stimulate the migration of various cancer types. These observations, together with the fact that Epo also promotes cancer cell proliferation (Lopez et al., 2011), may explain, at least in part, the negative effect of Epo administration on progression-free and overall survival times of cancer patients (Temkin et al., 2006).

In 2007, the European Organization for Research and Treatment of Cancer determined that additional clinical and basic studies were still needed in order to define the real benefits and consequences of using Epo in anemic cancer patients (Bokemeyer et al., 2007). Presently, as new data continue to accumulate, it is becoming clear that the use of Epo in cancer patients must be carefully considered. An ideal practice would be to evaluate the expression of EpoR in the tumor before deciding whether to treat the patients with Epo or a blood transfusion, with the ultimate purpose of improving rather than hampering the outcome of the cancer treatment.

The expression of an active c-Kit/SCF axis is associated with cell mobilization. SCF is produced and released by fibroblasts, keratinocytes, and endothelial cells (Li, Goodyer, Fellows, & Wang, 2006). Furthermore, mSFC is cleaved by MMP-9 (Heissig et al., 2002), releasing soluble SCF that can in turn stimulate migration of c-Kit-expressing tumor cells. MMP-9 is expressed and secreted by a variety of tumors (Roy, Yang, & Moses, 2009).

In addition, SCF-mediated activation of PI3K/AKT signaling induces the upregulation of MMP-9 expression (Guo et al., 2014), assembling an uncontrolled circuit of cell mobilization promotion.

Since RTKs such as c-Kit play a decisive role in cancer development, in recent years a number of tyrosine kinase inhibitors have been developed. Most of them are multitarget inhibitors, but dasatinib, nilotinib, and masitinib have shown an important ability to inhibit c-Kit (Lennartsson & Rönnstrand, 2012). Administration of this kind of molecules and also the development of novel approaches to inhibition of SCF/c-Kit cascades will contribute to blocking the interdependent cellular constituents that regulate cell motility, and may help to circumvent or at least slow down the advent of metastasis.

6. CONCLUDING REMARKS

Tumor cell motility is regulated by the activation of a complex group of molecular components; they induce the formation and maturation of invadopodia, which protrude into the extracellular matrix and degrade it, allowing the cell to migrate from the primary tumor and initiate the formation of metastasis in distant organs. Growing experimental evidence shows that Epo/EpoR and SCF/c-Kit systems participate in tumor cell migration as a result of their capacity to activate key elements of motility-associated signaling cascades, and invadopodia function (Fig. 3). It is known that Epo stimulates the expression of high levels of SDF-1, the ligand of CXCR4 (Liu et al., 2013). On the other hand, tumor cells can coexpress c-Kit and CXCR4, after SCF and SDF-1 binding the receptors cooperate to activate AKT and p70 S6 kinase, inducing the formation of cytoplasmic protrusions (Kijima et al., 2002).

The binding of Epo to its receptor induces a sustained activation of ERK1/2 (Aguilar et al., 2014) and the activation of the RhoA GTPase (Hamadmad & Hohl, 2008), which are key players in invadopodia formation. In addition, formation of invadopodia depends on the presence of primary extracellular stimuli that may in turn activate RTKs such as EGFGR, PDGFR, and c-Kit. Thus, the activity of the SCF/c-Kit axis in tumor cells may trigger the formation of invadopodia. Furthermore, c-Kit promotes the activation of the ERK and PKC kinases, which participate in invadopodia function regulation. Finally, c-Kit induces the upregulation of MMP-9

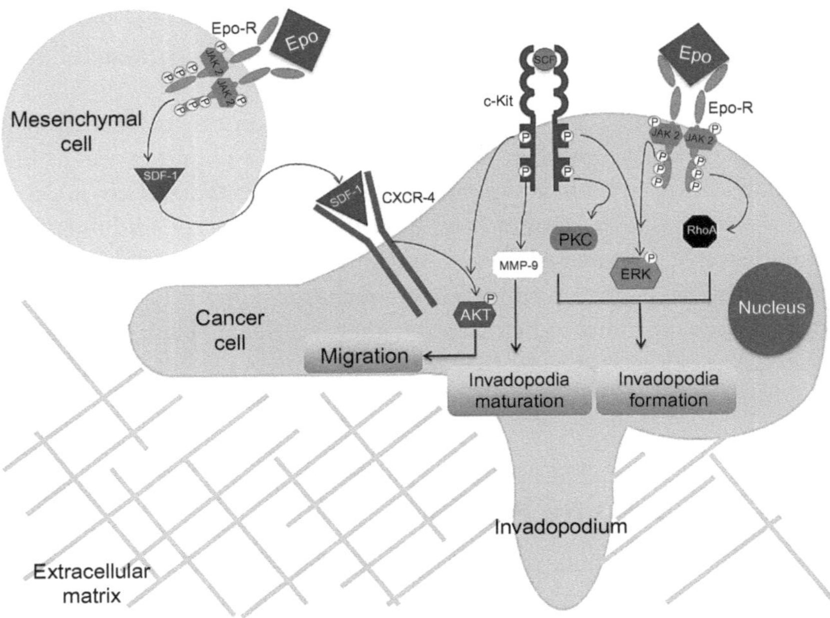

Fig. 3 Signaling molecules activated by erythropoietin (Epo) and stem cell factor (SCF) participate in cell migration. Epo induces the expression of SDF-1 in mesenchymal cells. SDF-1 and SCF promote migration of tumor cells coexpressing c-Kit and CXCR4. Activation of RhoA GTPase, PKC, and ERK1/2 by Epo and/or SCF triggers the formation of invadopodia. Upregulation of MMP-9 expression and activity by SCF allows the maturation of invadopodia, promoting tumor cell migration.

expression and activity (Guo et al., 2014), allowing the maturation of invadopodia and facilitating degradation of extracellular matrix components.

It is evident that Epo/EpoR and SCF/c-Kit participate in tumor cell migration. Thus, it would be important to identify which components of these signaling transduction pathways are crucial regulators of cell motility, in order to design novel therapeutic regimens including existing specific inhibitors, or testing new potential targets.

ACKNOWLEDGMENTS

This work is supported by grants from the Universidad Nacional Autónoma de México (PAPIIT-IN204816) and Fundación Miguel Alemán, A.C. M.J.V.-M. is a PhD student from the Doctorado en Ciencias Biomédicas, Universidad Nacional Autónoma de México (CONACyT: 239704; PAPIIT: IN209613), V.M.-E. is an MD student at the Universidad Panamericana, and L.R.-Z. is a Researcher at the Universidad Nacional Autónoma de México and Lecturer at the Universidad Panamericana.

REFERENCES

Abkowitz, J. L., Robinson, A. E., Kale, S., Long, M. W., & Chen, J. (2003). Mobilization of hematopoietic stem cells during homeostasis and after cytokine exposure. *Blood*, *102*, 1249–1253.

Aguilar, C., Aguilar, C., Lopez-Marure, R., Jimenez-Sanchez, A., & Rocha-Zavaleta, L. (2014). Co-stimulation with stem cell factor and erythropoietin enhances migration of c-Kit expressing cervical cancer cells through a sustained activation of ERK1/2. *Molecular Medicine Reports*, *9*, 1895–1902.

Alexeev, V., & Yoon, K. (2006). Distinctive role of the cKit receptor tyrosine kinase signaling in mammalian melanocytes. *Journal of Investigative Dermatology*, *126*, 1102–1110.

Bao, H., Jacobs-Helber, S. M., Lawson, A. E., Penta, K., Wickrema, A., & Sawyer, S. T. (1999). Protein kinase B (c-Akt), phosphatidylinositol 3-kinase, and STAT5 are activated by erythropoietin (EPO) in HCD57 erythroid cells but are constitutively active in an EPO-independent, apoptosis-resistant subclone (HCD57-SREI cells). *Blood*, *93*, 3757–3773.

Bokemeyer, C., Aapro, M. S., Courdi, A., Foubert, J., Link, H., Osterborg, A., et al. (2007). EORTC guidelines for the use of erythropoietic proteins in anaemic patients with cancer: 2006 update. *European Journal of Cancer*, *43*, 258–270.

Bredin, C. G., Liu, Z., Hauzenberger, D., & Klominek, J. (1999). Growth-factor-dependent migration of human lung-cancer cells. *International Journal of Cancer*, *82*, 338–345.

Brown, W. M., Maxwell, P., Graham, A. N. J., Yakkundi, A., Dunlop, E. A., Shi, Z., et al. (2007). Erythropoietin receptor expression in non-small cell lung carcinoma: A question of antibody specificity. *Stem Cells*, *25*, 718–722.

Damen, J. E., Wakao, H., Miyajima, A., Krosl, J., Humphries, R. K., Cutler, R. L., et al. (1995). Tyrosine 343 in the erythropoietin receptor positively regulates erythropoietin-induced cell proliferation and Stat5 activation. *The EMBO Journal*, *14*, 5557–5568.

Doleschel, D., Mundigl, O., Wessner, A., Gremse, F., Bachmann, J., Rodriguez, A., et al. (2012). Targeted near-infrared imaging of the erythropoietin receptor in human lung cancer xenografts. *Journal of Nuclear Medicine*, *53*, 304–311.

Doleschel, D., Rix, A., Arns, S., Palmowski, K., Gremse, F., Merkle, R., et al. (2015). Erythropoietin improves the accumulation and therapeutic effects of carboplatin by enhancing tumor vascularization and perfusion. *Theranostics*, *5*, 905–918.

Elliott, S., Busse, L., Bass, M. B., Lu, H., Sarosi, I., Sinclair, A. M., et al. (2006). Anti-Epo receptor antibodies do not predict Epo receptor expression. *Blood*, *107*, 1892–1895.

El-Serafi, M. M., Bahnassy, A. A., Ali, N. M., Eid, S. M., Kamel, M. M., Abdel-Hamid,- N. A., et al. (2010). The prognostic value of c-Kit, K-ras codon 12, and p53 codon 72 mutations in Egyptian patients with stage II colorectal cancer. *Cancer*, *116*, 4954–4964.

Feldman, L., Wang, Y., Rhim, J. S., Bhattacharya, N., Loda, M., & Sytkowski, A. J. (2006). Erythropoietin stimulates growth and STAT5 phosphorylation in human prostate epithelial and prostate cancer cells. *The Prostate*, *66*, 135–145.

Ferrario, E., Ferrari, L., Bidolio, P., De Candis, D., Del Vecchio, M., De Dosso, S., et al. (2004). Treatment of cancer-related anemia with epoetin alfa: A review. *Cancer Treatment Reviews*, *30*, 563–575.

Friederichs, J., von Weyhern, C. W., Rosenberg, R., Doll, D., Busch, R., Lordick, F., et al. (2010). Immunohistochemical detection of receptor tyrosine kinases c-kit, EGF-R, and PDGF-R in colorectal adenocarcinomas. *Langenbeck's Archives of Surgery*, *395*, 373–379.

Fu, P., Jiang, X., & Arcasoy, M. O. (2009). Constitutively active erythropoietin receptor expression in breast cancer cells promotes cellular proliferation and migration through a MAP-kinase dependent pathway. *Biochemical and Biophysical Research Communications*, *379*, 696–701.

Fuge, F., Doleschel, D., Rix, A., Gremse, F., Wessner, A., Winz, O., et al. (2015). In-vivo detection of the erythropoietin receptor in tumours using positron emission tomography. *European Radiology*, *25*, 472–479.

Gao, C., Li, S., Zhao, T., Chen, J., Ren, H., Zhang, H., et al. (2015). SCF, regulated by HIF-1α, promotes pancreatic ductal adenocarcinoma cell progression. *PLoS One*, *10*, e0121338.

Gerard, A., van der Kammen, R. A., Janssen, H., Ellenbroek, S. I., & Collard, J. G. (2009). The Rac activator Tiam1 controls efficient T-cell trafficking and route of trans-endothelial migration. *Blood*, *113*, 6138–6147.

Guo, J., Jie, W., Shen, Z., Li, M., Lan, Y., Kong, Y., et al. (2014). SCF increases cardiac stem cell migration through PI3K/AKT and MMP-2/-9 signaling. *International Journal of Molecular Medicine*, *34*, 112–118.

Hamadmad, S. N., & Hohl, R. J. (2008). Erythropoietin stimulates cancer cell migration and activates RhoA protein through a mitogen-activated protein kinase/extracellular signal-regulated kinase-dependent mechanism. *The Journal of Pharmacology and Experimental Therapeutics*, *324*, 1227–1233.

Han, C.-P., Lin, W.-L., Wang, P.-H., Yang, S.-F., Lewis, J. S., Chen, C.-K., et al. (2011). Overexpression of c-KIT (CD117) occurs infrequently in squamous cell carcinoma of the uterine cervix. *Histopathology*, *58*, 982–998.

Heeschen, C., Aicher, A., Lehmann, R., Fichtlscherer, S., Vasa, M., Urbich, C., et al. (2003). Erythropoietin is a potent physiologic stimulus for endothelial progenitor cell mobilization. *Blood*, *102*, 1340–1346.

Heinrich, M. C., Blanke, C. D., Druker, B. J., & Corless, C. L. (2002). Inhibition of KIT tyrosine kinase activity: A novel molecular approach to the treatment of KIT-positive malignancies. *Journal of Clinical Oncology*, *20*, 1692–1703.

Heissig, B., Hattori, K., Dias, S., Friedrich, M., Ferris, B., Hackett, N. R., et al. (2002). Recruitment of stem and progenitor cells from the bone marrow niche requires MMP-9 mediated release of kit-ligand. *Cell*, *109*, 625–637.

Hibi, K., Takahashi, T., Sekido, Y., Ueda, R., Hida, T., Ariyoshi, Y., et al. (1991). Coexpression of the stem cell factor and the c-kit genes in small-cell lung cancer. *Oncogene*, *6*, 2291–2296.

Hsu, Y. R., Wu, G. M., Mendiaz, E. A., Syed, R., Wypych, J., Toso, R., et al. (1997). The majority of stem cell factor exists as monomer under physiological conditions. Implications for dimerization mediating biological activity. *The Journal of Biological Chemistry*, *272*, 6406–6415.

Inbar, D., Cohen-Armon, M., & Neumann, D. (2012). Erythropoietin-driven signalling and cell migration mediated by polyADP-ribosylation. *British Journal of Cancer*, *107*, 1317–1326.

Inoue, M., Kyo, S., Fujita, M., Enomoto, T., & Kondoh, G. (1994). Coexpression of the c-kit receptor and the stem cell factor in gynecological tumors. *Cancer Research*, *54*, 3049–3053.

Jelkmann, W. (2003). Erythropoietin. *Journal of Endocrinological Investigation*, *26*, 832–837.

Jeong, J.-Y., Feldman, L., Solar, P., Szenajch, J., & Sytkowski, J. (2008). Characterization of erythropoietin receptor and erythropoietin expression and function in human ovarian cancer cells. *International Journal of Cancer*, *122*, 274–280.

Jeong, J.-Y., Hoxhaj, G., Socha, A. L., Sytkowski, A. J., & Feldman, L. (2009). An erythropoietin autocrine/paracrine axis modulates the growth and survival of human prostate cancer cells. *Molecular Cancer Research*, 7, 1150–1157.

Kijima, T., Maulik, G., Ma, P. C., Tibaldi, E. V., Turner, R. E., Rollins, B., et al. (2002). Regulation of cellular proliferation, cytoskeletal function, and signal transduction through CXCR4 and c-Kit in small cell lung cancer cells. *Cancer Research*, *62*, 6304–6311.

Kirito, K., Nakajima, K., Watanabe, T., Uchida, M., Tanaka, M., Ozawa, K., et al. (2002). Identification of the human erythropoietin receptor region required for Stat1 and Stat3 activation. *Blood*, *99*, 102–110.

Klingmuller, U., Bergelson, S., Hsiao, J. G., & Lodish, H. F. (1996). Multiple tyrosine residues in the cytosolic domain of the erythropoietin receptor promote activation of STAT5. *Proceedings of the National Academy of Sciences of the United States of America*, *93*, 8324–8328.

Knight, K., Wade, S., & Balducci, L. (2004). Prevalence and outcomes of anaemia in cancer: A systematic review of the literature. *American Journal of Medicine*, *116*, 11S–26S.

Krystal, G. W., Hines, S. J., & Organ, C. P. (1996). Autocrine growth of small cell lung cancer mediated by coexpression of c-kit and stem cell factor. *Cancer Research*, *56*, 370–376.

Kumar, S. M., Zhang, G., Bastian, B. C., Arcasoy, M. O., Karande, P., Pushparajan, A., et al. (2012). Erythropoietin receptor contributes to melanoma cell survival *in vivo*. *Oncogene*, *31*, 1649–1660.

Landuzzi, L., De Giovanni, C., Nicoletti, G., Rossi, I., Ricci, C., Astolfi, A., et al. (2000). The metastatic ability of Ewing's sarcoma cells is modulated by stem cell factor and by its receptor c-Kit. *American Journal of Pathology*, *157*, 2123–2131.

Langley, K. E., Bennett, L. G., Wypych, J., Yancik, S. A., Liu, X. D., Westcott, K. R., et al. (1993). Soluble stem cell factor in human serum. *Blood*, *81*, 656–660.

Lennartsson, J., & Rönnstrand, L. (2012). Stem cell factor receptor/c-Kit: From basic science to clinical implications. *Physiological Reviews*, *92*, 1619–1649.

Lester, R. D., Jo, M., Campana, W. M., & Gonias, S. L. (2005). Erythropoietin promotes MCF-7 breast cancer cell migration by an ERK/mitogen-activated protein kinase-dependent pathway and is primarily responsible for the increase in migration observed in hypoxia. *The Journal of Biological Chemistry*, *280*, 39273–39277.

Li, J., Goodyer, C. G., Fellows, F., & Wang, R. (2006). Stem cell factor/c-Kit interactions regulate human islet-epithelial cluster proliferation and differentiation. *The International Journal of Biochemistry & Cell Biology*, *38*, 961–972.

Liang, K., Qiu, S., Lu, Y., & Fan, Z. (2014). Autocrine/paracrine erythropoietin regulates migration and invasion potential and the stemness of human breast cancer cell. *Cancer Biology & Therapy*, *15*, 89–98.

Lin, J. S., Chen, Y. S., Chiang, H. S., & Ma, M. C. (2008). Hypoxic preconditioning protects rat hearts against ischaemia-reperfusion injury: Role of erythropoietin on progenitor cell mobilization. *The Journal of Physiology*, *586*, 5757–5769.

Linder, S., & Aepfelbacher, M. (2003). Podosomes: Adhesion hot-spots of invasive cells. *Trends in Cell Biology*, *13*, 376–385.

Liu, H., Chen, X., Focia, P. J., & He, X. (2007). Structural basis for stem cell factor-Kit signaling and activation of class III receptor tyrosine kinases. *The EMBO Journal*, *26*, 891–901.

Liu, N., Tian, J., Cheng, J., & Zhang, J. (2013). Effect of erythropoietin on the migration of bone marrow-derived mesenchymal stem cells to the acute kidney injury microenvironment. *Experimental Cell Research*, *319*, 2019–2027.

Lopez, T. V., Lappin, T. R. J., Maxwell, P., Shi, Z., Lopez-Marure, R., Aguilar, C., et al. (2011). Autocrine/paracrine erythropoietin signalling promotes JAK/STAT-dependent proliferation of human cervical cancer cells. *International Journal of Cancer*, *129*, 2566–2576.

Maiese, K., Chong, Z. Z., Hou, J., & Shang, Y. C. (2008). Erythropoietin and oxidative stress. *Current Neurovascular Research*, *5*, 125–142.

Martinho, O., Goncalves, A., Moreira, M. A. R., Ribeiro, L. F. J., Queiroz, G. S., Schmitt, F. C., et al. (2008). KIT activation in uterine cervix adenosquamous carcinomas by KIT/SCF autocrine/paracrine stimulation loops. *Gynecologic Oncology*, *111*, 350–355.

Matsui, J., Wakabayashi, T., Asada, M., Yoshimatsu, K., & Okada, M. (2004). Stem cell factor/c-kit signaling promotes the survival, migration, and capillary tube formation of human umbilical vein endothelial cells. *The Journal of Biological Chemistry*, *279*, 18600–18607.

Matsui, Y., Zsebo, K. M., & Hogan, B. L. (1990). Embryonic expression of a haematopoietic growth factor encoded by the Sl locus and the ligand for c-kit. *Nature*, *347*, 667–669.

Meininger, C. J., Yano, H., Rottapel, R., Bernstein, A., Zsebo, K. M., & Zetter, B. R. (1992). The c-kit receptor ligand functions as a mast cell chemoattractant. *Blood*, *79*, 958–963.

Mirkina, I., Hadzijusufovic, E., Krepler, C., Mikula, M., Mechtcheriakova, D., Strommer, S., et al. (2014). Phenotyping of human melanoma cells reveals a unique composition of receptor targets and a subpopulation co-expressing Erb4, EPO-R and NGF-R. *PLoS One*, *9*, e84417.

Mirmohammadsadegh, A., Marini, A., Gustrau, A., Delia, D., Nambiar, S., Hassan, M., et al. (2010). Role of erythropoietin receptor expression in malignant melanoma. *Journal of Investigative Dermatology*, *130*, 201–210.

Miyake, M., Goodison, S., Lawton, A., Zhang, G., Gomes-Giacoia, E., & Rosser, C. J. (2013). Erythropoietin is a JAK2 and ERK1/2 effector that can promote renal tumor cell proliferation under hypoxic conditions. *Journal of Hematology & Oncology*, *6*, 65.

Mort, R. L., Jackson, I. J., & Patton, E. (2015). The melanocyte lineage in development and disease. *Development*, *142*, 620–632.

Munugalavadla, V., & Kapur, R. (2005). Role of c-Kit and erythropoietin receptor in erythropoiesis. *Critical Reviews in Oncology/Hematology*, *54*, 63–75.

Paragh, G., Kumar, S. M., Rakosy, Z., Choi, S.-C., Xu, X., & Acs, G. (2009). RNA interference-mediated inhibition of erythropoietin receptor expression suppresses tumor growth and invasiveness in A2780 human ovarian carcinoma cells. *The American Journal of Pathology*, *174*, 1504–1514.

Paronetto, M. P., Farini, D., Sammarco, I., Maturo, G., Vespasiani, G., Geremia, R., et al. (2004). Expression of a truncated form of the c-Kit tyrosine kinase receptor and activation of Src kinase in human prostatic cancer. *American Journal of Pathology*, *164*, 1243–1251.

Paschka, P., Marcucci, G., Ruppert, A. S., Mrozek, K., Chen, H., Kittles, R. A., et al. (2006). Adverse prognostic significance of KIT mutations in adult acute myeloid leukemia with inv(16) and t(8;21): A cancer and leukemia group B study. *Journal of Clinical Oncology*, *24*, 3904–3911.

Peng, Y., Chen, Q., Gu, M., Chen, Y., Zhang, M., Zhou, J., et al. (2015). Human stromal cells in the peripheral zone of the prostate tumorigenesis of prostatic cancer stem cells through up-regulation of c-Kit expression. *Journal of Cancer*, *6*, 776–785.

Preto, A., Moutinho, C., Velho, S., Oliveira, C., Rebocho, A. P., Figueiredo, J., et al. (2007). A subset of colorectal carcinomas express c-Kit protein independently of BRAF and/or KRAS activation. *Virchows Archiv*, *450*, 619–626.

Reed, J., Ouban, A., Schickor, F. K., Muraca, P., Yeatman, T., & Coppola, D. (2002). Immunohistochemical staining for c-Kit (CD117) is a rare event in human colorectal carcinoma. *Clinical Colorectal Cancer*, *2*, 119–122.

Ronnstrand, L. (2004). Signal transduction via the stem cell factor receptor/c-Kit. *Cellular and Molecular Life Sciences*, *61*, 2535–2548.

Ropers, H. H., & Craig, I. W. (1989). Report of the committee on the genetic constitution of chromosomes 12 and 13. *Cytogenetics and Cell Genetics*, *51*, 259–279.

Roy, R., Yang, J., & Moses, M. A. (2009). Matrix metalloproteinases as novel biomarkers and potential therapeutic targets in human cancer. *Journal of Clinical Oncology*, *27*, 5287–5297.

Rozsas, A., Berta, J., Rojko, L., Horvath, L. Z., Keszthelyi, M., Kenessey, I., et al. (2013). Erythropoietin receptor expression is a potential prognostic factor in human lung adenocarcinoma. *PLoS One*, *8*, e77459.

Rygaard, K., Nakamura, T., & Spang-Thomsen, M. (1993). Expression of the protooncogenes c-met and c-kit and their ligands, hepatocyte growth factor/scatter factor and stem cell factor, in SCLC cell lines and xenografts. *British Journal of Cancer*, *67*, 37–46.

Sammarco, I., Capurso, G., Coppola, L., Bonifazi, A. P., Cassetta, S., Delle Fave, G., et al. (2004). Expression of the proto-oncogene c-Kit in normal and tumor tissues from colorectal carcinoma patients. *International Journal of Colorectal Disease*, *19*, 545–553.

Sasaki, R., Masuda, S., & Nagao, M. (2000). Erythropoietin: Multiple physiological functions and regulation of biosynthesis. *Bioscience, Biotechnology, and Biochemistry*, *64*, 1775–1793.

Satoh, K., Kagaya, Y., Nakano, M., Ito, Y., Ohta, J., Tada, H., et al. (2006). Important role of endogenous erythropoietin system in recruitment of endothelial progenitor cells in hypoxia-induced pulmonary hypertension in mice. *Circulation*, *113*, 1442–1450.

Sekido, Y., Takahashi, T., Ueda, R., Takahashi, M., Suzuki, H., Nishida, K., et al. (1993). Recombinant human stem cell factor mediates chemotaxis of small-cell lung cancer cell lines aberrantly expressing the c-kit protooncogene. *Cancer Research*, *53*, 1709–1714.

Shi, Z., Hodges, V. M., Dunlop, E. A., Percy, M. J., Maxwell, A. P., El-Tanani, M., et al. (2010). Erythropoietin-induced activation of the JAK2/STAT6, PI3K/Akt, and Ras/ERK pathways promotes malignant cell behaviour in a modified breast cancer cell line. *Molecular Cancer Research*, *8*, 615–626.

Shiozawa, Y., McGee, S., Pienta, M. J., McGregor, N., Jung, Y., Yumoto, K., et al. (2013). Erythropoietin supports the survival of prostate cancer, but not growth and bone metastasis. *Journal of Cellular Biochemistry*, *114*, 2471–2478.

Siemens, H., Jackstadt, R., Kaller, M., & Hermeking, H. (2013). Repression of c-Kit by p53 is mediated by miR-34 and is associated with reduced chemoresistance, migration and stemness. *Oncotarget*, *4*, 1399–1415.

Simak, R., Capodieci, P., Cohen, D. W., Fair, W. R., Scher, H., & Melamed, J. (2000). Expression of c-kit and kit-ligand in benign and malignant prostatic tissues. *Histology and Histopathology*, *15*, 365–374.

Solar, P., Hrckova, G., Varinska, L., Solarova, Z., Kriska, J., Uhrinova, I., et al. (2012). Location and the functionality of erythropoietin receptor(s) in A2780 cells. *Oncology Reports*, *28*, 141–146.

Spuul, P., Ciufici, P., Veillat, V., Leclercq, A., Daubon, T., Kramer, I. J., et al. (2014). Importance of Rho GTPases in formation, characteristics, and function of invadosomes. *Small GTPases*, *5*, e28195.

Temkin, S. M., Hellmann, M., Serur, E., Lee, Y. C., & Abulafia, O. (2006). Erythropoietin administration during primary treatment for locally advanced cervical carcinoma is associated with poor response to radiation. *International Journal of Gynecological Cancer*, *16*, 1855–1861.

Toyoda, T., Itai, T., Arakawa, T., Aoki, K. H., & Yamaguchi, H. (2000). Stabilization of human recombinant erythropoietin through interactions with the highly branched N-glycans. *The Journal of Biochemistry*, *128*, 731–737.

Um, M., Gross, A. W., & Lodish, H. F. (2007). A "classical" homodimeric erythropoietin receptor is essential for the antiapoptotic effects of erythropoietin on differentiated neuroblastoma SH-SY5Y and pheochromocytoma PC-12 cells. *Cellular Signalling*, *19*, 634–645.

Wang, H. L., & Lu, D. W. (2004). Overexpression of c-Kit protein is an infrequent event in small cell carcinomas of the uterine cervix. *Modern Pathology*, *17*, 732–738.

Wang, L., Wang, J. T., Li, Z., Liu, Y. Y., Jiang, M., Li, Y., et al. (2015). Silencing stem cell factor attenuates stemness and inhibits migration of cancer stem cells derived from Lewis lung carcinoma cells. *Tumor Biology*, *37*, 7213–7227.

Went, P. T., Dirnhofer, S., Bundi, M., Mirlacher, M., Schraml, P., Mangialaio, S., et al. (2004). Prevalence of KIT expression in human tumors. *Journal of Clinical Oncology*, *22*, 4514–4522.

Wiesner, C., Nabha, S. M., Dos Santos, E. B., Yamamoto, H., Meng, H., Melchior, S. W., et al. (2008). C-Kit and its ligand stem cell factor: Potential contribution to prostate cancer bone metastasis. *Neoplasia*, *10*, 996–1003.

Witthuhn, B. A., Quelle, F. W., Silvennoinen, O., Yi, T., Tang, B., Miura, O., et al. (1993). JAK2 associates with the erythropoietin receptor and is tyrosine phosphorylated and activated following stimulation with erythropoietin. *Cell*, *74*, 227–236.

Xu, C., Zhou, T., He, M., & Sun, Y. (2010). Differential up-regulation of erythropoietin and its receptor in benign and malignant prostatic tissue. *Urologic Oncology*, *28*, 314–319.

Yarden, Y., Kuang, W. J., Yang-Feng, T., Coussens, L., Munemitsu, S., Dull, T. J., et al. (1987). Human proto-oncogene c-kit: A new cell surface receptor tyrosine kinase for an unidentified ligand. *The EMBO Journal*, *6*, 3341–3351.

Yasuda, Y., Fujita, Y., Musha, T., Tanaka, H., Shiokawa, S., Nakamatsu, K., et al. (2001). Expression of erythropoietin in human female reproductive organs. *Italian Journal of Anatomy and Embryology*, *106*, 215–222.

Yasuda, Y., Masuda, S., Chikuma, M., Inoue, K., Nagao, M., & Sasaki, R. (1998). Estrogen-dependent production of erythropoietin in uterus and its implication in uterine angiogenesis. *The Journal of Biological Chemistry*, *273*, 25381–25387.

Yasuda, A., Sawai, H., Takahashi, H., Ochi, N., Matsuo, Y., Funahashi, H., et al. (2007). Stem cell factor/c-kit receptor signaling enhances the proliferation and invasion of colorectal cancer cells through the PI3K/Akt pathway. *Digestive Diseases and Sciences*, *52*, 2292–2300.

Yokomizo, R., Matsuzaki, S., Uehara, S., Murakami, T., Yaegashi, N., & Okamura, K. (2002). Erythropoietin and erythropoietin receptor expression in human endometrium throughout the menstrual cycle. *Molecular Human Reproduction*, *8*, 441–446.

Yorke, R., Chirala, M., & Younes, M. (2003). c-kit proto-oncogene product is rarely detected in colorectal adenocarcinoma. *Journal of Clinical Oncology*, *21*, 3885–3886.

Yoshimura, A., Longmore, G., & Lodish, H. F. (1990). Point mutation in the exoplasmic domain of the erythropoietin receptor resulting in hormone-independent activation and tumorigenicity. *Nature*, *348*, 647–649.

CHAPTER FIFTEEN

Significance of Erythropoietin Receptor Antagonist EMP9 in Cancers

Yoshiko Yasuda*[,1], Mitsugu Fujita†[,1]
*Cancer Center, Kyoto Prefectural University of Medicine, Kyoto, Japan
†Kindai University Faculty of Medicine, Osaka-Sayama, Japan
[1]Corresponding authors: e-mail address: y1126yas@gmail.com; mfujita47@gmail.com

Contents

Abstract

We have clarified that cancer cells express their own erythropoietin (Epo) and its receptor (EpoR) mRNA levels, and the respective proteins, which are under the control of Epo–EpoR signaling. Then we explored to inhibit the Epo–EpoR signaling with an EpoR antagonist Epo mimetic peptide 9 (EMP9) that is a derivative of an Epo-mimicking peptide EMP1. In the study of the cancer cell lines in vitro, rhEpo accelerated the cancer cell growth, whereas the EMP9 inhibited the cell growth along with the inhibition of STAT5 tyrosine phosphorylation. Moreover, in vitro study of surgically resected histoculture of lung cancers revealed that EMP9 diminishes the expression of myoglobin in the cancer cells and destroys the feeding vessels. Additionally, in the xenografts of lung cancer histoculture, the EMP9 destroyed the xenografts by inducing apoptosis and suppressing proliferation of cancer cells in concomitant with macrophage accumulation. Furthermore, two types of perforations were detected in their cytoplasm: the one is mediated by nNOS in the cancer cells and the other one is by iNOS in the innate immune cells. These findings suggest that the inhibition of the Epo–EpoR signaling by EMP9 induces the cancer cell death that is mediated by the apoptosis and calcification of the cancer

Vitamins and Hormones, Volume 105
ISSN 0083-6729
http://dx.doi.org/10.1016/bs.vh.2017.03.001
© 2017 Elsevier Inc.
All rights reserved.

cells as well as the oxygen deficiency through the feeding vessels. Taken together, EMP9-based therapy may be a promising strategy to treat cancer patients.

1. INTRODUCTION

Until 2001, the main function of erythropoietin (Epo) has been considered as erythropoiesis. That is, Epo binds to its receptor (EpoR) expressed on erythroid precursor cells and stimulates them to proliferate and differentiate into hemoglobin (Hb)-containing erythrocytes (Jelkmann, 1992; Krantz, 1991; Youssoufian, Longmore, Neumann, Yoshimura, & Lodish, 1993). In this regard, we found that the EpoR mRNA and EpoR protein are expressed in murine embryo proper at the early postimplantation stage (Yasuda et al., 1993). During this developmental period (6.5–7.5 days of gestation), the growth of murine embryos extremely accelerates and their doubling time reaches 4.8–8.1 h (Snow, 1977). These findings directed us to address whether the Epo–EpoR signaling is involved in carcinogenesis. As expected, the Epo–EpoR signaling has been found in malignant female reproductive organs (Yasuda et al., 2001), breast cancers (Acs, Zhang, Rebbeck, Acs, & Verma, 2002; Arcasoy et al., 2002), and prostate cancers (Feldman et al., 2006). Subsequently, we have found that most cancers express Epo and EpoR regardless of their origin (Yasuda et al., 2003). From then on, we have been trying to dissect the carcinogenic mechanisms of the Epo–EpoR signaling using an anti-Epo antibody (clone R2) and/or a soluble form of EpoR, both of which bind to Epo and act as Epo antagonists (Yasuda et al., 2001, 2002). These studies demonstrated that the inhibition of the Epo–EpoR signal induces the destruction of the cancer cells as well as the cancer-feeding vessels.

In parallel, we and others have searched for Epo-mimicking synthetic peptides as Epo agonist (Wrighton et al., 1996). Among them, Epo mimetic peptide 9 (EMP9), a derivative of EMP1 (Epo agonist), has been found to bind weakly to EpoR but not to induce their proliferation (Table 1; Johnson et al., 1998). This finding led us to hypothesize that the EMP9 might act as an EpoR antagonist. Indeed, intraperitoneal injection of the EMP9 induced the destruction of melanoma cell line (P39) and stomach choriocarcinoma cell line (SCH) by suppressing the cancer cell growth, inducing the cancer cell death, and destructing the neovascular formation (Yasuda et al., 2003). These phenomena were associated with the alteration of STAT5

Table 1 Property of Epo Mimetic Peptides[a]

	EMP1	EMP9
Sequence	**GGTYSCHFGPLTWVCKPQGG**	**GGTYSCHFAPLTWVCKPQGG**
Binding IC_{50} (μM) in EBP beads	5	80
Binding IC_{50} (μM) in TF-1 cells	0.07	1.0
Epo-ED_{50} (μM) in tr. human receptor	0.1	Inactive
Epo-ED_{50} (μM) in human receptor	0.1	Inactive

[a]Based on Johnson, D. L., Farrell, F. X., Barbone, F. P., McMahon, F. J., Tullai, J., Hoey, K., et al. (1998). Identification of a 13 amino acid peptide mimetic of erythropoietin and description of amino acids critical for the mimetic activity of EMP1. Biochemistry, 37, 3699–3710.
EBP, Epo-binding protein; *ED_{50}*, median effective dose; *IC_{50}*, median inhibitory concentration; *tr*, truncated.

phosphorylation status. Then, we screened a line of cell lines to address the impacts of Epo–EpoR signal inhibition on cancers by using EMP9 in vitro (Yasuda, Maeda, et al., 2010) and in vivo (Yasuda et al., 2015). These data revealed how the Epo–EpoR signaling is involved in carcinogenesis and how the EMP9 promotes the destruction of cancer cells.

In this chapter, we provide the following: (1) most cancers express EpoR mRNA; (2) most cancers secrete Epo; (3) the cancer-derived Epo binds to the EpoR expressed on their cell surface in an autocrine and/or paracrine manner; (4) the cancer-derived Epo protects them from apoptotic cell death; (5) the cancer-derived Epo stimulates them to proliferate and produce monomeric myoglobin (Mb) and tetrameric hemoglobins along with their heme synthetic enzyme, all of which further promote tumor growth; and (6) EMP9 inhibits the downstream events of the Epo–EpoR signaling (Szenajch, Wcislo, Jeong, Szczylik, & Feldman, 2010), which primarily attributes to the cancer cell destruction due to the deficiency of oxygen reservation as well as cancer-feeding vessel regeneration.

2. STRUCTURE AND PROPERTIES OF EMP9

EMP9 is a synthetic peptide that is composed of 20 amino acid residues; it is a derivative of the EMP1 that exhibits erythropoietic activities (Johnson et al., 1998). The amino acid components of EMP9 differ from those of EMP1 in that a glycine in the EMP1 is replaced with an alanine in the EMP9 (Table 1). EMP1 effectively binds to both monomeric and dimeric EpoR, whereas the binding affinities of the EMP9 to the EpoRs are 0.06 (to monomeric EpoR) and 0.07 (to dimeric EpoR) times lower than that of the EMP1 (Table 1). EMP1 effectively promotes the proliferation of EpoR-expressing cells at the dose of approximately 0.1 μM, whereas EMP9 does not induce the cell proliferation (Table 1).

3. BIOLOGICAL ROLE OF EMP9 AGAINST CANCERS

In parallel with the EMP9, we have clarified that the inhibition of Epo–EpoR signaling by anti-Epo antibody (R2) or soluble form of EpoR (sEpoR) effectively promotes the inhibition of cancer cell growth (Yasuda et al., 2001, 2002). These data led us to explore whether and how EMP9 is involved in the inhibition of carcinogenesis.

3.1 EMP9 Requires EpoR Expressions on Cancer Cells

We decided to address the therapeutic significance of Epo–EpoR signal inhibition in cancers. EMP1 is known to bind to the EpoR on cancer cells at the level of 0.07 μM, whereas EMP9 binds at the level of 1.0 μM, which is 14.3-fold higher concentration than EMP1 (Table 1; Johnson et al., 1998). This finding suggests that EMP9 requires EpoR expressions on the cancer cells at higher levels than EMP1 does. We first evaluated the mRNA expression levels of EpoR and Epo in 24 malignant human cancer cell lines, normal hepatocytes, and normal dermal fibroblasts under normoxic or anoxic conditions (Yasuda et al., 2003; Yasuda, Maeda, et al., 2010). The comparison of the EpoR mRNA levels of these cells revealed that EpoR mRNA is expressed on the malignant cells at 50- to 500-fold higher levels than the normal cells (Yasuda, Maeda, et al., 2010). Based on these findings, we speculated that EMP9 effectively destroys malignant cells but does not affect the growth of normal cells.

3.2 EMP9 Suppresses Cancer Cell Growth In Vitro

To confirm that EMP9 would effectively inhibit the growth of cancer cells expressing EpoR at higher levels than those with less, we first analyzed several cell lines (PC3, T98G, P39, G361, HepG2, SCH, and DLD1) for the expression levels of EpoR mRNA (Fig. 1A;

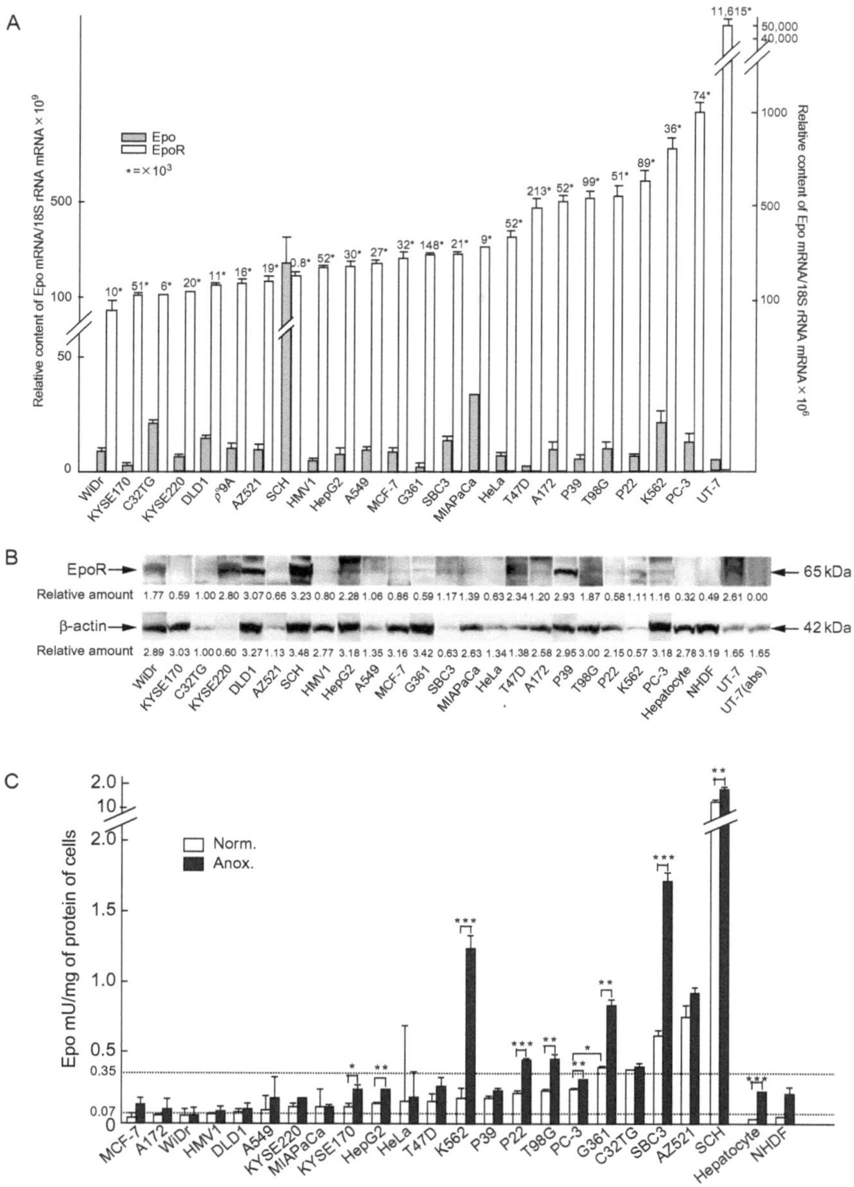

Fig. 1 Expression levels of Epo mRNA and EpoR mRNA with their proteins in 24 malignant cells. (A) Expression levels of Epo and EpoR mRNA compared with those of 18S rRNA mRNA. Each *bar* indicates means ± SEM of values. The numbers above each *bar* indicate the expression levels relative to that of Epo mRNA. (B) Western blot analyses of EpoR protein expressed in malignant and normal cell lines. The numbers under each blot indicate the band intensity relative to that of C32TG. (C) Secreted Epo protein levels in the culture supernatants of malignant and normal cells under normoxia and anoxia per 24 h. *Horizontal dotted lines* indicate the *upper* and *lower* ranges of Epo in normal serum. *$P < 0.05$, **$P < 0.01$, ***$P < 0.001$ based on Student's *t*-test.

Yasuda, Maeda, et al., 2010). Concomitantly, we analyzed the protein expression level of the EpoR by western blotting (Fig. 1B) and the secretion levels of Epo protein per 24 h by the enzyme-linked immunoassay (Fig. 1C; Yasuda et al., 2003). These data suggest that EMP9 suppresses the growth of cancer cells by the inhibition of the Epo secretion and/or EpoR protein expression.

We next sought to confirm that these cells would respond to an exogenous rhEpo. As a result, rhEpo significantly accelerated the growth of the cancer cells 1.2- to 1.7-fold faster than the saline control did (Fig. 2A and B). In contrast, EMP9 decelerated the growth of most cell lines dose dependently by 0.3- to 0.8-fold decrease compared with the control, except for SCH that expresses EpoR mRNA at lower levels than the other cell lines (Fig. 2C and D). Although no association was observed between the expression levels of EpoR mRNA and EpoR protein (Fig. 1), these data suggest that EMP9-induced suppression of cancer cell growth negatively correlates with the secretion levels of Epo protein.

To determine the effects of Epo protein and EMP9 on the Epo–EpoR signaling, we evaluated STAT5 phosphorylation status as a downstream event of the Epo–EpoR interaction. To this end, we used 5 U/mL of rhEpo, and 0.25 or 0.5 mg/mL of EMP9. Western blot analyses showed that the cancer cells exhibited the tyrosine phosphorylation of STAT5 even without STAT5 stimulators such as Epo (Fig. 2E). These cells exhibited the tyrosine phosphorylation of STAT5 in response to rhEpo (Fig. 2E). In contrast, EMP9 suppressed the STAT5 phosphorylation (Fig. 2E). Since these cells express EpoR mRNA at high levels, EMP9 effectively inhibits Epo–EpoR signaling in those with high EpoR expressions. These data suggest that Epo–EpoR signaling is triggered by Epo, suppressed by EMP9, and reactivated in an autocrine fashion in cancer cells.

3.3 EMP9 Suppresses Cancer Cell Growth In Vivo

We detected the monomeric hemoproteins (cytoglobin and Mb) and hemoglobins in normal and malignant lung tissues (Yasuda, Konishi, Koinuma, et al., 2006). Moreover, lung cancers (adenocarcinoma, squamous cell carcinoma, and large cell carcinoma including metastatic cancers) express the Epo and EpoR with the activation of MAP kinases (Yasuda, Hara, Hirohata, et al., 2010). We injected EMP9 into histoculture of lung cancers in vitro per our previous study based on anti-Epo antibody, soluble EpoR,

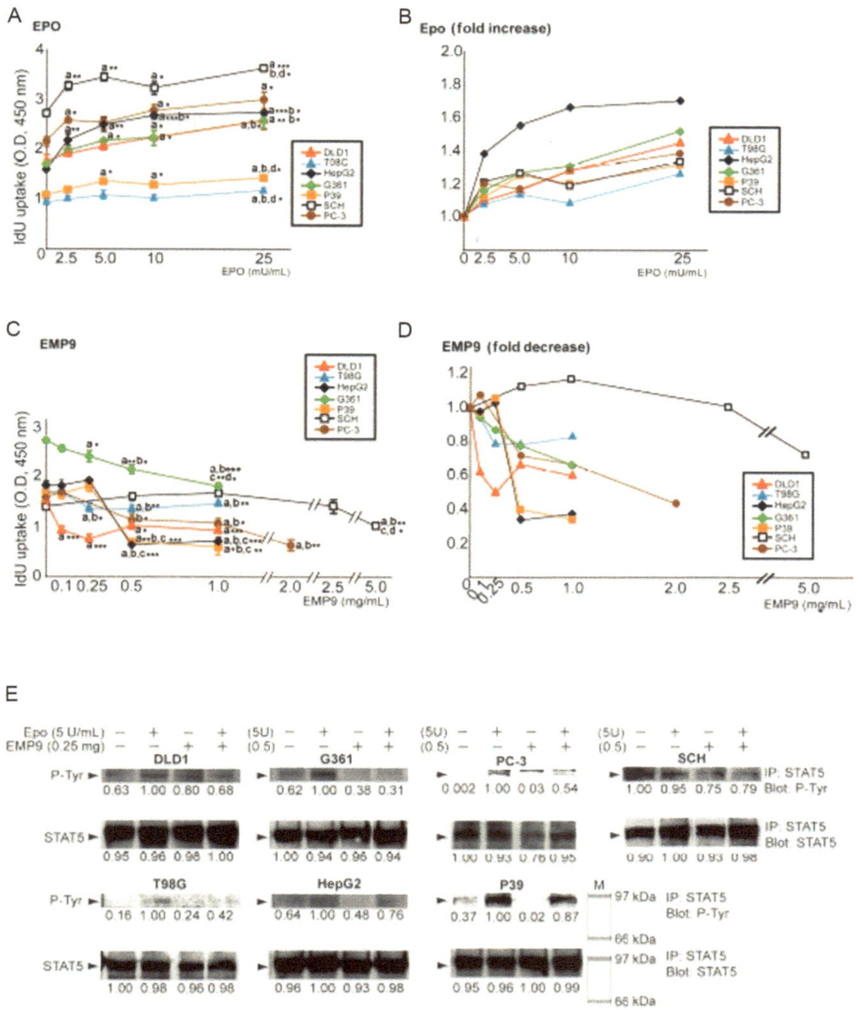

Fig. 2 Epo–EpoR signaling in cancer cell lines. (A) Absolute growth of each cell line in response to rhEpo (2.5–25 mU/mL) for 4 days based on IdU uptake. (B) Growth ratios of each cell line in response to rhEpo relative to the cell growth without rhEpo treatment. (C) Absolute growth of each cell lines in response to EMP9 (0.1–5 mg/mL) four times at 1-h intervals for 24 h based on IdU uptake. (D) Growth ratios of each cell lines in response to EMP9 relative to the cell growth without EMP9. Each point indicates mean ± SEM. (E) Tyrosine phosphorylation of STAT5 (pSTAT5) in the cell lines exposed to rhEpo and/or EMP9. Numbers under each blot indicate the intensities relative to the pSTAT5 level with the rhEpo treatment. *M*, molecular weight markers.

ovarian cancers, and endometrial cancers (Yasuda et al., 2002). We detected Mb-expressing cancer cells at high frequencies in the histoculture before the treatment (Fig. 3A). In contrast, no cancer cells expressed Mb after the treatment (Fig. 3B). These findings indicate that, in addition to the oxygen

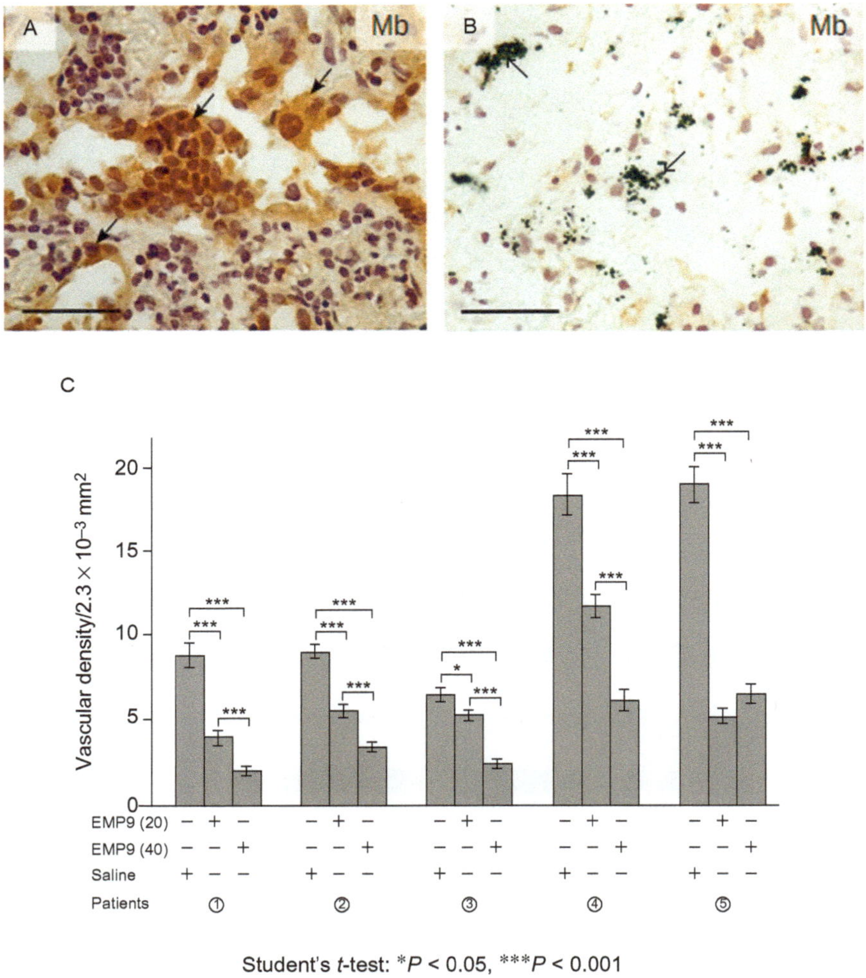

Fig. 3 Treatment with EMP9 in blocks of squamous cell lung cancer in vitro. (A) Accumulation of myoglobin (Mb)-expressing lung cancer cells without EMP9 treatment. (B) EMP9 treatment diminishes Mb-expressing cancer cells and leaves the deposits of charcoal particles (*arrows*). Bars: 50 μm. (C) Vascular densities in the lung cancer histocultures with or without the EMP9 treatment in vitro.

supply via feeding vessels, lung cancer cells synthesize Mb by themselves to reserve or store the oxygen in the form of oxy-Mb, MbO_2 through Epo–EpoR signaling.

We also evaluated the effects of the Epo–EpoR signaling inhibition on the feeding vessels in the lung cancers (Fig. 3C). Using the EMP9- and saline-injected samples derived from lung cancer patients, we stained the tissue sections with antifactor VIII antibody to evaluate the effects on cancer-feeding vessels. In all the samples treated with EMP9, the number of vessels significantly decreased compared with those of the controls ($P < 0.001$, $P < 0.01$). Four of the five samples showed a dose-dependent decrease in the vascular density. Consequently, in vitro treatment with EMP9 revealed that cancer cells die due to the loss of oxygen reservation mediated by Mb in the cancer cells themselves and the loss of oxygen supply via the cancer-feeding vessels.

Subsequently, we examined how EMP1 (Epo agonist) and EMP9 (EpoR antagonist) react to the transplanted lung cancer histoculture in nude mice. Four times injection of 1.0 mg EMP1 at 1-h intervals developed well-differentiated glandular structures (Fig. 4A and B), which included PCNA-positive epithelial cells at high frequencies (Fig. 4C) and macrophages at lower frequencies (Fig. 4D). In contrast, four times injection of 2.5 mg EMP9 induced the significant destruction of the glandular structures (Fig. 4E) with the apoptosis induction of the cancer cells (Fig. 4F) as well as the reduction of their proliferation (Fig. 4G). Of note, a substantial number of macrophages were recruited to the treated tumor sites in response to this dose of EMP9 treatment (Fig. 4H). Taken together, these data clearly suggest that EMP9 promotes the suppression and apoptosis of lung cancer cells.

4. MECHANISMS OF EMP9-MEDIATED INHIBITION OF CANCER CELL GROWTH

As mentioned above, during the last 16 years, we demonstrated how the Epo–EpoR signaling attributes to carcinogenesis using the EpoR antagonist EMP9, several cancer cell lines, and surgically resected specimens. Based on these data, the Epo–EpoR signaling is found to be involved in carcinogenesis through the protection for their survival, proliferation, and hemoprotein synthesis. In addition, we have demonstrated other pathological profiles recently as listed below.

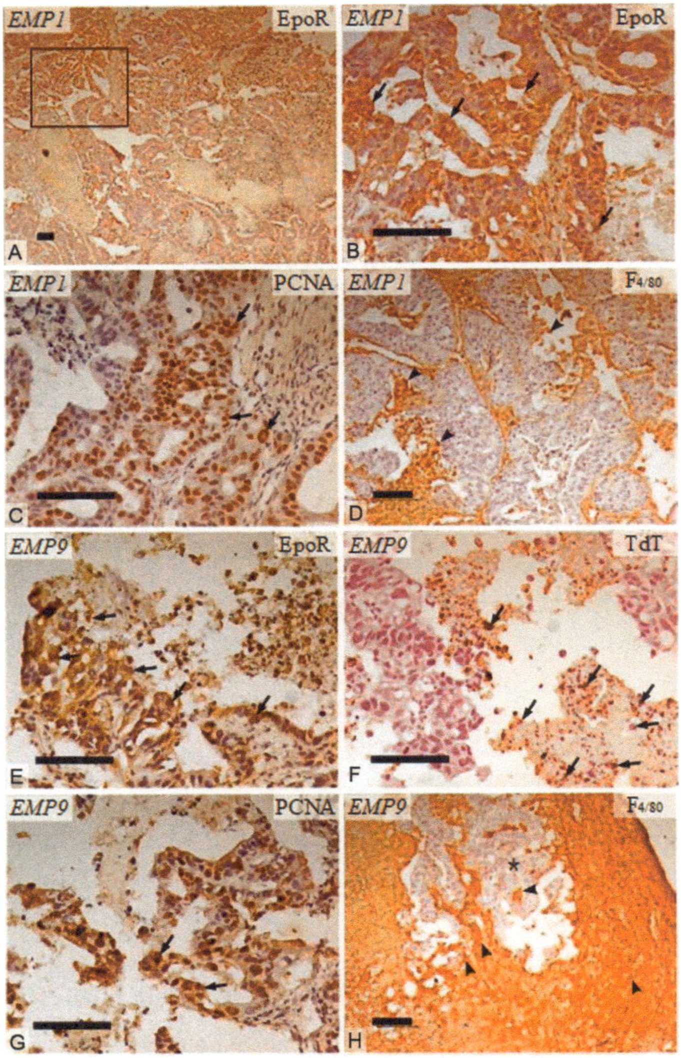

Fig. 4 Effects of EMP1 and EMP9 on xenografts of lung adenocarcinoma in nude mice. (A–D) Immunohistochemistry of lung cancer cell-based xenografts treated with EMP1. (A) and (B) Cancer cells express EpoR (*arrows*). (C) Cancer cells express PCNA (*arrows*). (D) Macrophages are detected in the treated cancer tissues (*arrowheads*). (E–H) Immunohistochemistry of lung cancer cell-based xenografts treated with EMP9. (E) Destructive regions exhibit dying cancer cells at high frequencies and EpoR-expressing cells sporadically (*arrows*). (F) Apoptotic death of cancer cells indicated by TdT assay (*arrows*). (G) Decrease in PCNA-expressing cancer cells in response to the EMP9 treatment. (H) Accumulation of $F_{4/80}$-positive macrophages (*arrowheads*) around the cancer foci (*asterisk*). Bars: 100 μm.

4.1 NO Flux

In xenografts of HeLa cells, we detected small pores at high frequencies in the cancer tissues in response to the EMP9 treatment (Yasuda et al., 2015). Two types of perforation are observed in the cancer cells: those with or without accompanying macrophages. When the cancer cells are accompanied with macrophages, the cancer cells exhibit pores in their cytoplasm at high levels (Fig. 5B). In contrast, cancer cells without accompanying macrophages exhibit pores in the cytoplasm at lower levels (Fig. 5C). These two types of pores originate from different processes. The former is mediated by iNOS from immune cells without Ca^{2+} dependency. The latter is derived from NO gas that is released from nNOS in the cytoplasm due to the high Ca^{2+}.

The Epo–EpoR signaling induces an increase in intracellular free Ca^{2+} [Ca^{2+}i] through a voltage-independent Ca^{2+} channels, which is a prerequisite for red blood cell formation (Cheung, Zhang, Bokvist, Tillotson, & Miller, 1997; Miller et al., 1999). Once EMP9 blocks the activation of the Epo–EpoR signaling, an excessive increase in the cytoplasmic influx of [Ca^{2+}i] occurs through a voltage-dependent Ca^{2+} channels. This excessive Ca^{2+} influx causes cell death and then is released into the extracellular space. Under these circumstances, nNOS in the cancer cells can be activated, and innate cells such as macrophages, NK cells, and neutrophils increase their tumor-killing activities.

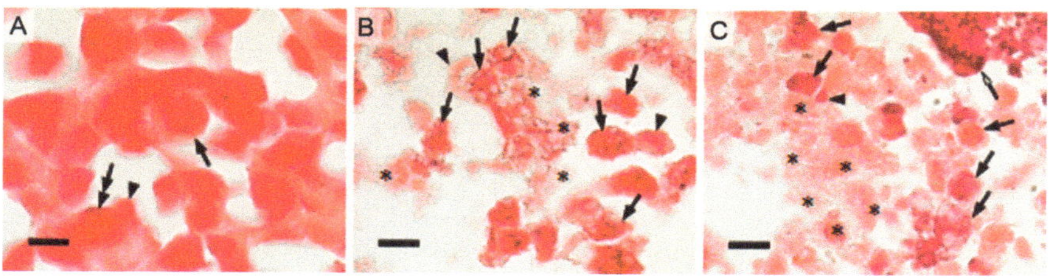

Fig. 5 Perforation of HeLa cells and calcification of the dying HeLa cells. (A) Histopathology of the HeLa cell-based xenografts treated with saline. *Arrow*, HeLa cells in mitosis; *double arrow*, HeLa cell attached to macrophages; *arrowhead*, macrophages. (B) The xenografts exposed to low-dose EMP9 exhibit perforated HeLa cells (*arrows*) along with macrophage accumulation (*arrowheads*). *Asterisks* indicate remnants of dying cells with perforations. (C) The xenografts exposed to high-dose EMP9 exhibit remnants of dying cells (*asterisks*), HeLa cells with perforation (*arrows*), and calcified dying cells (*open arrow*). Bars: 10 µm.

4.2 Calcification of Cancer Cells

EMP9 increases the intracellular and extracellular Ca^{2+} concentrations (Yasuda et al., 2015). In the cytoplasm of the cancer cells, the excessive Ca^{2+} binds easily to phosphate ions to form calcium deposits. In fact, the xenografts of breast cancer (Daitoh, Azumi, Yoshida, & Watatani, 2002), stomach carcinoma (Yoshida, Daitoh, Shigeoka, & Imamoto, 2003), and HeLa cells (Fig. 5C; Yasuda et al., 2015) have been shown to exhibit calcified masses in the destructing tumor foci.

4.3 Accumulation of Innate Immune Cells

Since EMP9 destructs the cancer-feeding vessels, the cancer foci become severe hypoxic in response to the EMP9 treatment. As a result, various immune cells and cancer cells express HIF-1α protein, and the population of macrophages, NK cells, and DC significantly increases in the xenografts exposed to EMP9 than that in the control mice ($P < 0.001$; Yasuda et al., 2015). This finding is associated with an increase in NK cells expressing IFN-γ in the cancer foci treated with EMP9. These data suggest that the EMP9-induced hypoxic and Ca^{2+}-rich environment significantly enhances innate immune activities.

5. CONCLUSION

In this chapter, we clarified how Epo–EpoR system is involved in carcinogenesis (Fig. 6). These data led us to realize that the blockade of Epo signaling through the EpoR antagonist EMP9 is a promising strategy to treat cancer patients from a multiple viewpoint described above.

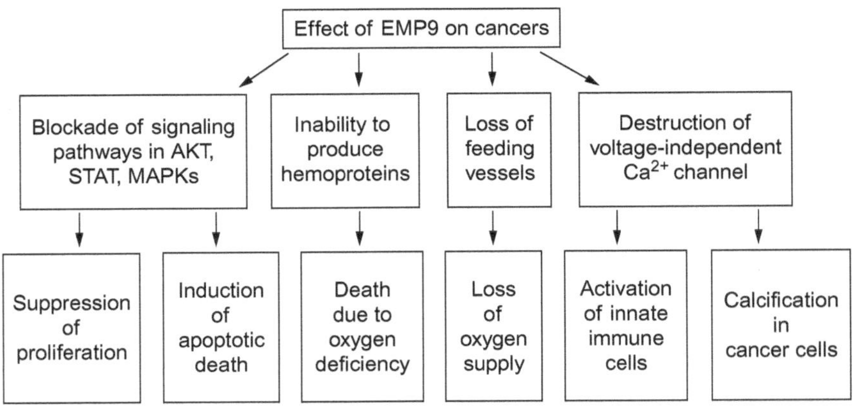

Fig. 6 Summarized view of functional activity of EMP9 in cancers.

REFERENCES

Acs, G., Zhang, P. J., Rebbeck, T. R., Acs, P., & Verma, A. (2002). Immunohistochemical expression of erythropoietin and erythropoietin receptor in breast carcinoma. *Cancer, 95*, 965–981.

Arcasoy, M. O., Amin, K., Karayal, A. F., Chou, S.-C., Raleigh, J. A., et al. (2002). Functional significance of erythropoietin receptor expression in breast cancer. *Laboratory Investigation, 82*, 911–918.

Cheung, J. Y., Zhang, X. Q., Bokvist, K., Tillotson, D. L., & Miller, B. A. (1997). Modulation of calcium channels in human erythroblasts by erythropoietin. *Blood, 89*, 92–100.

Daitoh, K., Azumi, T., Yoshida, T., & Watatani, M. (2002). A role of erythropoietin in breast cancers. *Medical Journal of Kinki University, 27*, 203–213. in Japanese.

Feldman, L., Wang, Y., Rhim, J. S., Bharttacharya, N., Loda, M., & Sytkowski, A. J. (2006). Erythropoietin stimulates growth and STAT5 phosphorylation in human prostate epithelial and prostate cancer cells. *Prostate, 66*, 135–145.

Jelkmann, W. (1992). Erythropoietin: Structure, control of production, and function. *Physiological Reviews, 72*, 449–489.

Johnson, D. L., Farrell, F. X., Barbone, F. P., McMahon, F. J., Tullai, J., Hoey, K., et al. (1998). Identification of a 13 amino acid peptide mimetic of erythropoietin and description of amino acids critical for the mimetic activity of EMP1. *Biochemistry, 37*, 3699–3710.

Krantz, S. B. (1991). Erythropoietin. *Blood, 77*, 419–434.

Miller, B. A., Barber, D. L., Bell, L. L., Beattie, B. K., Zhang, M. Y., et al. (1999). Identification of the erythropoietin receptor domain required for calcium channel activation. *The Journal of Biological Chemistry, 274*, 20465–20472.

Snow, M. H. L. (1977). Gastrulation in the mouse: Growth and regionalization of the epiblast. *Journal of Embryology and Experimental Morphology, 42*, 293–303.

Szenajch, J., Wcislo, G., Jeong, J. Y., Szczylik, C., & Feldman, L. (2010). The role of erythropoietin and its receptor in growth, survival and therapeutic response of human tumor cells. From clinic to bench—A critical review. *Biochimica et Biophysica Acta, 1806*, 82–95.

Wrighton, N. C., Farrell, F. X., Chang, R., Kashyap, A. K., Barbone, F. P., Mulcahy, L. S., et al. (1996). Small peptides as potent mimetics of the protein hormone erythropoietin. *Science, 273*, 458–464.

Yasuda, Y., Fujita, M., Koike, E., Obata, K., Shiota, M., Kotani, Y., et al. (2015). Erythropoietin receptor antagonist suppressed ectopic hemoglobin synthesis in xenografts of HeLa cells to promote their destruction. *PLoS One, 10,* e0122458.

Yasuda, Y., Fujita, Y., Masuda, S., Musha, T., Ueda, K., Tanaka, H., et al. (2002). Erythropoietin is involved in growth and angiogenesis in malignant tumours of female reproductive organs. *Carcinogenesis, 23*, 1797–1805.

Yasuda, Y., Fujita, Y., Matsuo, T., Koinuma, S., Hara, S., Tazaki, A., et al. (2003). Erythropoietin regulates tumour growth of human malignancies. *Carcinogenesis, 24*, 1021–1029.

Yasuda, Y., Hara, S., Hirohata, K., et al. (2010). Erythropoietin—Responsive sites in normal and malignant human lung tissues. *Anatomical Science International, 85*, 204–213.

Yasuda, Y., Konishi, H., Koinuma, S., et al. (2006). Distribution of globin proteins in normal lung tissues. *Acta anatomica Nipponica, 81*, 142.

Yasuda, Y., Maeda, Y., Koike, E., Watanabe, Y., Masuda, S., Yamasaki, H., et al. (2010). Cancer cell lines growth is promoted through individual responsiveness to autocrine and/or exogenous erythropoietin in vitro. In: *Recent Advances in Clinical Medicine* (pp. 337–348). ISBN: 978-960-474-165-6.

Yasuda, Y., Musha, T., Tanaka, H., Fujita, Y., Fujita, H., Utsumi, H., et al. (2001). Inhibition of erythropoietin signaling destroys xenografts of ovarian and uterine cancers in nude mice. *British Journal of Cancer, 84*, 836–843.

Yasuda, Y., Nagao, M., Okano, M., Masuda, S., Sasaki, R., et al. (1993). Localization of erythropoietin and erythropoietin-receptor in post-implantation mouse embryos. *Development, Growth & Differentiation, 35*, 711–722.

Yoshida, T., Daitoh, K., Shigeoka, H., & Imamoto, H. (2003). Expression of erythropoietin in stomach cancers. *Medical Journal of Kinki University, 28*, 359–367. in Japanese.

Youssoufian, H., Longmore, G., Neumann, D., Yoshimura, A., & Lodish, H. F. (1993). Structure, function, and activation of the erythropoietin receptor. *Blood, 81*, 2223–2236.

FSC
www.fsc.org
MIX
Papier | Fördert
gute Waldnutzung
FSC® C083411

Kolibri 360 GmbH
Ferdinand-Jühlke-Straße 7
99095 Erfurt, Deutschland
produktsicherheit@kolibri360.de